CHAPTERS IN GAME THEORY

THEORY AND DECISION LIBRARY

General Editors: W. Leinfellner (*Vienna*) and G. Eberlein (*Munich*)

Series A: Philosophy and Methodology of the Social Sciences

Series B: Mathematical and Statistical Methods

Series C: Game Theory, Mathematical Programming and Operations Research

Series D: System Theory, Knowledge Engineering an Problem Solving

SERIES C: GAME THEORY, MATHEMATICAL PROGRAMMING AND OPERATIONS RESEARCH

VOLUME 31

Scope: Particular attention is paid in this series to game theory and operations research, their formal aspects and their applications to economic, political and social sciences as well as to socio-biology. It will encourage high standards in the application of game-theoretical methods to individual and social decision making.

The titles published in this series are listed at the end of this volume.

CHAPTERS IN GAME THEORY

In honor of Stef Tijs

Edited by

PETER BORM

University of Tilburg,
The Netherlands

and

HANS PETERS

University of Maastricht,
The Netherlands

KLUWER ACADEMIC PUBLISHERS

BOSTON / DORDRECHT / LONDON

A C.I.P. Catalogue record for this book is available from the Library of Congress.

ISBN 1-4020-7063-2

Published by Kluwer Academic Publishers,
P.O. Box 17, 3300 AA Dordrecht, The Netherlands.

Sold and distributed in North, Central and South America
by Kluwer Academic Publishers,
101 Philip Drive, Norwell, MA 02061, U.S.A.

In all other countries, sold and distributed
by Kluwer Academic Publishers,
P.O. Box 322, 3300 AH Dordrecht, The Netherlands.

Printed on acid-free paper

Printed in the Netherlands

Preface

On the occasion of the 50th birthday of Stef Tijs in 1987 a volume of surveys in game theory in Stef's honor was composed[1]. All twelve authors who contributed to that book still belong to the twenty-nine authors involved in the present volume, published fifteen years later on the occasion of Stef's 65th birthday. Twenty-five of these twenty-nine authors wrote—or write, in one case—their Ph.D. theses under the supervision of Stef Tijs. The other four contributors are indebted to Stef Tijs to a different but hardly less decisive degree.

What makes a person deserve to be the honorable subject of a scientific *liber amicorum*, and that on at least two occasions in his life? If that person is called Stef Tijs then the answer includes at least the following reasons. First of all, until now Stef has supervised about thirty Ph.D. students in game theory alone. More importantly than sheer numbers, most of these students stayed in academics; for instance all those who contributed to the 1987 volume. It is beyond doubt that this fact has everything to do with the devotion, enthusiasm and deep knowledge invested by Stef in guiding students. Moreover, the number of his internationally published papers has increased from about sixty in 1987 to about two hundred now. His papers cover every field in game theory, and extend to related areas as social choice theory, mathematical economics, and operations research. Last but not least, Stef's numerous coauthors come from and live in all parts of this world: he has been a true missionary in game theory, and the contributors to this volume are proud to be among his apostles.

PETER BORM
HANS PETERS

Tilburg/Maastricht
February 2002

[1]H.J.M. Peters and O.J. Vrieze, eds., *Surveys in Game Theory and Related Topics*, CWI Tract 39, Amsterdam, 1987.

About Stef Tijs

The first work of Stef Tijs in game theory was his Ph.D. dissertation *Semi-infinite and infinite matrix games and bimatrix games* (1975). He took his Ph.D. at the University of Nijmegen, where he had held a position since 1960. His Ph.D. advisors were A. van Rooij and F. Delbaen. From 1975 on he gradually started building a game theory school in the Netherlands with a strong international focus. In 1991 he left Nijmegen to continue his research at Tilburg University. In 2000 he was awarded a doctorate honoris causa at the Miguel Hernandez University in Elche, Spain.

About this book

The authors of this book were asked to write on topics belonging to their expertise and having a connection with the work of Stef Tijs. Each contribution has been reviewed by two other authors. This has resulted in fourteen chapters on different subjects: some of these can be considered surveys while other chapters present new results. Most contributions can be positioned somewhere in between these categories. We briefly describe the contents of each chapter. For the references the reader should consult the list of references in each chapter under consideration.

Chapter 1, *Stochastic cooperative games: theory and applications* by Peter Borm and Jeroen Suijs, considers cooperative decision making under risk. It provides a brief survey on three existing models introduced by Charnes and Granot (1973), Suijs *et al.* (1999), and Timmer *et al.* (2000), respectively. It also compares their performance with respect to two applications: the allocation of random maintenance cost of a communication network tree to its users, and the division of a stochastic estate among the creditors in a bankruptcy situation.

Chapter 2, *Sequencing games: a survey* by Imma Curiel, Herbert Hamers, and Flip Klijn, gives an overview of the start and the main developments in the research area that studies the interaction between sequencing situations and cooperative game theory. It focuses on results related to balancedness and convexity of sequencing games.

In Chapter 3, *Game theory and the market* by Eric van Damme and Dave Furth, it is argued that both cooperative and non-cooperative game models can substantially increase our understanding of the functioning of actual markets. In the first part of the chapter, by going back to the work of the founding fathers von Neumann, Morgenstern, and Nash, a brief historical sketch of the differences and complementarities between the two types of models is provided. In the second part, the main point is illustrated by means of examples of bargaining, oligopolistic interaction and auctions.

In Chapter 4, *On the number of extreme points of the core of a transferable utility game* by Jean Derks and Jeroen Kuipers, it is derived from a more general result that the upper core and the core of a transferable utility game have at most $n!$ different extreme points, with n the number of players. This maximum number is attained by strict convex games but other games may have this property as well. It is shown that $n!$ different extreme core points can only be obtained by strict exact games, but that not all such games have $n!$ different extreme points.

In Chapter 5, *Consistency and potentials in cooperative TU-games: Sobolev's reduced game revived* by Theo Driessen, a consistency property for a wide class of game-theoretic solutions that possess a potential representation is studied. The consistency property is based on a modified reduced game related to Sobolev's. A detailed exposition of the developed theory is given for semivalues of cooperative TU-games and the Shapley and Banzhaf values in particular.

In Chapter 6, *On the set of equilibria of a bimatrix game: a survey* by Mathijs Jansen, Peter Jurg, and Dries Vermeulen, the methods used by different authors to write the set of equilibria of a bimatrix game as the union of a finite number of polytopes, are surveyed.

Chapter 7, *Concave and convex serial cost sharing* by Maurice Koster, introduces the concave and convex serial rule, two new cost sharing rules that are closely related to the serial cost sharing rule of Moulin and Shenker (1992). It is shown that the concave serial rule is the unique rule that minimizes the range of cost shares subject to the excess lower bounds. Analogous results are derived for the convex serial rule. In particular, these characterizations show that the serial cost sharing rule is consistent with diametrically opposed equity properties, depending on the nature of the cost function: the serial rule equals the concave (convex) serial rule in case of a concave (convex) cost function.

In Chapter 8, *Centrality orderings in social networks* by Herman Monsuur and Ton Storcken, a centrality ordering arranges the vertices in a social network according to their centrality position in that network.

Centrality addresses notions like focal points of communication, potential of communicational control, and being close to other network vertices. In social network studies they play an important role. Here the focus is on the conceptual issue of what makes a position in a network more central than another position. Characterizations of the cover, the median and degree centrality orderings are discussed.

In Chapter 9, *The Shapley transfer procedure for NTU-games* by Gert-Jan Otten and Hans Peters, the Shapley transfer procedure (Shapley, 1969) is extended in order to associate with every solution correspondence for transferable utility games satisfying certain regularity conditions, a solution for nontransferable utility games. An existence and a characterization result are presented. These are applied to the Shapley value, the core, the nucleolus, and the τ-value.

Chapter 10, *The nucleolus as equilibrium price* by Jos Potters, Hans Reijnierse, and Anita van Gellekom, studies exchange economies with indivisible goods and money. The notions of a stable equilibrium and regular prices are introduced. It is shown that the nucleolus concept for TU-games can be used to single out specific regular prices. Algorithms to compute the nucleolus can therefore be used to determine regular price vectors.

Chapter 11, *Network formation, costs, and potential games* by Marco Slikker and Anne van den Nouweland, studies strategic-form games of network formation in which an exogenous allocation rule is used to determine the players' payoffs in various networks. It is shown that such games are potential games if the cost-extended Myerson value is used as the exogenous allocation rule. The question is then studied which networks are formed according to the potential maximizer, a refinement of Nash equilibrium for potential games.

Chapter 12, *Contributions to the theory of stochastic games* by Frank Thuijsman and Koos Vrieze, presents an introduction to the history and the state of the art of the theory of stochastic games. Dutch contributions to the field, initiated by Stef Tijs, are addressed in particular. Several examples are provided to clarify the issues.

Chapter 13, *Linear (semi-)infinite programs and cooperative games* by Judith Timmer and Natividad Llorca, gives an overview of cooperative games arising from linear semi-infinite or infinite programs.

Chapter 14, *Population uncertainty and equilibrium selection: a maximum likelihood approach* by Mark Voorneveld and Henk Norde, introduces a general class of games with population uncertainty and, in line with the maximum likelihood principle, stresses those strategy profiles that are most likely to yield an equilibrium in the game selected by chance. Under mild topological restrictions, an existence result for maximum likelihood equilibria is derived. Also, it is shown how maximum likelihood equilibria can be used as an equilibrium selection device for finite strategic games.

About the authors

PETER BORM (p.e.m.borm@kub.nl) is affiliated with the Department of Econometrics of the University of Tilburg. He wrote his Ph.D. thesis, *On game theoretic models and solution concepts*, under the supervision of Stef Tijs.

IMMA CURIEL (curiel@math.umbc.edu) is affiliated with the Department of Mathematics and Statistics of the University of Maryland, Baltimore County. She wrote her Ph.D. thesis, *Cooperative game theory and applications*, under the supervision of Stef Tijs.

ERIC VAN DAMME (eric.vandamme@kub.nl) is affiliated with CentER, University of Tilburg. He wrote his master's thesis under the supervision of Stef Tijs and his Ph.D. thesis, *Refinements of the Nash equilibrium concept*, under the supervision of Jaap Wessels and Reinhard Selten.

JEAN DERKS (jean.derks@math.unimaas.nl) is affiliated with the Department of Mathematics of the University of Maastricht. His Ph.D. thesis, *On polyhedral cones of cooperative games*, was written under the supervision of Stef Tijs and Koos Vrieze.

THEO DRIESSEN (t.s.h.driessen@math.utwente.nl) is affiliated with the Department of Mathematical Sciences of the University of Twente. He wrote his Ph.D. thesis, *Contributions to the theory of cooperative games: the τ-value and k-convex games*, under the supervision of Stef Tijs and Michael Maschler.

DAVE FURTH (dfurth@fee.uva.nl) is affiliated with the Faculty of Economics and Econometrics of the University of Amsterdam. He wrote his Ph.D. thesis on oligopoly theory with Arnold Heertje and has been a regular guest of the game theory seminars organized since 1983 by Stef Tijs.

ANITA VAN GELLEKOM (anita.v.gellekom@mail.cadans.nl) works for a nonprofit institution. Her Ph.D. thesis, *Cost and profit sharing in a cooperative environment*, was written under the supervision of Stef Tijs.

HERBERT HAMERS (h.j.m.hamers@kub.nl) is affiliated with the Department of Econometrics of the University of Tilburg. Stef Tijs supervised his Ph.D. thesis, *Sequencing and delivery situations: a game theoretic approach.*

MATHIJS JANSEN (m.jansen@ke.unimaas.nl) is affiliated with the Department of Quantitative Economics of the University of Maastricht. His Ph.D. supervisors were Frits Ruymgaart and T.E.S. Raghavan, and his thesis *Equilibria and optimal threat strategies in two-person games* was written in close cooperation with Stef Tijs.

PETER JURG (peter@jurg.nl) works for a private software company. He wrote his thesis, *Some topics in the theory of bimatrix games*, under the supervision of Stef Tijs.

FLIP KLIJN (fklijn@uvigo.es) is affiliated with the Department of Statistics and Operations Research of the University of Vigo, Spain. His Ph.D. thesis, *A game theoretic approach to assignment problems*, was written under the supervision of Stef Tijs.

MAURIC KOSTER (mkoster@fee.uva.nl) is affiliated with the Department of Economics and Econometrics of the University of Amsterdam. Stef Tijs supervised his thesis *Cost sharing in production situations and network exploitation.*

JEROEN KUIPERS (jeroen.kuipers@math.unimaas.nl) is affiliated with the Department of Mathematics of the University of Maastricht. His Ph.D. thesis, *Combinatorial methods in cooperative game theory*, was supervised by Stef Tijs and Koos Vrieze.

NATIVIDAD LLORCA (nllorca@umh.es) is a Ph.D. student, under the supervision of Stef Tijs, at the Department of Statistics and Applied Mathematics of the University of Elche, Spain.

HERMAN MONSUUR (h.monsuur@kim.nl) is affiliated with the Royal Netherlands Naval College, section International Security Studies. His Ph.D. thesis, *Choice, ranking and circularity in asymmetric relations*, was supervised by Stef Tijs.

HENK NORDE (h.norde@kub.nl) is a member of the Department of Econometrics of the University of Tilburg, where his main research is in the area of game theory. He wrote a Ph.D. thesis in the field of differential equations, supervised by Leonid Frank at the University of Nijmegen.

ANNE VAN DEN NOUWELAND (annev@oregon.uoregon.edu) is affiliated with the Department of Economics of the University of Oregon, Eugene. She wrote her Ph.D. thesis, *Games and graphs in economic situations*, under the supervision of Stef Tijs.

GERT-JAN OTTEN (g.j.otten@kpn.com) works for KPN Telecom. His Ph.D. thesis, *On decision making in cooperative situations*, was supervised by Stef Tijs.

HANS PETERS (h.peters@ke.unimaas.nl) is affiliated with the Department of Quantitative Economics of the University of Maastricht. He wrote his Ph.D. thesis, *Bargaining game theory*, under the supervision of Stef Tijs.

JOS POTTERS (potters@sci.kun.nl) is affiliated with the Department of Mathematics of the University of Nijmegen. He wrote a Ph.D. thesis on a subject in geometry at the University of Leiden and cooperates with Stef Tijs in the area of game theory since the beginning of the eighties.

HANS REIJNIERSE (j.h.reijnierse@kub.nl) is affiliated with the Department of Econometrics of the University of Tilburg. He wrote his Ph.D. thesis, *Games, graphs, and algorithms*, supervised by Stef Tijs.

MARCO SLIKKER (m.slikker@tm.tue.nl) is affiliated with the Department of Technology Management of the Eindhoven University of Technology. Stef Tijs supervised his Ph.D. thesis *Decision making and cooperation restrictions.*

TON STORCKEN (t.storcken@ke.unimaas.nl) is affiliated with the Department of Quantitative Economics of the University of Maastricht. His Ph.D. thesis, *Possibility theorems for social welfare functions*, was supervised by Pieter Ruys, Stef Tijs, and Harrie de Swart.

JEROEN SUIJS (j.p.m.suijs@kub.nl) is affiliated with the CentER Accounting Research Group. He wrote his Ph.D. thesis, *Cooperative decision making in a stochastic environment*, under the supervision of Stef Tijs.

FRANK THUIJSMAN (frank@math.unimaas.nl) is affiliated with the Department of Mathematics of the University of Maastricht. He wrote his Ph.D. thesis, *Optimality and equilibria in stochastic games*, under the supervision of Stef Tijs and Koos Vrieze.

JUDITH TIMMER (j.b.timmer@math.utwente.nl) is afiliated with the Department of Mathematical Sciences of the University of Twente. Stef

Tijs supervised her Ph.D. thesis *Cooperative behaviour, uncertainty and operations research.*

DRIES VERMEULEN (d.vermeulen@ke.unimaas.nl) is affiliated with the Department of Quantitative Economics of the University of Maastricht. His Ph.D. thesis, *Stability in non-cooperative game theory*, was written under the supervision of Stef Tijs.

MARK VOORNEVELD (mark.voorneveld@hhs.se) works at the University of Stockholm and wrote a Ph.D. thesis, *Potential games and interactive decisions with multiple criteria*, supervised by Stef Tijs.

KOOS VRIEZE (o.j.vrieze@math.unimaas.nl) is affiliated with the Department of Mathematics of the University of Maastricht. His Ph.D. thesis, *Stochastic games with finite state and action spaces*, was written under the supervision of Henk Tijms and Stef Tijs.

Contents

Chapter 1

Stochastic Cooperative Games: Theory and Applications

BY PETER BORM AND JEROEN SUIJS

1.1 Introduction

Cooperative behavior generally emerges for the individual benefit of the people and organizations involved. Whether it is an international agreement like the GATT or the local neighborhood association, the main driving force behind cooperation is the participants' belief that it will improve their welfare. Although these believed welfare improvements may provide the necessary incentives to explore the possibilities of cooperation, it is not sufficient to establish and maintain cooperation. It is only the beginning of a bargaining process in which the coalition partners have to agree on which actions to take and how to allocate any joint benefits that possibly result from these actions. Any prohibitive objections in this bargaining process may eventually break up cooperation.

Since its introduction in von Neumann and Morgenstern (1944), cooperative game theory serves as a mathematical tool to describe and analyze cooperative behavior as mentioned above. The literature, however, mainly focuses on a deterministic setting in which the synergy between potential coalitional partners is known with certainty beforehand. An actual example in this regard is provided by the automobile

P. Borm and H. Peters (eds.), Chapters in Game Theory, 1–26.

industry, where some major car manufacturers collectively design new models so as to save on the design costs. Since they know the cost of designing a new car, they also know how much they will save on design expenditures by cooperating.

In every day life, however, not everything is certain and many of the decisions that people make are done so without the precise knowledge of the consequences. Moreover, the risks that people face as a result of their social and economic activities may affect their cooperative behavior in various instances. A typical example in this respect is a joint-venture. Investing in a new project is risky and therefore a company may prefer to share these risks by cooperating in a joint-venture with other companies. A joint-venture is thus arranged before the project starts when it is still unknown to what extent it will be a success. A similar argument applies for investment pools/funds, where investors pool their capital and make joint investments to benefit from risk sharing and risk diversification. As opposed to joint ventures and investment pools/funds, risk sharing need not be the primary incentive for cooperation. In many other cases, cooperation arises for other reasons and risk is just involved with the actions and decisions of the coalition partners. Small retailers, for instance, organize themselves in a buyers' cooperative to stipulate better prices when purchasing their inventory. Any economic risks, however, are not reduced by such a cooperation.

The first game theoretical literature on cooperative decision-making under risk dates from the early 70s with the introduction of chance-constrained games by Charnes and Granot (1973). Research on this subject was almost non-existent in the following decades until recently it was picked up again by Suijs *et al.* (1999) and Timmer *et al.* (2000). This paper provides a brief survey on the three existing models and compares their performance in two situations: the allocation of random maintenance costs of a communication network tree to its users, and the division of a stochastic estate among creditors in a bankruptcy situation.

Chance-constrained games were introduced in Charnes and Granot (1973) to encompass situations where the benefits obtained by the agents are random variables. Their attention is focused on dividing the benefits of the grand coalition. Although the benefits are random, the authors allocate a deterministic amount in two stages. In the first stage, before the realization of the benefits is known, payoffs are promised to the individuals. In the second stage, when the realization is known, the payoffs promised in the first stage are modified if needed. In several papers, Charnes and Granot introduce allocation rules for the first stage like the prior core, the prior Shapley value, and the prior nucleolus. To modify these so-called prior allocations in the second stage they define the two-stage nucleolus. We confine our discussion to the prior core and refer to Charnes and Granot (1976, 1977), and Granot (1977) for the other solution concepts.

Suijs *et al.* (1999) introduced stochastic cooperative games, which deal with the same kind of problems as chance-constrained games do, albeit in a completely different way. A drawback of the model introduced by Charnes and Granot (1973) is that it does not explicitly take into account the individuals' behavior towards risk. The effects of risk averse behavior, for example, are difficult to trace in this model. The model introduced in Suijs *et al.* (1999) explicitly includes the preferences of the individuals. Any kind of behavior towards risk, from risk loving behavior to risk averse behavior, can be expressed by these preferences. Another major difference is the way in which the benefits are allocated. As opposed to a two stage allocation, which assigns a deterministic payoff to each agent, an allocation in a stochastic cooperative game assigns a random payoff to each agent. Furthermore, for a two stage allocation the agents must come to an agreement twice. In the first stage, before the realization of the payoff is known, they have to agree on a prior allocation. In the second stage, once the realization is known, they have to agree on how the prior payoff is modified. In stochastic cooperative games the agents decide on the allocation before the realization is known. As a result, random payoffs are allocated so that no further decisions have to be taken once the realization of the payoff is known.

The model introduced by Timmer *et al.* (2000) is based on the model of stochastic cooperative games introduced by Suijs *et al.* (1999). The difference lies in the way random payoffs are allocated. Suijs *et al.* (1999) distinguishes two parts in an allocation. The first part concerns the allocation of the risk. In this regard, non-negative multiples of random payoffs are allocated to the agents. The second part then concerns deterministic transfer payments between the agents. The inclusion of deterministic transfer payments enables the agents to conclude mutual insurance deals. In exchange for a deterministic amount of money, i.e. an insurance premium, agents may be willing to bear a larger part of the risk. In order to exclude these insurance possibilities from the analysis, Timmer *et al.* (2000) does not allow for any deterministic transfer payments.

Besides a theoretical discussion of the abovementioned models, we will compare their performance in two possible applications. The focus of our analysis will be on the existence of core allocations.

The first application concerns the allocation of the random maintenance costs of a communication network tree that connects a service provider to its clients. Typical examples one can think of in this context are cities that are connected to a local powerplant, a cable TV network or computer workstations that are connected to a central server. Megiddo (1978) considered this cost allocation problem in a deterministic setting. It was shown that the corresponding TU-game has a nonempty core. Maintenance costs, however, are generally random of nature, for one does not know up front when connections are going to fail and what the resulting costs of repair will be. In this regard, random variables are more appropriate to describe the maintenance costs. In addition, by using random variables we are able to take the reliability or quality of a connection into account. Low quality connections may be cheap in construction, but they are also more likely to require repair and/or maintenance to keep them in operation. So, by assuming that these costs are deterministic one passes over the important aspect of a network's reliability.

The second application concerns the division of the estate of a bankrupt enterprise among its creditors. In this paper we assume that the exact value of the estate is uncertain. Generally, the value of a firm's assets (e.g. inventory, production facilities, knowledge) is ambiguous in case of bankruptcy as market values are no longer appropriate for valuation purposes. We assume the creditors all have a deterministic claim on this random estate. Since the value of the estate is insufficient to meet all claims, an allocation problem arises. O'Neill (1982) offers a game theoretical analysis of bankruptcy problems in a deterministic setting. It is shown that the core of bankruptcy games is nonempty. In fact, any allocation that gives each creditor at least zero and at most his claim is a core allocation.

This chapter is organized as follows. Section 1.2 provides a theoretical discussion of the three existing types of cooperative games that can deal with random payoffs. In particular, we focus on the core of these games and the corresponding requirements for it to be nonempty. Section 1.3 considers the cost allocation problem in a network tree while Section 1.4 considers the bankruptcy problem. Finally, Section 1.5 concludes.

1.2 Cooperative Decision-Making under Risk

For cooperative games with transferable utility, the payoff of a coalition is assumed to be known with certainty. In many cases though, the payoffs to coalitions can be uncertain. This would not raise a problem if the agents can await the realizations of the payoffs before deciding which coalitions to form and which allocations to settle on. But if the formation of coalitions and allocations has to take place before the payoffs are realized, the framework of TU-games is no longer appropriate.

This section presents three different models especially designed to deal with situations in which the benefits from cooperation are best described by random variables. The following models will pass in review consecutively: chance-constrained games (cf. Charnes and Granot, 1973), stochastic cooperative games with transfer payments (cf. Suijs *et al.*, 1999), and stochastic cooperative games without transfer payments (cf. Timmer *et al.*, 2000).

1.2.1 Chance-Constrained Games

Chance-constrained games as introduced by Charnes and Granot (1973) extend the theory of cooperative games in characteristic function form to situations where the benefits from cooperation are random variables. So, when several agents decide to cooperate, they do not exactly know the benefits that this cooperation generates. What they do know is the probability distribution function of these benefits. Let $\boldsymbol{V}(S)$ denote the random variable describing the benefits of coalition S. Furthermore, denote its probability distribution function by $F_{\boldsymbol{V}(S)}$. Thus,

$$F_{\boldsymbol{V}(S)}(t) \;=\; \mathbf{P}\{\boldsymbol{V}(S) \leq t\},$$

for all $t \in I\!R$. Then a chance-constrained game is defined by the pair $(N, \boldsymbol{V})$, where N is a finite set of agents and $\boldsymbol{V}$ is the characteristic function assigning to each coalition $S \subset N$ the nonnegative random benefits $\boldsymbol{V}(S)$. Note that chance-constrained games are based on the formulation (N, v) of TU-games with the deterministic benefits $v(S)$ replaced by stochastic benefits $\boldsymbol{V}(S)$.

For dividing the benefits of the grand coalition, the authors propose two stage allocations. In the first stage, when the realization of the benefits is still unknown, each agent is promised a certain payoff. These so-called prior payoffs are such that there is a fair chance that they are realized. Once the benefits are known, the total payoff allocated in the prior payoff can differ from what is actually available. In that case, we come to the second stage and modify the prior payoff in accordance with the realized benefits.

Let us start with discussing the prior allocations. A prior payoff is denoted by a vector $x \in I\!R^N$, with the interpretation that agent $i \in N$ receives the amount x_i. To comply with the condition that there is a reasonable probability that the promised payoffs can be kept, the prior payoff x must be such that

$$\underline{\alpha} \leq \mathbf{P}(\{\mathbf{V}(N) \leq \sum_{i \in N} x_i\}) = F_{\mathbf{V}(N)}(\sum_{i \in N} x_i) \leq \bar{\alpha}, \tag{1.1}$$

with $0 < \underline{\alpha} \leq \bar{\alpha} < 1$. This condition assures that the total amount $\sum_{i \in N} x_i$ that is allocated is not too low or too high. Note that expression (1.1) can also be written as

$$\xi_{\underline{\alpha}}(\mathbf{V}(N)) \leq \sum_{i \in N} x_i \leq \xi_{\bar{\alpha}}(\mathbf{V}(N)),$$

where ξ_α denotes the α-quantile. The α-quantile of a random payoff $\boldsymbol{X}$, denoted by $\xi_\alpha(\boldsymbol{X})$, is the largest value ξ such that the realization of $\boldsymbol{X}$ will be less than ξ with at most probability α. Formally, the α-quantile of $\boldsymbol{X}$ is defined by $\xi_\alpha(\boldsymbol{X}) = \sup\{t | \mathbf{P}(\boldsymbol{X} < t) \leq \alpha\}$ where $\alpha \in (0,1)$.

To come to a prior core for chance-constrained games, one needs to specify when a coalition S is satisfied with the amount $\sum_{i \in S} x_i$ it receives, so that it does not threaten to leave the grand coalition N. Charnes and Granot (1973) assumes that a coalition S is satisfied with what it gets, if the probability that they can obtain more on their own is small enough. This means that for each coalition $S \neq N$ there exists a number $\alpha(S) \in (0,1)$ such that coalition S is willing to participate in the coalition N given the proposed allocation $x \in I\!R^N$, whenever $\mathbf{P}(\{\mathbf{V}(S) > \sum_{i \in S} x_i\}) \leq 1 - \alpha(S)$. The number $\alpha(S)$ is a measure of assurance for coalition S. Note that the measures of assurance may vary over the coalitions. Furthermore, they reflect the coalitions' attitude towards risk, the willingness to bargain with other coalitions, and so on. The prior core of a chance-constrained game $(N, \mathbf{V})$ is then defined by

$$C(\mathbf{V}) = \left\{ x \in I\!R^N \;\middle|\; \begin{array}{l} \forall_{S \subset N: S \neq N} : F_{\mathbf{V}(S)}(\sum_{i \in S} x_i) \geq \alpha(S), \\ \underline{\alpha} \leq F_{\mathbf{V}(N)}(\sum_{i \in N} x_i) \leq \bar{\alpha} \end{array} \right\}. \tag{1.2}$$

Example 1.1 Consider the following three-person chance-constrained game $(N, \mathbf{V})$ defined by $\mathbf{V}(\{i\}) = 0$, $i = 1,2,3$, $\mathbf{V}(S) \sim U(0,2)$ if

$\#S = 2$, and $\mathbf{V}(N) \sim U(1,2)$. Furthermore, let $\alpha(S) = \frac{1}{2}$ for all $S \neq N$, $\underline{\alpha} = \frac{2}{5}$, and $\bar{\alpha} = \frac{3}{5}$. Next, let $x \in \mathbb{R}^3$ be a prior allocation. Since

$$F_{\mathbf{V}(N)}(t) = \begin{cases} 0, & \text{if } t \leq 1, \\ t-1, & \text{if } 1 < t < 2, \\ 1, & \text{if } 2 \leq t, \end{cases}$$

condition (1.1) implies that $1\frac{2}{5} \leq \sum_{i \in N} x_i \leq 1\frac{3}{5}$. For the one-person coalitions we have that $F_{\mathbf{V}(\{i\})}(x_i) \geq \alpha(\{i\}) = \frac{1}{2}$ if $x_i \geq 0$. Furthermore, since for all two-person coalitions S it holds that

$$F_{\mathbf{V}(S)}(t) = \begin{cases} 0, & \text{if } t \leq 0, \\ \frac{1}{2}t, & \text{if } 0 < t < 2, \\ 1, & \text{if } 2 \leq t, \end{cases}$$

it follows that $\sum_{i \in S} x_i \geq 1$ if $F_{\mathbf{V}(S)}(\sum_{i \in S} x_i) \geq \alpha(\{i\}) = \frac{1}{2}$. Hence, the prior core of this game is given by

$$C(\mathbf{V}) = \{x \in \mathbb{R}^3_+ | \ \forall_{S \subset N: \#S=2} : \sum_{i \in S} x_i \geq 1, 1\tfrac{2}{5} \leq \sum_{i \in N} x_i \leq 1\tfrac{3}{5}\}.$$

◁

A chance-constrained game $(N, \mathbf{V})$ with a nonempty core is called balanced. Furthermore, if the core of every subgame $(S, \mathbf{V}_{|S})$ is nonempty, then $(N, \mathbf{V})$ is called totally balanced. The subgame $(S, \mathbf{V}_{|S})$ is given by $rV_{|S}(U) = \mathbf{V}(U)$ for all $U \subset S$.

A necessary and sufficient condition for nonemptiness of the prior core is given by the following theorem, which can be found in Charnes and Granot (1973).

Theorem 1.2 *Let $(N, \mathbf{V})$ be a chance-constrained game. Then $C(\mathbf{V}) = \emptyset$ if and only if $\mu > \bar{\alpha}$ with*

$$\begin{aligned} \mu = \min_{x \in \mathbb{R}^N} \quad & F_{\mathbf{V}(N)}(\textstyle\sum_{i \in N} x_i) \\ s.t.: \quad & F_{\mathbf{V}(S)}(\textstyle\sum_{i \in S} x_i) \geq \alpha(S), \text{for all } S \subset N \\ & F_{\mathbf{V}(N)}(\textstyle\sum_{i \in N} x_i) \geq \bar{\alpha}. \end{aligned}$$

1.2.2 Stochastic Cooperative Games with Transfer Payments

A stochastic cooperative game with transfer payments is described by a tuple $(N, \mathcal{V}, \{\succsim_i\}_{i \in N})$, where $N = \{1, 2, \ldots, n\}$ is the set of agents, $\mathcal{V}$ a map assigning to each nonempty coalition S a collection $\mathcal{V}(S) \subset L^1(\mathbb{R}_+)$ of stochastic payoffs, and $\succsim_i$ the preference relation of agent i over the set $L^1(\mathbb{R})$ of random payoffs with finite expectation. In particular, it is assumed that the random payoffs are expressed in some infinitely divisible commodity like money. Benefits consisting of several different or indivisible commodities are excluded. The interpretation of $\mathcal{V}(S)$ is that each random payoff $\boldsymbol{X} \in \mathcal{V}(S)$ represents the benefit that results from one of the several actions that coalition S has at its disposal when cooperating.

Let $S \subset N$ and let $\boldsymbol{X} \in \mathcal{V}(S)$ be a stochastic payoff for coalition S. An allocation of $\boldsymbol{X}$ is represented by a pair $(d, r) \in \mathbb{R}^S \times \mathbb{R}^S$ with $\sum_{i \in S} d_i \leq 0$, $\sum_{i \in S} r_i = 1$, and $r_i \geq 0$ for all $i \in S$. Given a pair (d, r), agent $i \in S$ then receives the random payoff $d_i + r_i\boldsymbol{X}$. So, an allocation consists of two parts. The first part represents deterministic transfer payments between the agents in S. Note that the $\leq$-sign allows the agents to discard some of the money. The second part allocates a fraction of the random payoff to each agent in S. The class of stochastic cooperative games with agent set N is denoted by $SG(N)$ and its elements are denoted by Γ. Furthermore, $\mathcal{Z}_\Gamma(S)$ denotes the set of allocations that coalition S can obtain in $\Gamma \in SG(N)$, that is

$$\mathcal{Z}_\Gamma(S) = \{(d_i + r_i\boldsymbol{X})_{i \in S} |\, \boldsymbol{X} \in \mathcal{V}(S), (d, r) \in H^S \times \Delta^S\}, \tag{1.3}$$

where $H^S = \{d \in \mathbb{R}^S |\, \sum_{i \in S} d_i \leq 0\}$ and $\Delta^S = \{r \in \mathbb{R}^S_+ |\, \sum_{i \in S} r_i = 1\}$.

A core allocation is an allocation such that no coalition has an incentive to part company with the grand coalition because they can do better on their own. The core of a stochastic cooperative game $\Gamma \in SG(N)$ is thus defined by

$$\begin{aligned} \mathcal{C}(\Gamma) = \{&(d_i + r_i\boldsymbol{X})_{i \in N} \in \mathcal{Z}_\Gamma(N) | \\ &\forall_{S \subset N}\, \nexists_{(\hat{d}_i + \hat{r}_i\hat{\boldsymbol{X}})_{i \in S} \in \mathcal{Z}_\Gamma(S)} \forall_{i \in S} : \hat{d}_i + \hat{r}_i\hat{\boldsymbol{X}} \succ_i d_i + r_i\boldsymbol{X}\Big\}. \end{aligned} \tag{1.4}$$

A stochastic cooperative game $\Gamma \in SG(N)$ with a nonempty core is called balanced. Furthermore, if the core of every subgame $\Gamma_{|S}$ is nonempty, then Γ is called totally balanced. The subgame $\Gamma_{|S}$ is given by $(S, \mathcal{V}_{|S}, \{\succsim_i\}_{i \in S})$ where $\mathcal{V}_{|S}(U) = \mathcal{V}(U)$ for all $U \subset S$.

The core of a stochastic cooperative game can be empty. Necessary and sufficient conditions for nonemptiness of the core only exist for a specific subclass of stochastic cooperative games that we wil discuss below; they are still unknown for the general case.

Let $\Gamma \in SG(N)$ be a stochastic cooperative game with preferences $\{\succsim_i\}_{i \in N}$ such that for each $i \in N$ there exists a function $m_i : L^1(\mathbb{R}) \to \mathbb{R}$ satisfying

(M1) for all $\boldsymbol{X}, \boldsymbol{Y} \in L^1(\mathbb{R}) : \boldsymbol{X} \succsim_i \boldsymbol{Y}$ if and only if $m_i(\boldsymbol{X}) \geq m_i(\boldsymbol{Y})$;

(M2) for all $\boldsymbol{X} \in L^1(\mathbb{R})$ and all $d \in \mathbb{R}$: $m_i(d + \boldsymbol{X}) = d + m_i(\boldsymbol{X})$.

The interpretation is that $m_i(\boldsymbol{X})$ equals the amount of money for which agent i is indifferent between receiving the amount $m_i(\boldsymbol{X})$ with certainty and receiving the stochastic payoff $\boldsymbol{X}$, i.e. $\boldsymbol{X} \sim_i m_i(\boldsymbol{X})$. The amount $m_i(\boldsymbol{X})$ is called the certainty equivalent of $\boldsymbol{X}$. Condition (M1) states that agent i weakly prefers one stochastic payoff to another one if and only if the certainty equivalent of the former is greater than or equal to the certainty equivalent of the latter. Condition (M2) states that the certainty equivalent is linearly separable in the deterministic amount of money d.

Example 1.3 Let the preference $\succsim_\alpha$ with $\alpha \in (0,1)$ be such that for $\boldsymbol{X}, \boldsymbol{Y} \in L^1(\mathbb{R})$ it holds that $\boldsymbol{X} \succsim_\alpha \boldsymbol{Y}$ if $\xi_\alpha(\boldsymbol{X}) \geq \xi_\alpha(\boldsymbol{Y})$. With the certainty equivalent of $\boldsymbol{X} \in L^1(\mathbb{R})$ given by $m(\boldsymbol{X}) = \xi_\alpha(\boldsymbol{X})$, the conditions (M1) and (M2) are satisfied. That (M1) is fulfilled is straightforward. For (M2), note that

$$\begin{aligned}
\xi_\alpha(d + \boldsymbol{X}) &= \sup\{t \in \mathbb{R} |\ \mathbf{P}(\{d + \boldsymbol{X} < t\}) \leq \alpha\} \\
&= \sup\{t \in \mathbb{R} |\ \mathbf{P}(\{\boldsymbol{X} < t - d\}) \leq \alpha\} \\
&= \sup\{t + d \in \mathbb{R} |\ \mathbf{P}(\{\boldsymbol{X} < t\}) \leq \alpha\} \\
&= d + \sup\{t \in \mathbb{R} |\ \mathbf{P}(\{\boldsymbol{X} < t\}) \leq \alpha\} \ = \ d + \xi_\alpha(\boldsymbol{X}),
\end{aligned}$$

for all $d \in \mathbb{R}$ and all $\boldsymbol{X} \in L^1(\mathbb{R})$. Hence, $m(d + \boldsymbol{X}) = d + m(\boldsymbol{X})$. ◁

Let $\Gamma \in SG(N)$ be a stochastic cooperative game satisfying conditions (M1) and (M2). Take $S \subset N$. An allocation $(d_i + r_i\boldsymbol{X})_{i \in S} \in \mathcal{Z}_\Gamma(S)$ is Pareto optimal for coalition S if there exists no allocation $(\hat{d}_i + \hat{r}_i\hat{\boldsymbol{X}})_{i \in S} \in \mathcal{Z}_\Gamma(S)$ such that $\hat{d}_i + \hat{r}_i\hat{\boldsymbol{X}} \succ_i d_i + r_i\boldsymbol{X}$ for all $i \in S$. Pareto optimal allocations are characterized by the following proposition, which is due to Suijs and Borm (1999).

Proposition 1.4 *Let $\Gamma \in SG(N)$ satisfy conditions (M1) and (M2). Then $(d_i + r_i\boldsymbol{X})_{i \in S} \in \mathcal{Z}_\Gamma(S)$ is Pareto optimal if and only if*

$$\sum_{i \in S} m_i(d_i + r_i\boldsymbol{X}) \ = \ \max\left\{\sum_{i \in S} m_i(\hat{d}_i + \hat{r}_i\hat{\boldsymbol{X}}) |\ (\hat{d}_i + \hat{r}_i\hat{\boldsymbol{X}})_{i \in S} \in \mathcal{Z}_\Gamma(S)\right\} \tag{1.5}$$

For interpreting condition (1.5), consider a particular allocation for coalition S and let each member pay the certainty equivalent of the random payoff he receives. Acting in this way, the initial wealth of each member does not change and, instead of the random payoff, the coalition now has to divide the certainty equivalents that have been paid by its members. Since the preferences are strictly increasing in the deterministic amount of money one receives, the more money a coalition can divide, the better it is for all its members. So, the best way to allocate the random benefits, is to maximize the sum of the certainty equivalents. Furthermore, we can describe the random benefits of each coalition by the maximum sum of the certainty equivalents they can obtain, provided that this maximum exists, of course. This follows from the fact that for each $S \subset N$ it holds that

$$\begin{aligned}\{(m_i(d_i + r_i\mathbf{X}))_{i\in S} |\ (d_i + r_i\mathbf{X})_{i\in S} \in \mathcal{Z}_\Gamma(S)\} &= \\ &\{x \in I\!R^S |\ \textstyle\sum_{i\in S} x_i \leq v_\Gamma(S)\},\end{aligned} \tag{1.6}$$

where

$$v_\Gamma(S) = \max\left\{\sum_{i\in S} m_i(\hat{d}_i + \hat{r}_i\hat{\mathbf{X}})|\ (\hat{d}_i + \hat{r}_i\hat{\mathbf{X}})_{i\in S} \in \mathcal{Z}_\Gamma(S)\right\}. \tag{1.7}$$

Expression (1.6) means that it does not matter for coalition S whether they allocate a random payoff $\mathbf{X} \in \mathcal{V}(S)$ or the deterministic amount $v_\Gamma(S)$. To see that this equality does indeed hold, note that the inclusion '$\subset$' follows immediately from the definition of $v_\Gamma(S)$. For the reverse inclusion '$\supset$', let $y \in \{x \in I\!R^S |\ \sum_{i\in S} x_i \leq v_\Gamma(S)\}$. Next, let $(d_i + r_i\mathbf{X})_{i\in S} \in \mathcal{Z}_\Gamma(S)$ be such that $\sum_{i\in S} m_i(d_i + r_i\mathbf{X}) = v_\Gamma(S)$ and define $\delta_i = y_i - m_i(r_i\mathbf{X})$ for each $i \in S$. Since $\sum_{i\in S} \delta_i \leq 0$ and $m_i(\delta_i + r_i\mathbf{X}) = \delta_i + m_i(r_i\mathbf{X}) = y_i$ for all $i \in S$ it holds that $y \in \{(m_i(d_i + r_i\mathbf{X}))_{i\in S} |\ (d_i + r_i\mathbf{X})_{i\in S} \in \mathcal{Z}_\Gamma(S)\}$.

So, if for a stochastic cooperative game $\Gamma \in SG(N)$ the value $v_\Gamma(S)$ is well defined for each coalition $S \subset N$, we can also describe the game Γ by a TU-game (N, v_Γ) with $v_\Gamma(S)$ as in (1.7) for all $S \subset N$. Let $MG(N) \subset SG(N)$ denote the class of stochastic cooperative games for which the conditions (M1) and (M2) are satisfied and the game (N, v_Γ) is well defined. The following theorem is taken from Suijs and Borm (1999).

Theorem 1.5 *Let $\Gamma \in MG(N)$. If $(d_i + r_i\mathbf{X})_{i\in N} \in \mathcal{Z}_\Gamma(N)$ and $x \in I\!R^N$ are such that $m_i(d_i + r_i\mathbf{X}) = x_i$ for all $i \in N$, then $(d_i + r_i\mathbf{X})_{i\in N} \in C(\Gamma)$ if and only if $x \in C(v_\Gamma)$.*

An immediate consequence of this result is that for each $\Gamma \in MG(N)$ it holds true that $\mathcal{C}(\Gamma) = \emptyset$ if and only if $C(v_\Gamma) = \emptyset$. Furthermore, to check the nonemptiness of $C(v_\Gamma)$ we can rely on the well-known theorem by Bondareva (1963) and Shapley (1967).

1.2.3 Stochastic Cooperative Games without Transfer Payments

Stochastic cooperative games without transfer payments are introduced in Timmer *et al.* (2000). We denote the class of these games by $SG^*(N)$. Its description is similar to the one that includes transfer payments except for the following assumptions.

First, it is assumed that cooperation by coalition S generates a single, nonnegative random payoff $\boldsymbol{X}_S \in L^1(\mathbb{R}_+)$, i.e. $\mathcal{V}(S) = \{\boldsymbol{X}_S\}$ for all $S \subset N$. So, in this respect, the model follows the structure of chance-constrained games.

Second, deterministic transfer payments between agents are not allowed. As a result, allocations of the random payoff $\boldsymbol{X}_S$ are described by a vector $r \in \mathbb{R}^S_+$ such that $\sum_{i \in S} r_i = 1$.[1] The set of feasible allocations for coalition $S \subset N$ is thus described by $\mathcal{Z}^*_\Gamma(S) = \{(r_i \boldsymbol{X}_S)_{i \in S} | r \in \mathbb{R}^S_+ : \sum_{i \in S} r_i = 1\}$. Note that in the absence of deterministic transfer payments insurance is not possible.

Finally, the transitive and complete preferences $\succsim_i$ satisfy the following conditions

(P1) for any $\boldsymbol{X} \in L^1(\mathbb{R}_+)$ and $r, \hat{r} \in \mathbb{R}_+$ it holds that $r\boldsymbol{X} \succsim_i \hat{r}\boldsymbol{X}$ if and only if $r \geq \hat{r}$.

(P2) for all $\boldsymbol{X}, \boldsymbol{Y} \in L^1(\mathbb{R}_+)$ with $r\boldsymbol{Y} \succ_i \boldsymbol{X} \succ_i \hat{r}\boldsymbol{Y}$ for some $r, \hat{r} \in \mathbb{R}_+$ there exists $p \in \mathbb{R}_+$ such that $\boldsymbol{X} \sim_i p\boldsymbol{Y}$.

Condition (P1) is implied by first order stochastic dominance which is generally accepted to be satisfied for a rationally behaving individual. Condition (P2) is a kind of continuity condition.

The core of a stochastic cooperative game without transfer payments is defined in the usual way, that is it contains all allocations that induce stable cooperation of the grand coalition. Formally, the core of a stochastic cooperative game without transfer payments $\Gamma \in SG^*(N)$ is given

[1] Timmer *et al.* (2000) do not require nonnegativity of r. However, since this paper focuses on the core only, the nonnegativity conditions on r impose no additional restrictions as nonnegativity also follows from individual rationality and $\boldsymbol{X} \in L^1(\mathbb{R}_+)$.

by

$$\mathcal{C}^*(\Gamma) = \{(r_i\boldsymbol{X}_N)_{i\in N} \in \mathcal{Z}^*_\Gamma(N) | \forall_{S\subset N} \nexists_{(\hat{r}_i\boldsymbol{X}_S)_{i\in S}\in\mathcal{Z}^*_\Gamma(S)} \forall_{i\in S} : \hat{r}_i\boldsymbol{X}_S \succ_i r_i\boldsymbol{X}_N\}. \tag{1.8}$$

We can provide necessary and sufficient conditions for nonemptiness of the core for a specific subclass of stochastic cooperative games without transfer payments. To describe this subclass, let us take a closer look at the individuals' preferences.

Let $\Gamma \in SG^*(N)$ be a stochastic cooperative game without transfer payments. Define $\overline{\mathbb{R}} = \mathbb{R} \cup \{\infty\}$. Since the preference relation $\succsim_i$ satisfies conditions (P1) and (P2), there exists a function $f^i : \mathbb{R} \to \overline{\mathbb{R}}^{2^N}$ such that $f^i(0) = 0$ and, if $f^i_S(t), f^i_U(\hat{t}) < \infty$ then

$$f^i_S(t)\boldsymbol{X}_S \succsim_i f^i_U(\hat{t})\boldsymbol{X}_U \text{ if and only if } t \geq \hat{t}, \tag{1.9}$$

for any $S, U \subset N$. The proof is straightforward. Take $i \in N$. Let $f^i_{\{i\}} : \mathbb{R} \to \mathbb{R}$ be a strictly increasing continuous function with $f^i_{\{i\}}(0) = 0$. For each $S \subset N$, define $f^i_S(t) \in \mathbb{R}$ such that $f^i_S(t)\boldsymbol{X}_S \sim_i f^i_{\{i\}}(t)\boldsymbol{X}_{\{i\}}$ and $r\boldsymbol{X}_S \prec_i f^i_{\{i\}}(t)\boldsymbol{X}_{\{i\}}$ for all $r < f^i_S(t)$.[2] If such $f^i_S(t)$ does not exist, define $f^i_S(t) = \infty$. Note that (P1) implies that $f^i_S(\hat{t}) = \infty$ if $\hat{t} \geq t$ and $f^i_S(t) = \infty$. Furthermore, note that $f^i_S(t)$ is strictly increasing in t. Transitivity implies for any $S, U \subset N$ that $f^i_S(t)\boldsymbol{X}_S \sim_i f^i_U(t)\boldsymbol{X}_U$ if $f^i_S(t), f^i_U(t) < \infty$, so that (1.9) follows from the monotonicity condition (P1).

Example 1.6 Consider the quantile preferences as presented in Example 1.3, that is, $\boldsymbol{X} \succsim \boldsymbol{Y}$ if and only if $\xi_\alpha(\boldsymbol{X}) \geq \xi_\alpha(\boldsymbol{Y})$. These preferences satisfy conditions (P1) and (P2). In addition, f can be defined as follows. Let $\boldsymbol{X}_S \in L^1(\mathbb{R}_+)$, $S \subset N$, denote the nonzero random payoffs. For all $t \in \mathbb{R}$ and all $S \subset N$, define $f_S(t) = \frac{t}{\xi_\alpha(\boldsymbol{X}_S)}$ if $\xi_\alpha(\boldsymbol{X}_S) \neq 0$ and $f_S(t) = \infty$ otherwise. Since $\xi_\alpha(f_S(t)\boldsymbol{X}_S) = f_S(t)\xi_\alpha(\boldsymbol{X}_S) = t$ it holds that if $f_S(t), f_U(\hat{t}) < \infty$ then $f_S(t)\boldsymbol{X}_S \succsim f_U(\hat{t})\boldsymbol{X}_U$ if and only if $t \geq \hat{t}$.$\triangleleft$

The subclass $MG^*(N) \subset SG^*(N)$ of stochastic cooperative games without transfer payments is defined as follows. For each $\Gamma \in MG^*(N)$ the preferences $\{\succsim_i\}_{i\in N}$ are such that for each $i \in N$ the function $f^i : \mathbb{R} \to \overline{\mathbb{R}}^{2^N}$ is linear, that is $f^i(at) = af^i(t)$ for all $a, t \in \mathbb{R}$. This implies the following property for $\succsim_i$:

[2] The latter condition guarantees that $f^i_S(t)$ is unique and that $f^i_S(0) = 0$.

(P3) if $\boldsymbol{X} \succsim_i \boldsymbol{Y}$, then $a\boldsymbol{X} \succsim_i a\boldsymbol{Y}$ for any $a \in I\!R_+$.

This property states that in a way only the relative dispersion of the outcomes of a random payoff is taken into account. Note that this property is different from (M2), which states that preferences over random payoffs are independent of one's initial wealth. We will illustrate this difference in the following example.

Example 1.7 For simplicity, let us restrict our attention to nonnegative random payoffs defined on two states of the world that occur with equal probability. So, we can represent a random payoff $\boldsymbol{X}$ by a vector (x_1, x_2) with $\mathbb{P}(\boldsymbol{X} = x_i) = 0.5$, $i = 1, 2$.

First, consider the preference relation based on the utility function $U(x) = -e^{-x}$, that is $\boldsymbol{X} \succsim \boldsymbol{Y}$ if and only if $E(U(\boldsymbol{X})) \geq E(U(\boldsymbol{Y}))$. The corresponding certainty equivalent is defined by $m(\boldsymbol{X}) = U^{-1}(E(U(\boldsymbol{X})))$. It is a straightforward exercise to show that (M2) is satisfied, i.e. $m(d+\boldsymbol{X}) = d + m(\boldsymbol{X})$. These preferences, however, violate (P3). To see this, consider the random payoff $(0, 1)$ that pays 0 with probability 0.5 and 1 with probability 0.5. The certainty equivalent of this random payoff equals $-\log(0.5e^{-0} + 0.5e^{-1}) \approx 0.38$. So, this individual is indifferent between receiving the random payoff and receiving 0.38 with certainty. Now, consider a multiple of this lottery, e.g. $(0, 3)$, then the individual should be indifferent between receiving the lottery and receiving $3 \cdot 0.38 = 1.14$. However, the certainty equivalent for this lottery only equals $-\log(0.5e^{-0} + 0.5e^{-3}) \approx 0.64$.

Next, consider the preferences based on the utility function $U(\boldsymbol{X}) = \sqrt{x_1 x_2}$, where x_1 and x_2 are the respective outcomes of $\boldsymbol{X}$. Then $\boldsymbol{X} \succsim \boldsymbol{Y}$ if and only if $U(\boldsymbol{X}) = \sqrt{x_1 x_2} \geq \sqrt{y_1 y_2} = U(\boldsymbol{Y})$. Note that since $U(a\boldsymbol{X}) = aU(\boldsymbol{X})$ it holds that $a\boldsymbol{X} \succsim a\boldsymbol{Y}$ if $\boldsymbol{X} \succsim \boldsymbol{Y}$, that is (P3) is satisfied. The certainty equivalent can be defined by $m(\boldsymbol{X}) = \sqrt{x_1 x_2}$ because $U(m(\boldsymbol{X})) = \sqrt{m(\boldsymbol{X}) m(\boldsymbol{X})} = m(\boldsymbol{X}) = \sqrt{x_1 x_2} = U(\boldsymbol{X})$. This certainty equivalent violates condition (M2). To see this, consider the lottery $(1, 4)$. The corresponding certainty equivalent equals 2. So, if (M2) would be satisfied, this individual should be indifferent between the lottery $(3, 6)$ (i.e. the lottery $(1, 4)$ increased with 2) and its certainty equivalent 4. However, the certainty equivalent of the lottery $(3, 6)$ equals 4.24.

Finally, note that preferences based on quantiles (see Example 1.3 and Example 1.6) satisfy both (M2) and (P3). ◁

For the class $MG^*(N)$ we can derive necessary and sufficient conditions for the core to be nonempty. For this purpose, we introduce some notation. Take $\boldsymbol{X}, \boldsymbol{Y} \in L^1(\mathbb{R}_+)$, $i \in N$ and let $\rho_i(\boldsymbol{X}, \boldsymbol{Y}) \in \mathbb{R}$ be such that $\rho_i(\boldsymbol{X}, \boldsymbol{Y})\boldsymbol{X} \sim_i \boldsymbol{Y}$. For the moment, assume that such $\rho_i(\boldsymbol{X}, \boldsymbol{Y})$ exists. So, $\rho_i(\boldsymbol{X}, \boldsymbol{Y})$ is that multiple of the random payoff $\boldsymbol{X}$ for which agent $i \in N$ is indifferent between $\rho_i(\boldsymbol{X}, \boldsymbol{Y})\boldsymbol{X}$ and $\boldsymbol{Y}$. Since $f_S^i(t)\boldsymbol{X}_S \sim_i f_U^i(t)\boldsymbol{X}_U$ we obtain for $t = (f_U^i)^{-1}(1)$ that $f_S^i(t)\boldsymbol{X}_S \sim_i \boldsymbol{X}_U$. Hence, $\rho_i(\boldsymbol{X}_S, \boldsymbol{X}_U) = f_S^i((f_U^i)^{-1}(1))$. Furthermore, since f^i is linear it follows that

$$\begin{aligned} \rho_i(p\boldsymbol{X}_S, q\boldsymbol{X}_U) &= \frac{f_S^i((f_U^i)^{-1}(q))}{p} \\ &= \frac{q}{p} f_S^i((f_U^i)^{-1}(1)) = \frac{q}{p}\rho_i(\boldsymbol{X}_S, \boldsymbol{X}_U) \end{aligned} \tag{1.10}$$

for all $p, q \geq 0$. If $\rho_i(\boldsymbol{X}, \boldsymbol{Y})$ does not exist, define $\rho_i(\boldsymbol{X}, \boldsymbol{Y}) = \infty$.

Recall that the core of a stochastic cooperative game $\Gamma \in MG^*(N)$ is given by expression (1.8). Consider an allocation $(r_i\boldsymbol{X}_N)_{i\in N} \in \mathcal{Z}_\Gamma^*(N)$ and suppose that coalition $S \subset N$ does not agree with this allocation because they can do better on their own. This means that there exists an allocation $(\hat{r}_i\boldsymbol{X}_S)_{i\in S} \in \mathcal{Z}_\Gamma^*(S)$ such that $\hat{r}_i\boldsymbol{X}_S \succ_i r_i\boldsymbol{X}_N$ for all $i \in S$. If $\rho_i(\boldsymbol{X}_S, r_i\boldsymbol{X}_N) = \infty$ then agent i prefers $r_i\boldsymbol{X}_N$ to any multiple of $\boldsymbol{X}_S$. Hence, coalition S cannot strictly improve the payoff of agent $i \in S$. Consequently, coalition S has no incentive to part company with the grand coalition N. Furthermore, since $r_i \leq 1$ implies that $\boldsymbol{X}_N \succsim_i r_i\boldsymbol{X}_N$ so that $\rho_i(\boldsymbol{X}_S, \boldsymbol{X}_N) \geq \rho_i(\boldsymbol{X}_S, r_i\boldsymbol{X}_N) = \infty$, it follows that each coalition $S \notin \mathcal{N} = \{S \subset N | \forall_{i\in S} : \rho_i(\boldsymbol{X}_S, \boldsymbol{X}_N) < \infty\}$ will stay in the grand coalition N. If $\rho_i(\boldsymbol{X}_S, r_i\boldsymbol{X}_N) < \infty$, we know that that $\rho_i(\boldsymbol{X}_S, r_i\boldsymbol{X}_N)\boldsymbol{X}_S \sim_i r_i\boldsymbol{X}_N$. Then (P1) and (P3) imply that $\hat{r}_i > \rho_i(\boldsymbol{X}_S, r_i\boldsymbol{X}_N) = r_i\rho_i(\boldsymbol{X}_S, \boldsymbol{X}_N)$. Since $(\hat{r}_i\boldsymbol{X}_S)_{i\in S}$ is a feasible allocation it holds that $1 = \sum_{i\in S}\hat{r}_i > \sum_{i\in S} r_i\rho_i(\boldsymbol{X}_S, \boldsymbol{X}_N)$. Hence, coalition S cannot construct a better allocation if and only if $\sum_{i\in S} r_i\rho_i(\boldsymbol{X}_S, \boldsymbol{X}_N) \geq 1$. Consequently, we have that

$$\mathcal{C}^*(\Gamma) = \left\{ (r_i\boldsymbol{X}_N)_{i\in N} \in \mathcal{Z}_\Gamma^*(N) \,\middle|\, \forall_{S\in\mathcal{N}} : \sum_{i\in S} r_i\rho_i(\boldsymbol{X}_S, \boldsymbol{X}_N) \geq 1 \right\}. \tag{1.11}$$

Nonemptiness of the core is characterized by the following theorem. Its proof is stated in the Appendix.

Theorem 1.8 *Let $\Gamma \in MG^*(N)$. Then the core $\mathcal{C}^*(\Gamma) \neq \emptyset$ if and only if the following statement is true: for all $\lambda : \mathcal{N} \to \mathbb{R}_+$ such that $\sum_{S\in\mathcal{N}: i\in S}\lambda(S)\rho_i(\boldsymbol{X}_S, \boldsymbol{X}_N) \leq 1$ for each $i \in N$ it holds true that $\sum_{S\in\mathcal{N}}\lambda(S) \leq 1$.*

Note that for the deterministic case, that is TU-games, this condition is similar to Bondareva (1963) and Shapley (1967). This follows immediately from the fact that $\rho_i(v(S), v(N)) = \frac{v(N)}{v(S)}$ for TU-games and substituting $\hat{\lambda}(S) = \lambda(S)\frac{v(N)}{v(S)}$ for each $S \subset N$.

In the next two sections, we will apply the three different theoretical models to two specific situations, namely cost allocation in network trees and bankruptcy situations.

1.3 Cost Allocation in a Network Tree

With the geographical dispersion of customers and service providers the need arises for a communication network that connects the provider to its clients. Typical examples one can think of in this context are cities that are connected to the local powerplant or computer workstations that are connected to a central server. Megiddo (1978) considered this cost allocation problem in a deterministic setting, which we discuss first.

Let N denote the finite set of agents that are connected through a network with a single service provider denoted by $\star$. It is assumed that the network is a tree. This assumption is derived from Claus and Kleitman (1973), which also considers the construction problem of such networks. Using the total construction costs as the objective, it is shown that a minimum cost network is a tree. The network is represented by $T \subset \{\{k,l\}|k,l \in N \cup \{\star\}\}$ with the interpretation that the link $\{k,l\}$ between agents k and l exists if and only if $\{k,l\} \in T$. The cost of each link $\{k,l\} \in T$ is denoted by $\kappa_{\{k,l\}} \geq 0$. Furthermore, let $T_i \subset T$ be the path that connects agent i to the source $\star$. Note that since T is a tree this path T_i is unique. The corresponding fixed tree game (N, v) is defined as $v(S) = -\sum_{\{k,l\} \in \cup_{i \in S} T_i} \kappa_{\{k,l\}}$ for all $S \subset N$. So, $v(S)$ represents the total cost of all the links that the members of S use to reach the source. Megiddo (1978) showed that these games are totally balanced and that the Shapley value belongs to the core.

Given an existing communication network tree, the cost of each link may be considered to consist of two parts, namely the construction cost and the maintenance cost. Observe that the latter is generally random of nature as one does not know up front when connections are going to fail and what the resulting cost of repair will be. In this regard, random variables are more appropriate to describe the maintenance cost of each link. So, let $\boldsymbol{X}_{\{k,l\}} \in L^1(\mathbb{R}_+)$ denote the random cost of the link $\{k,l\} \in T$ and assume that these costs are mutually independent.

In the remainder of this section we model the cost allocation problem as a chance-constrained game, stochastic cooperative game with transfer payments, and a stochastic cooperative game without transfer payments, respectively.

Recall that a chance-constrained game is denoted by $(N, \boldsymbol{V})$ with $\boldsymbol{V}(S) \in L^1(I\!R)$ representing the random payoff to coalition S. In this case, $\boldsymbol{V}(S)$ equals the total random costs of the links that the members of S use to reach the source, that is $\boldsymbol{V}(S) = -\sum_{\{k,l\}\in\cup_{i\in S}T_i} \boldsymbol{X}_{\{k,l\}}$ for each $S \subset N$. As the following example shows, the prior core of such chance-constrained games can be empty.

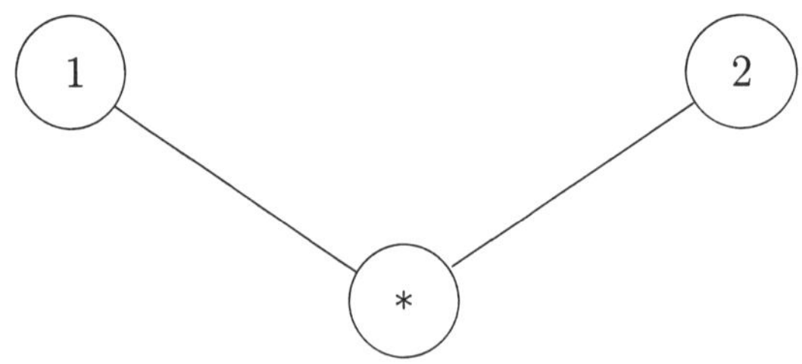

Figure 1.1: *A communication tree with two agents.*

Example 1.9 Consider the tree illustrated in Figure 1.1. There are only two agents and each agent has a direct connection to the source. The random costs $\boldsymbol{X}_{\{1,*\}}$ are uniformly distributed on $(0,1)$ and the random costs $\boldsymbol{X}_{\{2,*\}}$ are exponentially distributed with mean 1. Let the levels of assurance be $\alpha(\{1\}) = \alpha(\{2\}) = \bar{\alpha} = \alpha$ and $\underline{\alpha} < \alpha$. For an allocation (x_1, x_2) to belong to the prior core of the game, it must hold that $F_{\boldsymbol{V}(\{1\})}(x_1) \geq \alpha$, $F_{\boldsymbol{V}(\{2\})} \geq \alpha$, and $\underline{\alpha} \leq F_{\boldsymbol{V}(\{1,2\})}(x_1 + x_2) \leq \alpha$, with

$$F_{\boldsymbol{V}(\{1,2\})}(x) = \begin{cases} e^x(e-1), & \text{if } x \leq -1, \\ 2 + x - e^x, & \text{if } -1 < x \leq 0, \\ 1, & \text{if } x > 0, \end{cases}$$

the probability distribution function of $\boldsymbol{V}(\{1,2\}) = -\boldsymbol{X}_{\{1,*\}} - \boldsymbol{X}_{\{2,*\}}$. This implies that $x_1 \geq \xi_\alpha(\boldsymbol{V}(\{1\}))$, $x_2 \geq \xi_\alpha(\boldsymbol{V}(\{2\}))$, and $x_1 + x_2 \leq \xi_\alpha(\boldsymbol{V}(\{1,2\}))$. Hence, the prior core is empty if $\xi_\alpha(\boldsymbol{V}(\{1\})) + \xi_\alpha(\boldsymbol{V}(\{2\})) > \xi_\alpha(\boldsymbol{V}(\{1,2\}))$. As one can see in Figure 1.2, this is the case for $\alpha \in (0.4587, 1)$. $\triangleleft$

A stochastic cooperative game with transfer payments is described by a tuple $\Gamma = (N, \mathcal{V}, \{\succsim_i\}_{i\in N})$, with $\mathcal{V}(S) = \{\boldsymbol{V}(S)\}$ the total cost

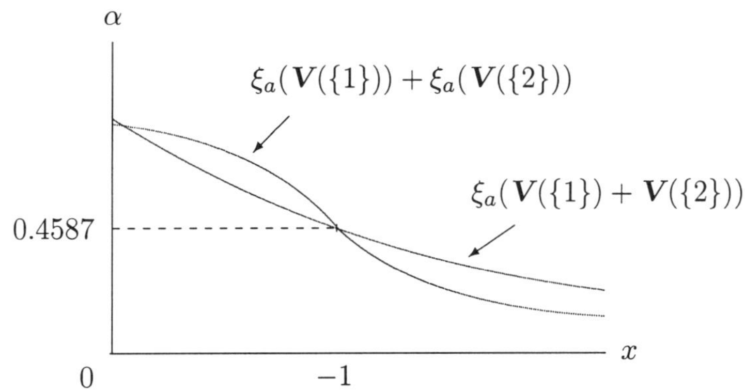

Figure 1.2: *The quantiles* $\xi_a(\boldsymbol{V}(\{1\})) + \xi_a(\boldsymbol{V}(\{2\}))$ *and* $\xi_\alpha(\boldsymbol{V}(\{1\}) + \boldsymbol{V}(\{2\}))$ *in relation to* α.

of the links that coalition S uses. In addition, we have to specify the preferences of the agents. We assume that the preference relation $\succsim_i$ of agent $i \in N$ can be represented by the α_i-quantile, that is $\boldsymbol{X} \succsim_i \boldsymbol{Y}$ if and only if $\xi_{\alpha_i}(\boldsymbol{X}) \geq \xi_{\alpha_i}(\boldsymbol{Y})$. Note that these preferences satisfy the properties (M1)–(M2). Hence, $\Gamma \in MG(N)$. The corresponding certainty equivalent equals $m_i(\boldsymbol{X}) = \xi_{\alpha_i}(\boldsymbol{X})$ so that

$$v_\Gamma(S) = \max \left\{ \sum_{i \in S} \xi_{\alpha_i}(d_i + r_i \boldsymbol{V}(S)) \,\middle|\, d \in H^S, r \in \Delta^S \right\} \tag{1.12}$$

for all $S \subset N$. Since $\xi_{\alpha_i}(d_i + r_i \boldsymbol{V}(S)) = d_i + r_i \xi_{\alpha_i}(\boldsymbol{V}(S))$ and $\sum_{i \in S} d_i \leq 0$ it easily follows that $v_\Gamma(S) = \max_{i \in S} \xi_{\alpha_i}(\boldsymbol{V}(S))$. Hence, a Pareto optimal allocation allocates the random costs $\boldsymbol{V}(S)$ to the member of S with the highest α_i, that is the agent that, relatively speaking, looks at the most optimistic outcomes of $\boldsymbol{V}(S)$.

From Theorem 1.5 we know that the core $\mathcal{C}(\Gamma)$ of the stochastic cooperative game is nonempty if and only if the core $C(v_\Gamma)$ of the corresponding TU-game (N, v_Γ) is nonempty. The following example shows that in this case the core can be empty.

Example 1.10 Consider the network allocation problem presented in Example 1.9. Let $\alpha_1 = \alpha_2 = \alpha$. Since $v_\Gamma(\{1\}) = \xi_\alpha(\boldsymbol{V}(\{1\}))$, $v_\Gamma(\{2\}) = \xi_\alpha(\boldsymbol{V}(\{2\}))$ and $v_\Gamma(\{1,2\}) = \xi_\alpha(\boldsymbol{V}(\{1\}) + \boldsymbol{V}(\{2\}))$ we have that the core of the game (N, v_Γ) is empty if and only if $\xi_\alpha(\boldsymbol{V}(\{1\})) + \xi_\alpha(\boldsymbol{V}(\{2\})) > \xi_\alpha(\boldsymbol{V}(\{1,2\}))$. From Figure 1.2 we know that this holds true for $\alpha \in (0.4587, 1)$. ◁

Finally, let us turn to stochastic cooperative games without transfer payments. Again, let us confine our attention to agents whose preferences can be represented by quantiles. Since these preferences satisfy conditions (P1)–(P3), we know that the core is given by expression (1.11). Recall that $\rho_i(\boldsymbol{X}, \boldsymbol{Y})$ is such that $\rho_i(\boldsymbol{X}, \boldsymbol{Y})\boldsymbol{X} \sim_i \boldsymbol{Y}$ if $\rho_i(\boldsymbol{X}, \boldsymbol{Y}) < \infty$. Since the latter holds if and only if $\xi_{\alpha_i}(\rho_i(\boldsymbol{X}, \boldsymbol{Y})\boldsymbol{X}) = \xi_{\alpha_i}(\boldsymbol{Y})$, it follows from $\xi_{\alpha_i}(\rho_i(\boldsymbol{X}, \boldsymbol{Y})\boldsymbol{X}) = \rho_i(\boldsymbol{X}, \boldsymbol{Y})\xi_{\alpha_i}(\boldsymbol{X})$ that $\rho_i(\boldsymbol{X}, \boldsymbol{Y}) = \xi_{\alpha_i}(\boldsymbol{Y})/\xi_{\alpha_i}(\boldsymbol{X})$. Hence,

$$\mathcal{C}^*(\Gamma) = \left\{ (r_i \boldsymbol{V}(N))_{i \in N} \in \mathcal{Z}^*_\Gamma(N) \,\middle|\, \forall_{S \in \mathcal{N}} : \sum_{i \in S} r_i \frac{\xi_{\alpha_i}(\boldsymbol{V}(N))}{\xi_{\alpha_i}(\boldsymbol{V}(S))} \geq 1 \right\}.$$

Notice that since $S \in \mathcal{N}$ implies $\rho_i(\boldsymbol{V}(S), \boldsymbol{V}(N)) < \infty$ for all $i \in S$, it follows that $\xi_{\alpha_i}(\boldsymbol{V}(S)) > 0$.

Example 1.11 Consider again the network allocation problem of Example 1.9. Let $\alpha_1 = \alpha_2 = \alpha$. A core allocation $(r_1 \boldsymbol{V}(\{1,2\}), r_2 \boldsymbol{V}(\{1,2\}))$ satsfies the condition
$r_i \xi_\alpha(\boldsymbol{V}(\{1,2\}))/\xi_\alpha(\boldsymbol{V}(\{i\})) \geq 1$ for $i = 1, 2$. This is equivalent to $r_i \geq \xi_\alpha(\boldsymbol{V}(\{i\}))/\xi_\alpha(\boldsymbol{V}(\{1,2\}))$ for $i = 1, 2$. Using $\boldsymbol{V}(\{1,2\}) = \boldsymbol{V}(\{1\}) + \boldsymbol{V}(\{2\})$, $r_1 + r_2 = 1$, and $\xi_\alpha(\boldsymbol{V}(\{1,2\})) < 0$ we obtain that the core is empty if and only if $\xi_\alpha(\boldsymbol{V}(\{1\})) + \xi_\alpha(\boldsymbol{V}(\{2\})) < \xi_\alpha(\boldsymbol{V}(\{1,2\}))$. This is the case if $\alpha \in (0, 0.4587)$ (see Figure 1.2). $\triangleleft$

Summarizing, stable allocations of the total random network costs need not exist for the three models under consideration. In fact, the example we provided concerned only two agents. This means that even a coalition consisting of two agents may not benefit from cooperation. This seems counterintuitive, for two agents can pretend that they cooperate and allocate the total costs as if they were standing alone. This, however, is not possible because of the allocation structure that we imposed. When cooperating, agents 1 and 2 cannot allocate the random costs $\boldsymbol{X}_{\{1,\star\}} + \boldsymbol{X}_{\{2,\star\}}$ in such a way that agent 1 pays $\boldsymbol{X}_{\{1,\star\}}$ and agent 2 pays $\boldsymbol{X}_{\{2,\star\}}$. Both agents must bear a proportional part of the total costs. So, for allocating network costs the allocation structure that is chosen in the models might be too restrictive. Therefore, it may be more appropriate to allow the agents to allocate the random costs $\boldsymbol{X}_{\{k,l\}}$ of each connection proprotionally among each other instead of the total costs $\sum_{\{k,l\} \in T} \boldsymbol{X}_{\{k,l\}}$.

1.4 Bankruptcy Problems with Random Estate

A bankruptcy petition is presented against a firm when it is not able to meet the claims of its creditors. Since the firm's assets have insufficient value to pay off the debts, not every individual creditor can receive his claim back in full. So, what would be a fair allocation of the firm's estate? O'Neill (1982) modeled this allocation problem by means of the following cooperative game. Let $E \in \mathbb{R}_+$ denote the estate of the firm and let N denote the finite set of creditors, each creditor having a claim $\delta_i \in \mathbb{R}_+$ on the firm's estate. Since the firm is in bankruptcy, it must hold that the total number of claims exceeds the estate, that is $\sum_{i \in N} \delta_i > E$. The value of a coalition $S \subset N$ is defined as the remains of the estate E when coalition S fulfills the claims of the noncooperating creditors. Thus, $v(S) = \max\{E - \sum_{i \in N \setminus S} \delta_i, 0\}$. To simplify notation, define $\delta_S = \sum_{i \in S} \delta_i$ and $(x)_+ = \max\{x, 0\}$. Then $v(S) = (E - \delta_{N \setminus S})_+$ for all $S \subset N$. O'Neill (1982) showed that the core of bankruptcy games is nonempty. Furthermore, it was shown that $C(v) = \{x \in \mathbb{R}^N \,|\, \sum_{i \in N} x_i = E,\ \forall_{i \in N} : 0 \leq x_i \leq \delta_i\}$. Thus, any allocation that gives each creditor no more than his claim is a core allocation.

Next, let us assume that the exact value of the estate is uncertain. For instance, the value of a firm's assets (e.g. inventory, production facilities, knowledge) is ambiguous in bankruptcy as market values are no longer appropriate. We assume that creditors have a deterministic claim on this random estate and that the total claims exceed the estate for all possible realizations, i.e. $\mathbf{P}(\boldsymbol{E} < \sum_{i \in N} \delta_i) = 1$.

First, we model this bankruptcy problem as a chance-constrained game. Similar to O'Neill (1982) we define the value of a coalition $S \subset N$ as the remains of the estate after they paid back the claims of the other creditors, that is $\boldsymbol{V}(S) = (\boldsymbol{E} - \delta_{N \setminus S})_+$ for all $S \subset N$. Notice that $\mathbf{P}(\rho(S) < \sum_{i \in S} \delta_i) = 1$. The following example shows that the prior core of this game can be empty.

Example 1.12 Let $\boldsymbol{E}$ be uniformly distributed between 0 and 10 and let the claims of the two creditors be equal to 6. Further, take $\alpha(\{1\}) = \alpha(\{2\}) = 0.9$, $\bar{\alpha} = 0.5$ and $\underline{\alpha} = 0.3$. The value of coalition $\{i\}$ is given by the random variable $\boldsymbol{V}(\{i\}) = (\boldsymbol{E} - 6)_+$ with probability distribution function

$$F_{\boldsymbol{V}(\{i\})}(x) = \begin{cases} 0, & \text{if } x < 0, \\ \frac{6+x}{10}, & \text{if } 0 \leq x \leq 4, \\ 1, & \text{if } x > 4. \end{cases}$$

Note that $\boldsymbol{V}(\{i\})$ equals zero with probability 0.6. In that case, the estate is insufficient to pay off the claim of creditor j, so that nothing remains for creditor i. Since $\boldsymbol{V}(\{1,2\}) = \boldsymbol{E}$, we have that (x_1, x_2) belongs to the prior core if $F_{\boldsymbol{V}(\{i\})}(x_i) \geq 0.7$ for $i = 1,2$ and $0.3 \leq F_{\boldsymbol{V}(\{1,2\})}(x_1 + x_2) \leq 0.5$. This implies that $x_i \geq 3$ for $i = 1,2$ and $x_1 + x_2 \leq 5$. Obviously, such allocations do not exist. $\triangleleft$

The main reason why the game in Example 1.12 has a nonempy core is because $\alpha(\{i\}) > \bar{\alpha}$. Given the interpretation of $\alpha(S)$, this inequality might be considered counterintuitive. Since individual i finds an allocation acceptable if the probability that he cannot do better on his own is at least $\alpha(\{i\})$, one might expect that individual i finds an allocation for coalition $S \ni i$ acceptable if the probability that this coalition cannot do better on its own is also at least $\alpha(\{i\})$. In other words, $\alpha(\{i\}) \leq \alpha(S)$. Imposing a monotonicity condition on $\alpha(S)$ is sufficient for nonemptiness of the prior core:

Theorem 1.13 *A bankruptcy game $(N, \boldsymbol{V})$ has a nonempty prior core if $\alpha(S) \leq \bar{\alpha}$ for all $S \subset N$.*

Second, we model the allocation problem as a stochastic cooperative game with transfer payments. Let $\Gamma = (N, \mathcal{V}, \{\succsim_i\}_{i \in N})$ be a stochastic bankruptcy game with $\mathcal{V}(S) = \{\boldsymbol{V}(S)\} = \{(\boldsymbol{E} - \delta_{N\setminus S})_+\}$ for all $S \subset N$ and $\succsim_i$ the preference relation based on the α_i-quantile, $i \in N$. Since the creditors' preferences satisfy conditions (M1) and (M2) we can define the corresponding TU-game (N, v_Γ) by

$$\begin{aligned} v_\Gamma(S) &= \max\left\{\sum_{i \in S} \xi_{\alpha_i}(d_i + r_i \boldsymbol{V}(S)) \,\middle|\, d_i \in H^S, r \in \Delta^S\right\} \\ &= \max\left\{\sum_{i \in S} r_i \xi_{\alpha_i}(\boldsymbol{V}(S)) \,\middle|\, r \in \Delta^S\right\}, \end{aligned}$$

for all $S \subset N$. Using that $\xi_{\alpha_i}(\boldsymbol{V}(S)) = (\xi_{\alpha_i}(\boldsymbol{E}) - \delta_{N\setminus S})_+$ it directly follows that $v_\Gamma(S) = \max_{i \in S}(\xi_{\alpha_i}(\boldsymbol{E}) - \delta_{N\setminus S})_+$ for all $S \subset N$. Now we can easily prove that the core $C(v_\Gamma)$ is nonempty. Define $\bar{\alpha} = \max_{i \in N} \alpha_i$ and define the TU-game $(N, \bar{v}_\Gamma)$ by $\bar{v}_\Gamma(S) = (\xi_{\bar{\alpha}}(\boldsymbol{E}) - \delta_{N\setminus S})_+$ for all $S \subset N$. Since $\bar{v}_\Gamma(S) \geq v_\Gamma(S)$ for all $S \subset N$ and $\bar{v}_\Gamma(N) = v_\Gamma(N)$ it follows that $C(\bar{v}_\Gamma) \subset C(v_\Gamma)$. Moreover, the game $(N, \bar{v}_\Gamma)$ is a bankruptcy game in the sense of O'Neill (1982) with estate $E = \xi_{\bar{\alpha}}(\boldsymbol{E})$. Hence, the core $C(\bar{v}_\Gamma) \neq \emptyset$ and thus $C(v_\Gamma) \neq \emptyset$. In particular it holds that

$\{x \in I\!R^N | \sum_{i \in N} x_i = \xi_{\bar{\alpha}}(\boldsymbol{E}),\ \forall_{i \in N} : 0 \leq x_i \leq \delta_i\} \subset C(v_\Gamma)$. Applying Theorem 1.5 then yields the following result.

Theorem 1.14 *Let Γ be a stochastic bankruptcy game and let $\bar{\alpha} = \max_{i \in N} \alpha_i$. Then*

$$\left\{(d_i + r_i\boldsymbol{E})_{i\in N} \in \mathcal{Z}_\Gamma(N) \,\middle|\, \begin{array}{l} \forall_{i\in N:\alpha_i<\bar{\alpha}} : r_i = 0,\ 0 \leq d_i \leq \delta_i \\ \forall_{i\in N:\alpha_i=\bar{\alpha}} : 0 \leq d_i + r_i\xi_{\alpha_i}(\boldsymbol{E}) \leq \delta_i \end{array}\right\} \subset \mathcal{C}(\Gamma).$$

Note that if all creditors have the same preferences, that is $\alpha_i = \alpha$ for all $i, j \in N$, then the game (N, v_Γ) is a bankruptcy game with estate $E = \xi_\alpha(\boldsymbol{E})$. Hence, equality holds in Theorem 1.14.

Next, let us turn to the case without transfer payments. Using that $\rho_i(\boldsymbol{V}(S), \boldsymbol{V}(N)) = \xi_{\alpha_i}(\boldsymbol{V}(N))/\xi_{\alpha_i}(\boldsymbol{V}(S))$ we have that

$$\mathcal{C}^*(\Gamma) = \left\{(r_i\boldsymbol{E})_{i\in N} \in \mathcal{Z}^*_\Gamma(N) \,\middle|\, \forall_{S\in\mathcal{N}} : \sum_{i\in S} r_i \frac{\xi_{\alpha_i}(\boldsymbol{E})}{\xi_{\alpha_i}(\boldsymbol{E}) - \delta_{N\setminus S}} \geq 1\right\}.$$

The following proposition states that the core is nonempty. The proof is provided in the Appendix.

Theorem 1.15 *Let $\Gamma \in MG^*(N)$ be a stochastic bankruptcy game. Then the core $\mathcal{C}^*(\Gamma)$ is nonempty.*

Bankruptcy games have a nonempty core independent of whether or not we allow for transfer payments between the agents. So how do the core allocations of these two bankruptcy games compare to each other? We can compare the core allocations by comparing the corresponding certainty equivalents. Therefore, take $(\hat{r}_i\boldsymbol{X}_N) \in \mathcal{C}^*(\Gamma)$. Since ξ_{α_i} satisfies conditions (M1)–(M2), the certainty equivalent of a random payoff $\boldsymbol{X}$ equals $\xi_{\alpha_i}(\boldsymbol{X})$. So, does there exists an allocation $(d_i + r_i\boldsymbol{X}_N)_{i\in N} \in \mathcal{C}(\Gamma)$ such that the certainty equivalent coincides with $(\hat{r}_i\boldsymbol{X}_N)_{i\in N}$, that is $\xi_{\alpha_i}(d_i + r_i\boldsymbol{X}_N) = \xi_{\alpha_i}(\hat{r}_i\boldsymbol{X}_N)$ for all $i \in N$? The answer is no. In fact, by allowing transfer payments, the agents can strictly improve upon the allocation $(\hat{r}_i\boldsymbol{X}_N)_{i\in N}$ if $\xi_{\alpha_i}(\boldsymbol{E}) \neq \xi_{\alpha_j}(\boldsymbol{E})$ for some $i, j \in N$. To see this, let $j \in N$ be such that $\alpha_j = \max_{i\in N} \alpha_i$. Then $r_j = 1$ and $r_i = 0$ for all $i \in N\backslash\{j\}$ is a Pareto optimal risk allocation. Further, take $d_i = \hat{r}_i\xi_{\alpha_i}(\boldsymbol{E}) + \frac{1}{|N|}\,\text{exc}$ for $i \in N\backslash\{j\}$ and $d_j = (\hat{r}_j - 1)\xi_{\alpha_j}(\boldsymbol{E}) + \frac{1}{|N|}\,\text{exc}$, where $\text{exc} = (1 - \hat{r}_j)\xi_{\alpha_j}(\boldsymbol{E}) - \sum_{i\in N\backslash\{j\}} \hat{r}_i\xi_{\alpha_i}(\boldsymbol{E})$. Note that

$$\text{exc} \;=\; (1 - \hat{r}_j)\xi_{\alpha_j}(\boldsymbol{E}) - \sum_{i\in N\backslash\{j\}} \hat{r}_i\xi_{\alpha_i}(\boldsymbol{E})$$

$$
\begin{aligned}
&> (1-\hat{r}_j)\xi_{\alpha_j}(\boldsymbol{E}) - \sum_{i\in N\backslash\{j\}} \hat{r}_i\xi_{\alpha_j}(\boldsymbol{E}) \\
&= (1-\hat{r}_j)\xi_{\alpha_j}(\boldsymbol{E}) - (1-\hat{r}_j)\xi_{\alpha_j}(\boldsymbol{E}) \;=\; 0.
\end{aligned}
$$

Since

$$
\begin{aligned}
\sum_{i\in N} d_i &= (\hat{r}_j - 1)\xi_{\alpha_j}(\boldsymbol{E}) + \frac{1}{|N|}\,\text{exc} + \sum_{i\in N\backslash\{j\}} \left(\hat{r}_i\xi_{\alpha_i}(\boldsymbol{E}) + \frac{1}{|N|}\,\text{exc}\right) \\
&= \text{exc} + (\hat{r}_j - 1)\xi_{\alpha_j}(\boldsymbol{E}) + \sum_{i\in N\backslash\{j\}} \hat{r}_i\xi_{\alpha_i}(\boldsymbol{E}) \;=\; 0
\end{aligned}
$$

$(d_i + r_i\boldsymbol{E})_{i\in N}$ is a feasible allocation. Moreover, $\xi_{\alpha_i}(d_i + r_i\boldsymbol{E}) = d_i + r_i\xi_{\alpha_i}(\boldsymbol{E}) = d_i = \hat{r}_i\xi_{\alpha_i}(\boldsymbol{E}) + \frac{1}{|N|}\,\text{exc} > \hat{r}_i\xi_{\alpha_i}(\boldsymbol{E})$ for all $i \in N\backslash\{j\}$ and $\xi_{\alpha_j}(d_j + r_j\boldsymbol{E}) = d_j + r_j\xi_{\alpha_j}(\boldsymbol{E}) = (\hat{r}_j - 1)\xi_{\alpha_j}(\boldsymbol{E}) + \frac{1}{|N|}\,\text{exc} + \xi_{\alpha_j}(\boldsymbol{E}) = \hat{r}_j\xi_{\alpha_j}(\boldsymbol{E}) + \frac{1}{|N|}\,\text{exc} > \hat{r}_j\xi_{\alpha_j}(\boldsymbol{E})$. Hence, all agents $i \in N$ prefer the allocation $(d_i + r_i\boldsymbol{E})_{i\in N}$ to the allocation $(\hat{r}_i\boldsymbol{E})_{i\in N}$.

Summarizing, core allocations exist for stochastic bankruptcy games whereas they need not exist for chance-constrained bankruptcy games. Furthermore, core allocations without transfer payments are strictly Pareto dominated by allocations with transfer payments. Hence, from the viewpoint of the creditors, transfer payments are preferable.

1.5 Concluding Remarks

This paper surveyed three existing models on cooperative behavior under risk. We discussed the differences and similarities between these models and examined their performance in two applications. The cooperative models under consideration were chance-constrained games, stochastic cooperative games with transfer payments, and stochastic cooperative games without transfer payments. The main difference between the first and the latter is that stochastic cooperative games explicitly incorporate the preferences of the agents. The main difference between the latter two is, as their names imply, in the deterministic transfer payments. Allowing for deterministic transfer payments allows for insurance as agents can transfer risk in exchange for a deterministic payment, i.e. an insurance premium. One reason to exclude insurance possibilities from the analysis when examining cooperative behavior is the following. Since the possibility to insure risks provides an incentive to cooperate (see, for instance, Suijs *et al.*, 1998), allowing for insurance may bias the results in the sense that it may not be clear whether the incentive to cooperate arises from the characteristics of the particular setting under consideration or from the insurance opportunities.

In the analysis of the applications, we focused on stability of cooperation. More precisely, for each of the three models we examined whether the core was nonempty. For the cost allocation problem in a communication network tree, all three models need not yield stable cooperation. This 'deficiency' is attributed to the different restrictions that the three models impose on the allocation space. For the bankruptcy case, conditions could be derived such that stable cooperation arises in all three models.

Appendix

Proof of Theorem 1.8. Let $\Gamma \in MG^*(N)$. The core $\mathcal{C}^*(\Gamma)$ is nonempty if the following system of linear equations has a nonnegative solution

$$-\sum_{i \in S} r_i \rho_i(\boldsymbol{X}_S, \boldsymbol{X}_N) \leq -1, \quad \text{for all } S \in \mathcal{N},$$
$$\sum_{i \in N} r_i \leq 1.$$

Using a variant of Farkas' Lemma, a nonnegative solution exists if and only if there exists no $\mu \geq 0$, $\lambda(S) \geq 0$, $S \in \mathcal{N}$ satisfying

$$\sum_{S \in \mathcal{N}: i \in S} \lambda(S) \rho_i(\boldsymbol{X}_S, \boldsymbol{X}_N) \leq \mu, \quad \text{for all } i \in N$$
$$\sum_{S \in \mathcal{N}} \lambda(S) > \mu.$$

Without loss of generality we may assume that $\mu > 0$, otherwise we can consider $\mu + \text{exc}$ with $\text{exc} > 0$ sufficiently small. Hence, the above is equivalent to there exists no $\lambda(S) \geq 0, S \in \mathcal{N}$ such that

$$\sum_{S \in \mathcal{N}: i \in S} \lambda(S) \rho_i(\boldsymbol{X}_S, \boldsymbol{X}_N) \leq 1, \quad \text{for all } i \in N$$
$$\sum_{S \in \mathcal{N}} \lambda(S) > 1.$$

Rewriting yields the statement: if for all $\lambda(S) \geq 0, S \in \mathcal{N}$ satisfying $\sum_{S \in \mathcal{N}: i \in S} \rho_i(\boldsymbol{X}_S, \boldsymbol{X}_N) \leq 1$ for each $i \in N$, then $\sum_{S \in \mathcal{N}} \lambda(S) \leq 1$. □

Proof of Theorem 1.13. Notice that $\xi_{\alpha(S)}((\boldsymbol{E} - d_{N \setminus S})_+) = (\xi_{\alpha(S)}(\boldsymbol{E}) - d_{N \setminus S})_+$ for all $S \subset N$. The prior core is nonemtpy if and only if there exists an allocation $x \in \mathbb{R}^N$ such that $\xi_{\underline{\alpha}}(\boldsymbol{E}) \leq \sum_{i \in N} x_i \leq \xi_{\bar{\alpha}}(\boldsymbol{E})$ and $\sum_{i \in S} x_i \geq (\xi_{\alpha(S)}(\boldsymbol{E}) - \delta_{N \setminus S})_+$.

If $\alpha(S) \leq \bar{\alpha}$ for all $S \subset N$ then $(\xi_{\alpha(S)}(\boldsymbol{E}) - \delta_{N \setminus S})_+ \leq (\xi_{\bar{\alpha}(N)}(\boldsymbol{E}) - \delta_{N \setminus S})_+$ for all $S \subset N$. Since the bankruptcy game with estate $E =$

$\xi_{\bar{\alpha}(N)}(\boldsymbol{E})$ and claims δ_i, $i \in N$, has a nonempty core, there exists an allocation $x \in I\!R^N$ such that $\sum_{i \in x_i} = \xi_{\bar{\alpha}(N)}(\boldsymbol{E})$ and $\sum_{i \in S} x_i \geq (\xi_{\bar{\alpha}(N)}(\boldsymbol{E}) - \delta_{N \backslash S})_+$ for all $S \subset N$. Hence, x is a core-allocation for the bankruptcy game $(N, \boldsymbol{V})$. □

Proof of Theorem 1.15. Define $\overline{E}_S = \max_{i \in S} \xi_{\alpha_i}(\boldsymbol{E})$ for each $S \subset N$. Notice, that since

$$\frac{\xi_{\alpha_i}(\boldsymbol{E})}{\xi_{\alpha_i} - \delta_{N\backslash S}} = \frac{\overline{E}_{\{i\}}}{\overline{E}_{\{i\}} - \delta_{N\backslash S}} \geq \frac{\overline{E}_S}{\overline{E}_S - \delta_{N\backslash S}},$$

an allocation $(r_i\boldsymbol{E})_{i \in N} \in \mathcal{Z}^*_\Gamma(N)$ is a core-allocation if $\sum_{i \in S} r_i \geq \frac{\overline{E}_S - \delta_{N\backslash S}}{\overline{E}_S}$ for all $S \subset \mathcal{N}$.

We will show that

$$\left\{ (r_i\boldsymbol{E})_{i \in N} \in \mathcal{Z}^*_\Gamma(N) \,\middle|\, \forall_{i \in N} : 0 \leq r_i \leq \frac{\delta_i}{\overline{E}_{N\backslash\{i\}}} \right\} \subset \mathcal{C}^*(\Gamma).$$

Let $j \in N$ be such that $\xi_{\alpha_j} = \overline{E}_N$. First, we consider the case that j is unique. Take $S \in \mathcal{N}$. Hence, $\overline{E}_S - \delta_{N\backslash S} > 0$. If $j \notin S$ then

$$\begin{aligned}
\sum_{i \in S} r_i &= 1 - \sum_{i \notin S} r_i \geq 1 - \sum_{i \notin S} \frac{\delta_i}{\overline{E}_{N\backslash\{i\}}} \\
&= 1 - \sum_{i \notin S, i \neq j} \frac{\delta_i}{\overline{E}_{N\backslash\{i\}}} - \frac{\delta_j}{\overline{E}_{N\backslash\{j\}}} = 1 - \sum_{i \notin S, i \neq j} \frac{\delta_i}{\overline{E}_N} - \frac{\delta_j}{\overline{E}_{N\backslash\{j\}}} \\
&\geq 1 - \sum_{i \notin S, i \neq j} \frac{\delta_i}{\overline{E}_{N\backslash\{j\}}} - \frac{\delta_j}{\overline{E}_{N\backslash\{j\}}} = 1 - \frac{\delta_{N\backslash S}}{\overline{E}_{N\backslash\{j\}}} \\
&= \frac{\overline{E}_{N\backslash\{j\}} - \delta_{N\backslash S}}{\overline{E}_{N\backslash\{j\}}} \geq \frac{\overline{E}_S - \delta_{N\backslash S}}{\overline{E}_S},
\end{aligned}$$

where the second inequality follows from $\overline{E}_N \geq \overline{E}_{N\backslash\{j\}}$, the third equality from $\overline{E}_{N\backslash\{i\}} = \overline{E}_N$ for $i \neq j$, and the third inequality from $\overline{E}_{N\backslash\{j\}} \geq \overline{E}_S$.

If $j \in S$ then

$$\begin{aligned}
\sum_{i \in S} r_i &= 1 - \sum_{i \notin S} r_i \geq 1 - \sum_{i \notin S} \frac{\delta_i}{\overline{E}_{N\backslash\{i\}}} \\
&= 1 - \sum_{i \notin S} \frac{\delta_i}{\overline{E}_N} = \frac{\overline{E}_N - \delta_{N\backslash S}}{\overline{E}_N}
\end{aligned}$$

$$\geq \frac{\overline{E}_S - \delta_{N\backslash S}}{\overline{E}_S},$$

where the second equality follows from $\overline{E}_{N\backslash\{i\}} = \overline{E}_N$ and the second inequality follows from $\overline{E}_N \geq \overline{E}_S$.

If agent j is not unique, then the proof is similar to the case $j \in S$ above. Hence, we leave that as an exercise to the reader. □

References

Bondareva, O. (1963): "Some applications of linear programming methods to the theory of cooperative games," *Problemi Kibernet*, 10, 119–139. In Russian.

Charnes, A., and D. Granot (1973): "Prior solutions: extensions of convex nucleolus solutions to chance-constrained games," *Proceedings of the Computer Science and Statistics Seventh Symposium at Iowa State University*, 323–332.

Charnes, A., and D. Granot (1976): "Coalitional and chance-constrained solutions to n-person games I," *SIAM Journal on Applied Mathematics*, 31, 358–367.

Charnes, A., and D. Granot (1977): "Coalitional and chance-constrained solutions to n-person games II," *Operations Research*, 25, 1013–1019.

Claus, A., and D. Kleitman (1973): "Cost allocation for a spanning tree," *Networks*, 3, 289–304.

Granot, D. (1977): "Cooperative games in stochastic characteristic function form," *Management Science*, 23, 621–630.

Megiddo, N. (1978): "Computational complexity of the game theory approach to cost allocation for a tree," *Mathematics of Operations Research*, 3, 189–196.

O'Neill, B. (1982): "A problem of rights arbitration from the Talmud," *Mathematical Social Sciences*, 2, 345–371.

Shapley, L. (1967): "On balanced sets and cores," *Naval Research Logistics Quarterly*, 14, 453–460.

Suijs, J. and P. Borm (1999): "Stochastic cooperative games: superadditivity, convexity, and certainty equivalents," *Games and Economic Behavior*, 27, 331–345.

Suijs, J., O. Borm, A. De Waegenaere, and S. Tijs (1999): "Cooperative games with stochastic payoffs," *European Journal of Operational Research*, 113, 193–205.

Suijs, J., A. De Waegenaere, and P. Borm (1998): "Stochastic cooperative games in insurance," *Insurance: Mathematics & Economics*, 22, 209–228.

Timmer, J., P. Borm, and S. Tijs (2000): "Convexity in stochastic cooperative situations," *CentER Discussion Paper Series 2000-04*, Tilburg University.

von Neumann, J., and O. Morgenstern (1944): *Theory of Games and Economic Behavior.* Princeton: Princeton University Press.

Chapter 2

Sequencing Games: a Survey

BY IMMA CURIEL, HERBERT HAMERS, AND FLIP KLIJN

2.1 Introduction

During the last three decades there have been many interesting interactions between linear and combinatorial optimization and cooperative game theory. Two problems meet here: On the one hand the problem of minimizing the costs or maximizing the revenues of a project, on the other hand the problem of allocating these costs or revenues among the participants in the project. The first problem is dealt with using techniques from linear and combinatorial optimization theory, the second problem falls in the realm of cooperative game theory. We mention minimum spanning tree games (cf. Granot and Huberman, 1981), linear production games (cf. Owen, 1975), traveling salesman games (cf. Potters *et al.*, 1992), Chinese postman games (cf. Granot *et al.*, 1999) and assignment games (cf. Shapley and Shubik, 1972). An overview of this type of games can be found in Tijs (1991) and Curiel (1997).

Another fruitful topic in this area has been and still is that of sequencing games. This paper gives an overview of the developments of the interaction between sequencing situations and cooperative games.

In operations research, sequencing situations are characterized by a finite number of jobs lined up in front of one (or more) machine(s) that have to be processed on the machine(s). A single decision maker wants to determine a processing order of the jobs that minimizes total costs.

P. Borm and H. Peters (eds.), Chapters in Game Theory, 27–50.

This single decision maker problem can be transformed into a multiple decision makers problem by taking agents into account who own at least one job. In such a model a group of agents (coalition) can save costs by cooperation. For the determination of the maximal cost savings of a coalition one has to solve the combinatorial problem corresponding to this coalition. In particular, the maximal cost savings for each coalition can be modeled by a cooperative transferable utility game, which is an ordered pair (N, v) where N denotes a non-empty, finite set of players (agents) and v is a mapping from the power set 2^N of N to the real numbers with $v(\emptyset) = 0$. The questions that arise are: which coalition(s) will form, and how should the maximal cost savings be allocated among the members of this coalition. One way to answer these questions is to look at solution concepts and properties of the game. One of the most prominent solution concepts in cooperative game theory is the core of a game. It consists of all vectors which distribute $v(N)$, i.e., the revenues incurred when all players in N cooperate, among the players in such a way that no subset of players can be better off by seceding from the rest of the players and acting on their own behalf. That is, a vector x is in the core of a game (N, v) if $\sum_{j \in N} x_j = v(N)$ and $\sum_{j \in S} x_j \geq v(S)$, for all $S \subset N$. A cooperative game whose core is not empty is said to be balanced.

A well-known class of balanced games is the class of convex games. A game (N, v) is called *convex* if for any $i \in N$ and any $S, T \subseteq N \backslash \{i\}$ it holds

$$v(T \cup \{i\}) - v(T) \geq v(S \cup \{i\}) - v(S).$$

Convex (or submodular) games are known to have nice properties, in the sense that some solution concepts for these games coincide and others have intuitive descriptions. For example, for convex games the core is equal to the convex hull of all marginal vectors (cf. Shapley, 1971, and Ichiishi, 1981), and, as a consequence, the Shapley value is the barycentre of the core (Shapley, 1971). Moreover, the bargaining set and the core coincide, the kernel coincides with the nucleolus (Maschler *et al.*, 1972) and the τ-value can be easily calculated (Tijs, 1981).

In this paper we will focus on balancedness and convexity of the several classes of sequencing games that will be discussed.

Sequencing games were introduced in Curiel *et al.* (1989). They considered the class of one-machine sequencing situations in which no restrictions like due dates and ready times are imposed on the jobs and the weighted completion time criterion was chosen as the cost criterion. It was shown for the corresponding sequencing games that they are convex and, thus, that the games are balanced. Hamers *et al.* (1995) extended the class of one-machine sequencing situations considered by Curiel *et al.* (1989) by imposing ready times on the jobs. In this case the corresponding sequencing games are balanced, but are not necessarily convex. For a special subclass of sequencing games with ready times, however, convexity could be established. Similar results are also obtained in Borm *et al.* (1999), in which due dates are imposed on the jobs.

Instead of imposing restrictions on the jobs, Hamers *et al.* (1999) and Calleja *et al.* (2001) extended the number of machines. Hamers *et al.* (1999) consider sequencing situations with m parallel and identical machines in which no restrictions on the jobs are imposed. Again, the weighted completion time criterion is used. They proved balancedness in case there are two machines, and show balancedness for special classes in case there are more than two machines. Calleja *et al.* (2001) established balancedness for a special class of sequencing games that arise from 2 machine sequencing situations in which a maximal weighted cost criterion is considered.

Van Velzen and Hamers (2001) consider some classes of sequencing games that arise from the same sequencing situations as used in Curiel *et al.* (1989). A difference, however, is that the coalitions in their games have more possibilities to maximize their profit. They show that some of these classes are balanced.

This chapter is organized as follows. We start in Section 2.2 by recalling permutation games and σ_0-component additive games, two classes of games that are closely related to sequencing games. Section 2.3 deals with the sequencing situations and games studied in Curiel *et al.* (1989). Section 2.4 discusses the sequencing games that arise if ready times or due dates are imposed on the jobs. Multiple-machine sequencing games are discussed in Section 2.5. Section 2.6 considers sequencing games that arise when the agents have more possibilities to maximize their profit.

2.2 Games Related to Sequencing Games

In this section we consider two classes of games that are closely related to sequencing games: permutation games, introduced by Tijs *et al.* (1984) and σ-component additive games, introduced by Curiel *et al.* (1993).

The main reason to start with these games is that they play an important role in the investigation of the balancedness of sequencing games.

Permutation games describe a situation in which n persons all have one job to be processed and one machine on which each job can be processed. No machine is allowed to process more than one job. Side-payments between the players are allowed. If player i processes his job on the machine of player j the processing costs are a_{ij}. Let $N = \{1, \ldots, n\}$ be the set of players. The *permutation game* (N, v) with costs a_{ij} is the cooperative game defined by

$$v(S) \quad = \quad \sum_{i \in S} a_{ii} - \min_{\pi_S \in \Pi_S} \sum_{i \in S} a_{i \pi_S(i)}$$

for all $S \subset N \backslash \{\emptyset\}$ and $v(\emptyset) = 0$. The class of all S-permuations is denoted by Π_S. The number $v(S)$ denotes the maximal cost savings a coalition can obtain by processing its jobs according to an optimal schedule compared to the situation in which every player processes his job on his own machine.

Theorem 2.1 *Permutation games are totally balanced.*

For Theorem 2.1 several proofs are presented in literature. We mention Tijs *et al.* (1984), using the Birkhoff-von Neumann theorem on doubly stochastic matrices. Curiel and Tijs (1986) gave another proof of the balancedness of permutation games. They used an equilibrium existence theorem of Gale (1984) for a discrete exchange economy with money, thereby showing a relation between assignment games (cf. Shapley and Shubik, 1972) and permutation games. Klijn *et al.* (2000) use the existence of envy-free allocations in economies with indivisible objects, quasi-linear utility functions, and an amount of money to prove the balancedness of permutation games.

Let σ_0 be an order on the player set. Then a game (N, v) is called a σ_0-component additive game if it satisfies the following three conditions:
(i) $v(\{i\}) = 0$ for all $i \in N$
(ii) (N, v) is superadditive
(iii) $v(S) = \sum_{T \in S/\sigma_0} v(T)$ for any $S \subset N$, where S/σ_0 is the set of maximally connected components of S. Here, a coalition is called connected with respect to σ_0 if for all $i, j \in S$ and k such that $\sigma_0(i) < \sigma_0(k) < \sigma(j)$ it holds that $k \in S$.

The next result of Curiel *et al.* (1994), shows that σ_0-component additive games have a non-empty core.

Theorem 2.2 *σ_0-component additive games are balanced.*

The proof shows that a specific vector, the β-value, which is the average of two specific marginal vectors, is in the core.

Moreover, Potters and Reijnierse (1995) showed that for Γ-component additive games, a class of games that contain σ_0-component additive games, the core is equal to the bargaining set and the nucleolus coincides with the kernel.

2.3 Sequencing Situations and Sequencing Games

In this section we describe the class of one-machine sequencing situations and the corresponding class of sequencing games as introduced in Curiel *et al.* (1989). Furthermore, we will discuss the EGS rule and the split core, solution concepts that generate allocations that are in the core of a sequencing games. Finally, we show that sequencing games are convex.

In a one-machine sequencing situation there is a queue of agents, each with one job, before a machine (counter). Each agent (player) has to have his job processed on this machine. The finite set of agents is denoted by $N = \{1, ..., n\}$. By a bijection $\sigma : N \to \{1, ..., n\}$ we can describe the position of the agents in the queue. Specifically, $\sigma(i) = j$ means that player i is in position j. We assume that there is an initial order $\sigma_0 : N \to \{1, ..., n\}$ of the jobs before the processing of the machine starts. The set of all possible processing orders is denoted by Π_N. The processing time p_i of the job of agent i is the time the machine takes to handle this job.

For each agent $i \in N$ the costs of spending time in the system can be described by a linear cost function $c_i : [0, \infty) \to I\!R$ defined by $c_i(t) = \alpha_i t$ with $\alpha_i > 0$. So $c_i(t)$ is the cost for agent i if he has spent t units of time in the system.

A sequencing situation as described above is denoted by (N, σ_0, p, α), where $N = \{1, ..., n\}$ is the set of players, $\sigma_0 : N \to \{1, ..., n\}$, $p = (p_i)_{i \in N} \in (0, \infty)^N$ and $\alpha = (\alpha_i)_{i \in N} \in (0, \infty)^N$.

The starting time $t_{\sigma,i}$ of the job of agent i if processed in a semi-active way according to a bijection σ is

$$t_{\sigma,i} = \begin{cases} t_{\sigma,j} + p_j & \text{if } \sigma(i) > 1 \\ 0 & \text{if } \sigma(i) = 1 \end{cases}$$

where $j \in N$ is such that $\sigma(j) = \sigma(i) - 1$. Here, a processing order is called semi-active if there does not exist a job which could be processed earlier without altering the processing order. In other words, there are no unnecessary delays in the processing order. Note that we may restrict our attention to semi-active processing orders since for each agent the cost function is weakly increasing. Consequently, the completion time $C(\sigma, i)$ of the job of agent i with respect to σ is equal to $t_{\sigma,i} + p_i$.

By reordering the jobs, the total costs of the agents will change. Clearly, there exists an ordering for which the total costs are minimized. A processing order that minimizes total costs, and thus maximizes total cost savings, is an order in which the players are processed in decreasing order with respect to the urgency index u_i defined by $u_i = \frac{\alpha_i}{p_i}$. This result is due to Smith (1956) and is formally presented (without proof) in the following proposition.

Proposition 2.3 *Let (N, σ_0, p, α) be a sequencing situation. Then for $\hat{\sigma} \in \Pi_N$*

$$\sum_{i \in N} \alpha_i(C(\sigma_0, i) - C(\hat{\sigma}, i)) = \max_{\sigma \in \Pi_N} \{\sum_{i \in N} \alpha_i(C(\sigma_0, i) - C(\sigma, i))\}$$

if and only if

$$u_{\hat{\sigma}^{-1}(1)} \geq u_{\hat{\sigma}^{-1}(2)} \geq \ldots \geq u_{\hat{\sigma}^{-1}(n)}.$$

Note that an optimal order can be obtained from the initial order by consecutive switches of neighbours i, j with i directly in front of j and $u_i < u_j$.

The problem of allocating the maximal surplus back to the players is tackled using game theory. First, we define a class of cooperative games that arises from the above described sequencing situations. Second, some allocation rules will be discussed and related to sequencing games.

For a sequencing situation (N, σ_0, p, α) the costs $C_S(\sigma)$ of coalition S with respect to a processing order σ are equal to $C_S(\sigma) = \sum_{i \in S} \alpha_i C(\sigma, i)$. Let $\hat{\sigma}$ be an optimal order of N. Then the maximal cost savings for coalition N are equal to $C_N(\sigma_0) - C_N(\hat{\sigma})$. We want to determine the maximal cost savings of a coalition S that decides to cooperate. For this, we have to define which rearrangements of the coalition S are admissible with respect to the initial order. A bijection $\sigma : N \rightarrow \{1, ..., n\}$ is called *admissible* for S if it satisfies the following condition:

$$P(\sigma_0, j) = P(\sigma, j) \text{ for all } j \in N \backslash S$$

where for any $\tau \in \Pi_N$ the set of predecessor of a player $j \in N$ with respect to τ is defined as $P(\tau, j) = \{k \mid \tau(k) < \tau(j)\}$.

This condition implies that the starting time of each agent outside the coalition S is equal to his starting time in the initial order, and the agents of S are not allowed to jump over players outside S. The set of admissible rearrangements for a coalition S is denoted by Σ_S.

By defining the worth of a coalition S as the maximum cost savings coalition S can achieve by means of an admissible rearrangement we obtain a cooperative game called a sequencing game. Formally, for a sequencing situation (N, σ_0, p, α) the corresponding *sequencing game* (N, v) is defined by

$$v(S) = \max_{\sigma \in \Sigma_S} \{\sum_{i \in S} \alpha_i [C(\sigma_0, i) - C(\sigma, i)]\} \tag{2.1}$$

for all $S \in 2^N$.
We will refer to the games defined in (2.1), introduced in Curiel *et al.* (1989), as standard sequencing games or s-sequencing games.

Expression (2.1) can be rewritten in terms of $g_{ij} = \max\{0, \alpha_j p_i - \alpha_i p_j\}$, the cost savings attainable by players i and j when i is directly in front of j. Then for any S that is connected with respect to σ_0 it holds that

$$v(S) = \sum_{i,j \in S : \sigma_0(i) < \sigma_0(j)} g_{ij}. \tag{2.2}$$

For a coalition T that is not connected with respect to σ_0 it follows that

$$v(T) = \sum_{S \in T \backslash \sigma_0} v(S)$$

where $T \backslash \sigma_0$ is the set of components of T, a component of T being a maximal, connected subset of T.

Example 2.4 Let $N = \{1, 2, 3\}, \sigma_0(i) = i$ for all $i \in N, p = (2, 2, 1)$ and $\alpha = (4, 6, 5)$. It follows that $g_{12} = g_{23} = 4$ and $g_{13} = 6$. Then $v(\{i\}) = 0$ for all $i \in N$, $v(\{1, 2\}) = v(\{2, 3\}) = 4$, $v(\{1, 3\}) = v(\{1\}) + v(\{3\}) = 0$ and $v(N) = 14$. $\triangleleft$

We can conclude that s-sequencing games are σ_0-component additive games. Hence, s-sequencing games are balanced. Nevertheless, we show that two context specific rules, the EGS (Equal Gain Splitting) rule and the split core, provide allocations that are in the core of a sequencing game.

From (2.2) it follows immediately that for an s-sequencing game that arises from a sequencing situation (N, σ_0, p, α):

$$v(N) = \sum_{i,j \in N: \sigma_0(i) < \sigma_0(j)} g_{ij}.$$

Recall that the set of predecessors of player $i \in N$ with respect to the processing order σ is given by $P(\sigma, i) = \{j \mid \sigma(j) < \sigma(i)\}$. We define the set of followers of $i \in N$ with respect to σ to be $F(\sigma, i) = \{j \mid \sigma(j) > \sigma(i)\}$. The Equal Gain Splitting, introduced in Curiel *et al.* (1989), is a map that assigns to each sequencing situation (N, σ_0, p, α) a vector in $\mathbb{R}^N$, and is defined by

$$EGS_i(N, \sigma_0, p, \alpha) = \frac{1}{2} \sum_{j \in F(\sigma_0, i)} g_{ij} + \frac{1}{2} \sum_{k \in P(\sigma_0, i)} g_{ki} \tag{2.3}$$

for all $i \in N$. Note that the EGS-rule is independent of the chosen optimal order and that the EGS-rule assigns to each player half of the gains of all neighbour switches he is actually involved in reaching an optimal order from the initial order.

From (2.3) it readily follows that the EGS-rule allocates the maximal cost savings that coalition N can obtain, i.e.,

$$\sum_{i \in N} EGS_i(N, \sigma_0, p, \alpha) = \sum_{i,j \in N: \sigma_0(i) < \sigma_0(j)} g_{ij}.$$

Example 2.5 Let $N = \{1, 2, 3\}$, $\sigma_0(i) = i$ for all $i \in N$, $p = (2, 2, 1)$ and $\alpha = (4, 6, 5)$. From Example 1 we have that $g_{12} = g_{23} = 4$ and $g_{13} = 6$. Then $EGS_1(N, \sigma_0, p, \alpha) = \frac{1}{2}(4+6) = 5, EGS_2(N, \sigma_0, p, \alpha) = \frac{1}{2}(4+4) = 4$ and $EGS_3(N, \sigma_0, p, \alpha) = \frac{1}{2}(6+4) = 5$. Moreover, we have $\sum_{i \in N} EGS_i(N, \sigma_0, p, \alpha) = 14 = v(N)$. $\triangleleft$

The EGS-rule divides the gain of each neighbour switch equally among both players involved. Generalizing the EGS-rule we consider Gain Splitting (GS) rules in which each player obtains a non-negative part of the gain of all neighbour switches he is actually involved in to reach the optimal order. The total gain of a neighbour switch is divided among both players that are involved. Formally, we define for all $i \in N$ and all $\lambda \in \Lambda$

$$GS_i^{\lambda}(N, \sigma_0, p, \alpha) = \sum_{j \in F(\sigma_0, i)} \lambda_{ij} g_{ij} + \sum_{k \in P(\sigma_0, i)} (1 - \lambda_{ki}) g_{ki}$$

where $\Lambda = \{\{\lambda_{ij}\}_{i,j\in N,\sigma_0(i)<\sigma_0(j)} \mid 0 \leq \lambda_{ij} \leq 1\}$. Note that for each $\lambda \in \Lambda$ we possibly obtain another allocation. Moreover, $GS^\lambda(N,\sigma_0,p,\alpha) = EGS(N,\sigma_0,p,\alpha)$ in case $\lambda_{ij} = \frac{1}{2}$ for all $i,j \in \{1,...,n\}, \sigma_0(i) < \sigma_0(j)$.

Example 2.6 If we take $\lambda_{12} = \frac{3}{4}, \lambda_{13} = \frac{1}{3}$ and $\lambda_{23} = 1$ in the sequencing situation of Example 2.5, then $GS^\lambda(N,\sigma_0,p,\alpha) = (5,5,4)$. $\triangleleft$

The *split core*, introduced in Hamers *et al.* (1996), of a sequencing situation (N,σ_0,p,α) is defined by

$$SPC(N,\sigma_0,p,\alpha) = \{GS^\lambda(N,\sigma_0,p,\alpha) \mid \lambda \in \Lambda\}.$$

The following theorem, due to Hamers *et al.* (1996), shows that the split core of a sequencing situation is a subset of the core of the corresponding s-sequencing game.

Theorem 2.7 *Let (N,σ_0,p,α) be a sequencing situation and let (N,v) be the corresponding s-sequencing game. Then $SPC(N,\sigma_0,p,\alpha) \subset C(v)$.*

Proof. It is sufficient to show $\sum_{i\in S} GS_i^\lambda(N,\sigma_0,p,\alpha) \geq v(S)$ for all connected coalitions $S \in 2^N$ with equality for $S = N$. Let $\lambda \in \Lambda$ and let S be a connected set. Then

$$\begin{aligned}
&\sum_{i\in S} GS_i^\lambda(N,\sigma_0,p,\alpha) \\
= &\sum_{i\in S}\Big[\sum_{j\in N:\sigma_0(i)<\sigma_0(j)} g_{ij}\lambda_{ij} + \sum_{k\in N:\sigma_0(k)<\sigma_0(i)} g_{ki}(1-\lambda_{ki})\Big] \\
\geq &\sum_{i\in S}\Big[\sum_{j\in S:\sigma_0(i)<\sigma_0(j)} g_{ij}\lambda_{ij} + \sum_{k\in S:\sigma_0(k)<\sigma_0(i)} g_{ki}(1-\lambda_{ki})\Big] \\
= &\sum_{i,j\in S:\sigma_0(i)<\sigma_0(j)} g_{ij} \\
= &\ v(S).
\end{aligned}$$

In case $S = N$ the inequality becomes an equality. Hence, $GS^\lambda(N,\sigma_0,p,\alpha) \in C(v)$. $\square$

Because $EGS(N,\sigma_0,p,\alpha)$ is the gain splitting rule in which all weights are equal to $\frac{1}{2}$, we have the following corollary.

Corollary 2.8 *Let (N,σ_0,p,α) be a sequencing situation and let (N,v) be the corresponding s-sequencing game. Then $EGS(N,\sigma_0,p,\alpha) \in C(v)$.*

The following theorem, due to Curiel *et al.* (1989), shows that s-sequencing games are convex games.

Theorem 2.9 *Let (N, σ_0, α, p) be a sequencing situation. Then the corresponding s-sequencing game (N, v) is convex.*

Proof. Let $i \in N$ and let $S \subset T \subset N \backslash \{i\}$. Then it follows that there exists $U_1, V_1 \in S \backslash \sigma_0 \cup \{\emptyset\}$ and $U_2, V_2 \in T \backslash \sigma_0 \cup \{\emptyset\}$ with $U_1 \subseteq U_2$ and $V_1 \subseteq V_2$ such that

$$\begin{aligned} v(S \cup \{i\}) - v(S) &= \sum_{k \in U_1} g_{ki} + \sum_{j \in V_1} g_{ij} \\ v(T \cup \{i\}) - v(T) &= \sum_{k \in U_2} g_{ki} + \sum_{j \in V_2} g_{ij} \end{aligned}$$

Since $g_{ij} \geq 0$ for all $i, j \in N, i \neq j$ we have that (N, v) is convex. □

2.4 On Sequencing Games with Ready Times or Due Dates

In this section we consider one-machine sequencing games with different types of restrictions on the jobs. First we study the situations in which jobs are available at different moments in time. Put differently, we impose ready times (release dates) on the jobs. We show that these games are balanced games, and that a special subclass is convex. Second, we will investigate situations in which jobs have a due date.

The description of a one-machine sequencing game in which the jobs have ready times is similar to that of the one-machine sequencing games in the previous section. We only have to include the notion of the ready time of a job and put an extra assumption on the initial order. The ready time r_i of the job of agent i is the earliest time the processing of his job can begin. Furthermore, it is assumed that there is an initial order $\sigma_0 : N \to \{1, ..., n\}$ such that

(*A*1) $r_i \leq r_j$ for all $i, j \in N$ with $\sigma_0(i) < \sigma_0(j)$

A sequencing situation as described above is denoted by $(N, \sigma_0, r, p, \alpha)$, where N is the set of players, $\sigma_0 : N \to \{1, ..., n\}$, $r = (r_i)_{i \in N} \in [0, \infty)^N$, $p = (p_i)_{i \in N} \in (0, \infty)^N$ and
$\alpha = (\alpha_i)_{i \in N} \in (0, \infty)^N$.

The starting time $t_{\sigma,i}$ of the job of agent i if processed according to a bijection σ (in a semi-active way) in the situation with ready times is

$$t_{\sigma,i} = \begin{cases} \max(r_i, t_{\sigma,j} + p_j) & \text{if } \sigma(i) > 1 \\ r_i & \text{if } \sigma(i) = 1 \end{cases}$$

where $j \in N$ is such that $\sigma(j) = \sigma(i) - 1$. Hence, the completion time $C(\sigma, i)$ of the job of agent i with respect to σ is equal to $t_{\sigma,i} + p_i$. The total costs $c_\sigma(S)$ of a coalition $S \subset N$ is given by

$$c_\sigma(S) = \sum_{i \in S} \alpha_i(C(\sigma, i)).$$

The set of admissible rearrangements of a coalition S is identical to the set defined in the previous section, i.e., Σ_S. Consequently, given a sequencing situation $(N, \sigma_0, r, p, \alpha)$ the corresponding sequencing game is analogously defined as in the previous section, i.e.,

$$v(S) \quad = \quad \max_{\sigma \in \Sigma_S} \{\sum_{i \in S} \alpha_i [C(\sigma_0, i) - C(\sigma, i)]\} \text{ for all } S \in 2^N. \tag{2.4}$$

for all $S \in 2^N$. We refer to the games defined in (2.4) as r-sequencing games. Because the set of admissible rearrangements is identical to the one in s-sequencing games we have again that for any coalition T it holds that

$$v(T) = \sum_{S \in T \backslash \sigma_0} v(S).$$

Because r-sequencing games are σ_0-component additive games, we can conclude that r-sequencing games have a non-empty core.

Theorem 2.10 *Let $(N, \sigma_0, r, p, \alpha)$ be a sequencing situation. Then the corresponding r-sequencing game (N, v) is balanced.*

The following example shows that convexity need not be satisfied.

Example 2.11 Let $N = \{1, 2, 3\}$, $\sigma_0 = (1, 2, 3), r = (0, 0, 1)$, $p = (1, 2, 3)$ and $\alpha = (1, 3, 12)$. The costs according to the initial order equal $1 \cdot 1 + 3 \cdot 3 + 6 \cdot 12 = 82$. The optimal rearrangement $\hat{\sigma}_N$ equals $(1, 3, 2)$ with corresponding costs of $1 \cdot 1 + 4 \cdot 12 + 6 \cdot 3 = 67$. Consequently, we have that $v(N) = 15$. Furthermore, $v(\{2, 3\}) = 15$ since $\hat{\sigma}_{\{2,3\}} = (1, 3, 2)$

and $v(\{1,2\}) = 1$ since $\hat{\sigma}_{\{1,2\}} = (2,1,3)$. From this we can infer that (N, v) is not a convex game:

$$v(N) - v(\{2,3\}) = 0 < 1 = v(\{1,2\}) - v(\{2\}).$$

$\triangleleft$

However, convexity can be established for a special subclass. More precisely, we restrict our attention to sequencing situations $(N, \sigma_0, r, p, \alpha)$ where there are no time gaps in the job processing according to the initial order σ_0, i.e.,

$$(A2) \qquad t_{\sigma_0,j} = t_{\sigma_0,i} + p_i \text{ for all } i,j \in N \text{ with } \sigma_0(j) = \sigma_0(i) + 1,$$

and

$$(A3) \qquad r_i \in \mathbf{N} \text{ and } p_i = 1 \text{ for all } i \in N.$$

Now, we state a convexity result, due to Hamers *et al.* (1995).

Theorem 2.12 *Let $(N, \sigma_0, r, p, \alpha)$ be a sequencing situation satisfying (A1)-(A3) and let (N, v) be the corresponding r-sequencing game. Then (N, v) is convex.*

In the second part of this section we focus on one-machine sequencing games in which due dates are imposed on the jobs.

The description of a one-machine sequencing game in which the jobs have due dates is similar to that of the one-machine sequencing games in which ready times are involved. We only have to replace the ready times by due dates and put an extra assumption on the initial order. The job of agent i has to be finished by the due date d_i. Furthermore, it is assumed that there is an initial order $\sigma_0 : N \to \{1, ..., n\}$ such that

$$(B1) \qquad d_i \leq d_j \text{ for all } i,j \in N \text{ with } \sigma_0(i) < \sigma_0(j).$$

A sequencing situation as described above is denoted by $(N, \sigma_0, d, p, \alpha)$, where N is the set of players, $\sigma_0 : N \to \{1, ..., n\}$, $d = (d_i)_{i \in N} \in (0, \infty)^N$, $p = (p_i)_{i \in N} \in (0, \infty)^N$, $\alpha = (\alpha_i)_{i \in N} \in (0, \infty)^N$.

The starting time of a job is defined identically to that in Section 2.3, and consequently the completion time is also defined identically. The set of admissible rearrangements of a coalition is in the same spirit: jobs outside the coalition cannot jump over jobs inside the coalition and their starting time is not changed, i.e., $P(\sigma_0, j) = P(\sigma, j)$ for all $j \in N \backslash S$. Moreover, we impose the restriction that a rearrangement is admissible only if all jobs are processed before their due dates. Formally, an admissible rearrangement σ satisfies

$(B2)\quad t_{\sigma,j} + p_j \leq d_j \quad \text{for all } j \in N.$

We denote the set of admissible rearrangements of a coalition S by $\mathcal{D}(S)$. The corresponding sequencing game is defined as follows

$$v(S) = \max_{\sigma \in \mathcal{D}(S)} \{\sum_{i \in S} \alpha_i [C(\sigma_0, i) - C(\sigma, i)]\}, \tag{2.5}$$

for all $S \in 2^N \backslash \{\emptyset\}$. We will refer to the game defined in (2.5) as d-sequencing game.

Because d-sequencing games are σ_0-component additive games, we have the following theorem.

Theorem 2.13 *Let $(N, \sigma_0, d, p, \alpha)$ be a sequencing situation satisfying (B1) and let σ_0 satisfy (B2). Then the corresponding d-sequencing game (N, v) is balanced.*

Convexity can be established for a special subclass. More precisely, we restrict attention to sequencing situations $(N, \sigma_0, r, p, \alpha)$ with

$(B3)\quad d_i \in \mathbf{N} \text{ and } p_i = 1 \text{ for all } i \in N.$

Now, we state a convexity result, due to Borm *et al.* (1999).

Theorem 2.14 *Let $(N, \sigma_0, d, p, \alpha)$ be a sequencing situation satisfying (B1) and (B3) and σ_0 satisfy (B2). Let (N, v) be the corresponding d-sequencing game. Then (N, v) is convex.*

The proof of this result is shown by establishing the equivalence between this class of d-sequencing games and the class of r-sequencing games described in Theorem 2.12.

Other convex classes of sequencing games that arise from sequencing situations in which due dates are involved can also be found in Borm *et al.* (1999). The related sequencing games arise from the same sequencing situations, but with a different cost criterion.

2.5 On Sequencing Games with Multiple Machines

In this section we consider multiple-machine sequencing games. First, we discuss m-machine sequencing situations in which the weighted completion time criterion is taken into account. Second, we deal with 2-machine sequencing situation with some maximal weighted completion criterion.

The first model in this section deals with multiple-machine sequencing situations with m parallel and identical machines. The weighted completion time criterion is used. Furthermore, each agent has one job that has to be processed on precisely one machine. These sequencing situations, which will be referred to as m-machine sequencing situations, give rise to the class of m-sequencing games. Formally, in an m-machine sequencing situation each agent has one job that has to be processed on precisely one machine. Each job can be processed on any machine. The finite set of machines is denoted by $M = \{1, ..., m\}$ and the finite set of agents is denoted by $N = \{1, ..., n\}$. We assume that each machine starts processing at time 0 and that the processing time of each job is independent of the machine the job is processed on. The processing time of the job of agent i is denoted by $p_i \geq 0$. We assume that every agent has a linear monetary cost function $c_i : [0, \infty) \rightarrow \mathbb{R}$ defined by $c_i(t) = \alpha_i t$ where $\alpha_i > 0$ is a (positive) cost coefficient.

We can use a one to one map $b : N \rightarrow \{1, ..., m\} \times \{1, ..., n\}$ to describe on which machine and in which position on that machine the job of an agent will be processed. Specifically, $b(i) = (r, j)$ means that agent i is assigned to machine r and that (the job of) agent i is in position j on machine r. Such a map b will be called a (processing) schedule.

In the following, an m-machine sequencing situation will be described by (M, N, b^0, p, α), where $M = \{1, ..., m\}$ is the set of machines, $N = \{1, ..., n\}$ the set of agents, b^0 the initial schedule, $p \in [0, \infty)^N$ the processing times, and $\alpha = (\alpha_i)_{i \in N} \in (0, \infty)^N$ the cost coefficients.

The starting time $t(b, i)$ of the job of agent i if processed in a semi-active way according to a schedule b equals

$$t(b, i) = \sum_{j \in N : b(j) \prec b(i)} p_j,$$

where $b(j) \prec b(i)$ if and only if the job of the agents j and i are on the same machine (i.e., $b(j)_1 = b(i)_1$) and j precedes i (i.e., $b(j)_2 < b(i)_2$). Consequently, the completion time $C(b, i)$ of the job of agent i with respect to b is equal to $t(b,i)+p_i$. The total costs $c_b(S)$ of a coalition $S \subseteq N$ with respect to the schedule b is given by

$$c_b(S) = \sum_{i \in S} \alpha_i(C(b,i)).$$

We will restrict our attention to m-machine sequencing situations (M, N, b^0, p, α) that satisfy the following condition: the starting time of a job that is in the last position on a machine with respect to b^0 is smaller than or equal to the completion time of each job that is in the last position with respect to b^0 on the other machines. Formally, for $k \in M$ let i_k be the last agent on machine k with respect to b^0, then we demand for all $k \in M$ that

$$t(b^0, i_k) \leq C(b^0, i_s) \quad \text{for all} \quad s \in M.$$

This condition states that each job that is in the last position of a machine cannot make any profit by joining the end of a queue of any other machine.

The (maximal) cost savings of a coalition S depend on the set of admissible rearrangements of this coalition. We call a schedule $b : N \to \{1, ..., m\} \times \{1, ..., n\}$ admissible for S with respect to b^0 if it satisfies the following two conditions:
(i) Two agents $i, j \in S$ that are on the same machine can only switch if all agents in between i and j on that machine are also members of S;
(ii) Two agents $i, j \in S$ that are on different machines can only switch places if the tail of i and the tail of j are contained in S. The tail of an agent i is the set of agents that follow agent i on his machine, i.e., the set of agents $k \in N$ with $b(i) \prec b(k)$.
The set of admissible schedules for a coalition S is denoted by B_S. An admissible schedule for coalition N will be called a schedule.

By defining the worth of a coalition as the maximum cost savings a coalition can achieve by means of admissible schedules we obtain a cooperative game called an m-sequencing game. Formally, for an m-machine sequencing situation (M, N, b^0, p, α) the corresponding m-*sequencing game* (N, v) is defined by

$$v(S) = \max_{b \in B_S} \{\sum_{i \in S} \alpha_i[C(b^0, i) - C(b, i)]\}$$

for all coalitions $S \in 2^N \backslash \{\emptyset\}$.

Now, we will focus on the balancedness of m-sequencing games. Because 1-machine sequencing situations coincide with s-sequencing implies that 1-sequencing games are balanced. The following theorem, due to Hamers *et al.* (1999), shows that 2-machine sequencing games are balanced.

Theorem 2.15 *Let (M, N, b^0, p, α) be such that $|M| = 2$. Then the corresponding 2-sequencing game (N, v) is balanced.*

Proof. Let $i_1, i_2, ..., i_{m_1}$ be the jobs on machine 1 such that $b^0(i_x) \prec b^0(i_y)$ if $x < y$ and let $i_n, ..., i_{m_1+1}$ be the jobs on machine 2 such that $b^0(i_x) \prec b^0(i_y)$ if $x > y$. Take $\sigma \in \Pi_N$ such that $\sigma(i_j) = j$ for all $j \in N$. Now, it is easy to see that (N, v) is a σ-component additive game. □

An open problem is the balancedness of m-machine sequencing games with $m \geq 3$. However, balancedness results, due to Hamers *et al.* (1999), are obtained for two special classes.

Theorem 2.16 *Let (N, v) be the m-sequencing game that arises from an m-machine sequencing situation (M, N, b^0, p, α) in which $\alpha_i = 1$ for all $i \in N$. Then (N, v) is balanced.*

In Theorem 2.16 we assumed that all cost coefficients are equal to one. This implies that the class of m-sequencing games generated by the unweighted completion time criterion is a subclass of the class of balanced games. Clearly, the balancedness result also holds true in the case that all cost coefficients are equal to some positive constant $c > 0$. Furthermore, a similar result, due to Hamers *et al.* (1999), holds for m-sequencing situations with identical processing times instead of identical cost coefficients.

Theorem 2.17 *Let (N, v) be the m-sequencing game that arises from an m-machine sequencing situation (M, N, b^0, p, α) in which $p_i = 1$ for all $i \in N$. Then (N, v) is balanced.*

The second model that will be discussed in this section considers sequencing situations with two parallel machines. Contrary to previous models in this paper, it is assumed that each agent owns two jobs to be processed, one on each machine. The costs of an agent depend linearly on the final completion time of his jobs. In other words, it depends on the time an agent has to wait until both his jobs have been processed.

Now, the formal description of the model is provided. The set of the two machines is denoted by $M = \{1, 2\}$. There is a finite set of agents $N = \{1, ..., n\}$. We assume that each agent has 2 jobs to be processed, one on each machine. Moreover, we assume that each machine starts processing at time 0 and by the vector $(p_i, q_i)_{i \in N}$ we denote the processing times of the jobs of every agent i, $p_i \geq 0$ for the job to be processed on machine 1 and $q_i \geq 0$ for the job to be processed on machine 2. We also assume that there is an initial scheme (π, φ) of the jobs on the machines where π and φ are the initial orders for the first and the second machine, respectively. Formally, π and φ are bijections from N to $\{1, ..., n\}$ where $\pi(i) = s$ and $\varphi(i) = t$ mean that initially, player i has a job in position s on machine 1 and a job in position t on machine 2 in the initial queues before the machines. Let Π_N be the set of orders of N, i.e., bijections from N to $\{1, ..., n\}$. Then $\Pi_N \times \Pi_N$ denotes the set of possible schemes.

Every agent has a linear cost function $c_i : [0, \infty) \to I\!R$ defined by $c_i(t) = \alpha_i t$ where $\alpha_i > 0$, and where t represents the time player i has to wait to have both his jobs processed.

A 2 parallel machines sequencing situation is a 5-tuple $(M, N, (\pi, \varphi), (p_i, q_i)_{i \in N}, (\alpha_i)_{i \in N})$ and we will refer to it as a 2–PS situation.

Let $(\sigma, \tau) \in \Pi_N \times \Pi_N$ be a scheme. We denote by $C_i(\sigma) = \sum_{j \in N : \sigma(j) \leq \sigma(i)} p_j$ the completion time $C_i(\sigma)$ of the job of agent i on the first machine with respect to the order σ. Similarly, $C_i(\tau) = \sum_{j \in N : \tau(j) \leq \tau(i)} q_i$ denotes the completion time of the job of agent i on the second machine with respect to τ. For every player i we consider the final completion time with respect to (σ, τ), that is $C_i(\sigma, \tau) = \max\{C_i(\sigma), C_i(\tau)\}$. Then the total costs of the agents with respect to (σ, τ) can be written as

$$c_N(\sigma, \tau) = \sum_{i \in N} \alpha_i C_i(\sigma, \tau).$$

A scheme $(\hat{\sigma}, \hat{\tau}) \in \Pi_N \times \Pi_N$ is called optimal for N if total costs are minimized, i.e.,

$$c_N(\hat{\sigma}, \hat{\tau}) = \min_{(\sigma, \tau) \in \Pi_N \times \Pi_N} c_N(\sigma, \tau).$$

Let $(M, N, (\pi, \varphi), (p_i, q_i)_{i \in N}, (\alpha_i)_{i \in N})$ be a 2–PS situation. The maximal cost savings of a set of players $S \subseteq N$ depend on the set of admissible rearrangements of this set of agents S. We call a scheme (σ, τ) an admissible rearrangement for S with respect to (π, φ) if it satisfies the following two properties: two agents $i, j \in S$ can only switch on one machine if all agents in between i and j on that machine with respect to the initial order on that machine are also members of S. Formally, given the initial order $\pi \in \Pi_N$, an admissible order for S on machine 1, is a bijection $\sigma \in \Pi_N$ such that $P(\sigma, i) = P(\pi, i)$ for all $i \in N \backslash S$. Similarly, an admissible order for S on machine 2 is a bijection $\tau \in \Pi_N$ such that $P(\tau, i) = P(\varphi, i)$ for all $i \in N \backslash S$.

Let $\mathcal{A}^1(S)$ and $\mathcal{A}^2(S)$ denote the set of admissible rearrangements of coalition S on machine 1 and machine 2, respectively. The set $\mathcal{A}^1(S) \times \mathcal{A}^2(S)$ is called the set of admissible schemes for S. In other words, we consider an scheme to be admissible for S if each agent outside S has the same completion time on each machine as in the initial scheme. Moreover, the agents of S are not allowed to jump over players outside S.

Then, given a 2–PS situation, the corresponding 2–PS game (N, v) is defined in such a way that the worth of a coalition $S \subseteq N$ is equal to the maximal cost savings the coalition can achieve by means of admissible schemes. Formally,

$$v(S) = \max_{(\sigma,\tau)\in\mathcal{A}^1(S)\times\mathcal{A}^2(S)} \sum_{i\in S} \alpha_i \left(C_i(\pi, \varphi) - C_i(\sigma, \tau)\right)$$

for all $S \in 2^N \backslash \{\emptyset\}$. The following example illustrates that a 2–PS game need not be convex.

Example 2.18 Consider the 2–PS situation with
$N = \{1, 2, 3, 4, 5, 6, 7, 8, 9\}$, $p = (1, ..., 1)$, $\alpha = (1, ..., 1)$ and the initial scheme given by:

π	5	6	7	1	2	3	8	9	4
φ	7	8	3	4	5	6	1	9	2

Take $S = \{1, 3\}$, $T = \{1, 3, 4, 5, 6\}$ and $i = 2$. Optimal schemes are:

$\hat{\sigma}$	×	×	×	3	1	2	×	×	×	for $S \cup \{i\}$,
$\hat{\tau}$	×	×	3	×	×	×	1	×	2	

$\hat{\sigma}$	5	6	×	1	×	3	×	×	4	for T and
$\hat{\tau}$	×	×	5	6	1	3	4	×	×	

$\hat{\sigma}$	5	6	×	3	1	2	×	×	4	for $T \cup \{i\}$.
$\hat{\tau}$	×	×	5	3	6	1	4	×	2	

Let (N, v) be the corresponding 2-PS sequencing game, then $v(T \cup \{i\}) - v(T) = 6 - 6 < 2 - 0 = v(S \cup \{i\}) - v(S)$. Hence, v is not convex. ◁

However, Calleja *et al.* (2001) show that simple 2-PS games, i.e., games that arise from situations in which all processing times and cost coefficients are equal to one, are balanced.

Theorem 2.19 *Let $(M, N, (\pi, \varphi), (p_i, q_i)_{i \in N}, (\alpha_i)_{i \in N})$ be a simple 2−PS situation and let (N, v) be the corresponding 2-PS game. Then (N, v) is balanced.*

Another class of sequencing games that arises from multiple machine sequencing situations are the FD-sequencing games, introduced in van den Nouweland *et al.* (1992). These games arise from flow shops with a dominant machine. They present an algorithm that provides an optimal order for these sequencing situations. If the first machine is the dominant machine the class of FD-sequencing games coincides with the class of s-sequencing games. In all other cases, the FD-sequencing games need not be balanced.

2.6 On Sequencing Games with more Admissible Rearrangements

This section discusses some classes of one machine sequencing situations in which the set of admissible rearrangements is enlarged. First, we consider one-relaxed sequencing situations, introduced in van Velzen and Hamers (2001), a restriction of relaxed sequencing situations, discussed in Curiel *et al.* (1993), and the corresponding games. Next, we consider rigid sequencing situations and related games, also introduced in van Velzen and Hamers (2001).

One-relaxed sequencing situations are similar to the one-machine sequencing situations discussed in Section 2.3, i.e., a one-relaxed sequencing situation is described by $(N, j, \sigma_0, p, \alpha)$, where $N = \{1, ..., n\}$ is the set of players, $j \in N$ the relaxed player, $\sigma_0 : N \to \{1, ..., n\}$, $p = (p_i)_{i \in N} \in (0, \infty)^N$ and $\alpha = (\alpha_i)_{i \in N} \in (0, \infty)^N$. Also starting time and completion time are defined identically.

Now, let the player set $N = \{1, 2, ..., n\}$ and the relaxed player $j \in N$ be fixed. We want to determine the maximal cost savings of a coalition S whose members decide to cooperate. For this, we have to define which

rearrangements of a coalition S are admissible with respect to the initial order. At this point the relaxed player will be used, which will create the difference between s-sequencing games. To define *admissible* rearrangements we distinguish between two sets of coalitions: $S \subset N\backslash\{j\}$, i.e., coalitions that do not contain the relaxed player, and $S, j \in S \subset N$, i.e., coalitions that contain the relaxed player. A bijection $\sigma : N \to \{1, ..., n\}$ is called *admissible* for $S \subset N\backslash\{j\}$ if it satisfies the conditions that the starting time of each agent outside the coalition S is equal to his starting time in the initial order, and the agents of S are not allowed to jump over players outside S, i.e., $P(\sigma_0, i) = P(\sigma, i)$ for all $i \in N\backslash S$. Hence, a coalition S that does not contain the relaxed player has the same set of admissible rearrangements as a coalition in a s-sequencing game.
A bijection $\sigma : N \to \{1, ..., n\}$ is called *admissible* for S with $j \in S$ if it satisfies the following two conditions:
(i) The starting time of each agent outside the coalition S is less than or equal to his starting time in the initial order: $t_{\sigma,i} \leq t_{\sigma_0,i}$ for all $i \in N\backslash S$.
(ii) If there exists an $m \in S$, such that $p_j \leq p_m$, then the agents of $S\backslash\{j, m\}$ are not allowed to jump over players outside S:

$$P(\sigma_0, i)\backslash\{j, m\} = P(\sigma, i)\backslash\{j, m\} \quad \text{for all} \quad i \in N\backslash S,$$

m has the same predecessors with respect to $N\backslash S$ in σ_0 as j in σ:

$$P(\sigma_0, m) \cap (N\backslash S) = P(\sigma, j) \cap (N\backslash S),$$

and j has the same predecessors with respect to $N\backslash S$ in σ_0 as m in σ:

$$P(\sigma_0, j) \cap (N\backslash S) = P(\sigma, m) \cap (N\backslash S).$$

Hence, a rearrangement is admissible if player j is the only player that can select another player in S and switch with this player (even if these players have to jump over players outside S), as long as the starting times of players outside S do not increase. Moreover, the jobs in S that are not job j, can only switch positions in the connected parts of S, except the player that is selected by j. The set of admissible rearrangements of a coalition S is denoted by $\mathcal{B}(S)$.

By defining the worth of a coalition S as the maximum cost savings coalition S can achieve by means of an admissible rearrangement we obtain again a sequencing game. Formally, for a 1-relaxed sequencing situation $(N, j, \sigma_0, p, \alpha)$ the corresponding *1-relaxed sequencing game* (N, v) is defined by

$$v(S) = \max_{\sigma \in \mathcal{B}(S)} \{\sum_{i \in S} \alpha_i [C(\sigma_0, i) - C(\sigma, i)]\}$$

for all $S \in 2^N \backslash \{\emptyset\}$.

Now, it can be shown, see van Velzen and Hamers (2001), that a specific marginal vector is in the core of a 1-relaxed sequencing game.

Theorem 2.20 *Let $(N, j, \sigma_0, p, \alpha)$ be a 1-relaxed sequencing situation and let (N, v) be the corresponding game. Then (N, v) is balanced.*

However, the following example shows that in general 1-relaxed sequencing games need not be convex.

Example 2.21 Consider the 1-relaxed sequencing situation $(N^4, \sigma_0, p, \alpha)$, $N^4 = \{1, 2, 3, 4\}$, $\sigma_0 = (1, 2, 3, 4)$, $p = (1, 2, 1, 1)$ and $\alpha = (1, 1, 3, 4)$. Take $S = \{4\}$, $T = \{3, 4\}$ and $i = \{1\}$. Let (N, v) be the corresponding 1-relaxed sequencing game. Then $v(S \cup \{i\}) - v(S) = v(\{1, 4\}) - v(\{4\}) = v(\{1, 4\}) = 12$, and $v(T \cup \{i\}) - v(T) = v(\{1, 3, 4\}) - v(\{3, 4\}) = 12 - 1 = 11$. Because $v(T \cup \{i\}) - v(T) = 11 < 12 = v(S \cup \{i\}) - v(S)$, we conclude that (N, v) is not convex. $\triangleleft$

Other possible relaxations of the set of admissible rearrangements are discussed in Curiel *et al.* (1993).

Rigid sequencing situations are described similarly to the one-machine sequencing situations discussed in Section 2.3, i.e., a rigid sequencing situation is described by (N, σ_0, p, α), where $N = \{1, ..., n\}$ is the set of players, $\sigma_0 : N \to \{1, ..., n\}$, $p = (p_i)_{i \in N} \in (0, \infty)^N$ and $\alpha = (\alpha_i)_{i \in N} \in (0, \infty)^N$. Also starting time and completion time are defined analogously.

Now, we want to determine the maximal cost savings of a coalition S whose members decide to cooperate. For this, we have to define which rearrangements of coalition S are admissible with respect to the initial order. A bijection $\sigma : N \to \{1, ..., n\}$ is called *admissible* for S if it satisfies the following two conditions:

(i) All players outside S have the same starting time: $t_{\sigma_0, j} = t_{\sigma, j}$ for all $j \in N \backslash S$,

(ii) In both orders the starting times coincide: $t_{\sigma_0, i} = t_{\sigma, \sigma^{-1}(\sigma_0(i))}$ for all $i \in N$.

Hence, players in S can only switch if their processing times are equal. The set of admissible rearrangements of a coalition S is denoted by $\mathcal{R}(S)$.

By defining the worth of a coalition S as the maximum cost savings coalition S can achieve by means of an admissible rearrangement we obtain a cooperative game called a rigid sequencing game. Formally, for a rigid sequencing situation (N, σ_0, p, α) the corresponding *rigid sequencing game* (N, v) is defined by

$$v(S) = \max_{\sigma \in \mathcal{R}(S)} \{\sum_{i \in S} \alpha_i [C(\sigma_0, i) - C(\sigma, i)]\}$$

for all $S \in 2^N \backslash \{\emptyset\}$.

Van Velzen and Hamers (2001) show that rigid sequencing games are balanced.

Theorem 2.22 *Let (N, σ_0, p, α) be a sequencing situation and let (N, v) be the corresponding rigid sequencing game. Then (N, v) is balanced.*

The proof is based on the fact that rigid sequencing games are a subclass of the class of permutation games. The latter one, introduced in Tijs *et al.* (1984), is a class of (totally) balanced games. In particular, if all processing times are equal, rigid games coincide with the class of permutation games. This implies immediately that rigid games in general are not convex, because not all permutation games are convex.

References

Borm, P., G. Fiestras-Janeiro, H. Hamers, E. Sánchez E., and M. Voorneveld (1999): "On the convexity of games corresponding to sequencing situations with due dates," CentER Discussion Paper 1999-49. To appear in: *European Journal of Operational Research.*

Calleja, P., P. Borm, H. Hamers, F. Klijn, and M. Slikker (2001): "On a new class of parallel sequencing situations and related games," CentER Discussion Paper 2001-3.

Curiel, I. (1997): *Cooperative game theory and applications.* Dordrecht: Kluwer Acacemic Publishers.

Curiel, I., G. Pederzoli, and S. Tijs (1989): "Sequencing games," *European Journal of Operational Research*, 40, 344–351.

Curiel, I., J. Potters, V. Rajendra Prasad, S. Tijs, and B. Veltman (1993): "Cooperation in one machine scheduling," *Methods of Operations Research*, 38, 113–131.

Curiel, I., J. Potters, V. Rajendra Prasad, S. Tijs, and B. Veltman (1994): "Sequencing and cooperation," *Operations Research*, 42, 566–568.

Curiel, I., and S. Tijs (1986): "Assignment games and permutation games," *Methods of Operations Research*, 54, 323–334.

Gale, D. (1984): "Equilibrium in a discrete exchange economy with money," *International Journal of Game Theory*, 13, 61–64.

Granot, D., H. Hamers, and S. Tijs (1999): "On some balanced, totally balanced and submodular delivery games," *Mathematical Programming*, 86, 355–366.

Granot, D., and G. Huberman (1981): "Minimum cost spanning tree games," *Mathematical Programming*, 21, 1–18.

Hamers, H., P. Borm, and S. Tijs (1995): "On games corresponding to sequencing situations with ready times," *Mathematical Programming*, 70, 1–13.

Hamers, H., F. Klijn, and J. Suijs (1999): "On the balancedness of multi-machine sequencing games," *European Journal of Operational Research* 119, 678–691.

Hamers, H., J. Suijs, S. Tijs, and P. Borm (1996): "The split core for sequencing games," *Games and Economic Behavior*, 15, 165–176.

Ichiishi, T. (1981): "Super-modularity: applications to convex games and the greedy algorithm for LP," *Journal of Economic Theory*, 25, 283–286.

Klijn, F., S. Tijs, and H. Hamers (2000): "Balancedness of permutation games and envy-free allocations in indivisible good economies, *Economics Letters*, 69, 323–326.

Maschler, M., B. Peleg, and L. Shapley (1972): "The kernel and bargaining set of convex games," *International Journal of Game Theory*, 2, 73–93.

Owen, G. (1975): "On the core of linear production games," *Mathematical Programming*, 9, 358–370.

Potters, J., I. Curiel, and S. Tijs (1992): "Traveling Salesman Games," *Mathematical Programming*, 53, 199–211.

Potters, J., and H. Reijnierse (1995): "Γ-component additive games," *International Journal of Game Theory*, 24, 49–56.

Shapley, L. (1971): "Cores of convex games," *International Journal of Game Theory*, 1, 11–26.

Shapley, L., and M. Shubik (1972): "The assignment game I: The core," *International Journal of Game Theory*, 1, 111–130.

Smith, W. (1956): "Various optimizers for single-stage production," *Naval Research Logistics Quarterly*, 3, 59–66.

Tijs, S. (1981): "Bounds for the core and the τ-value," in: O. Moeschlin and D. Pallaschke (eds.), *Game Theory and Mathematical Economics*. Amsterdam: North-Holland, 123–132.

Tijs, S. (1991): "LP-games and combinatorial optimization games," *Cahiers du Centre d'Etudes de Recherche Operationnelle*, 34, 167–186.

Tijs, S., T. Parthasarathy, J. Potters, and V. Rajendra Prasad (1984): "Permutation games: another class of totally balanced games," *OR Spektrum*, 6, 119–123.

van den Nouweland, A., M. Krabbenborg, and J. Potters (1992): "Flowshops with a dominant machine," *European Journal of Operational Research* 62, 38–46.

van Velzen, B., and H. Hamers (2001): "Relaxations on sequencing games," *Working paper Tilburg University.*

Chapter 3

Game Theory and the Market

BY ERIC VAN DAMME AND DAVE FURTH

3.1 Introduction

Based on the assumption that players behave rationally, game theory tries to predict the outcome in interactive decision situations, i.e. situations in which the outcome is determined by the actions of all players and no player has full control. The theory distinguishes between two types of models, cooperative and non-cooperative. In models of the latter type, emphasis is on individual players and their strategy choices, and the main solution concept is that of Nash equilibrium (Nash, 1951). Since the concept as originally proposed by Nash is not completely satisfactory—it does not adequately take into account that certain threats are not credible, many variations have been proposed, see van Damme (2002b), but in their main idea these all remain faithful to Nash's original insight. The cooperative game theory models, instead, focus on coalitions and outcomes, and, for cooperative games, a wide variety of solution concepts have been developed, in which few unifying principles can be distinguished. (See other chapters in this volume for an overview.) The terminology that is used sometimes gives rise to confusion; it is not the case that in non-cooperative games players do not wish to cooperate and that in cooperative games players automatically do so. The difference instead is in the level of detail of the model; non-cooperative models assume that all possibilities for cooperation have

P. Borm and H. Peters (eds.), Chapters in Game Theory, 51–81.

been included as formal moves in the game, while cooperative models are "incomplete" and allow players to act outside of the detailed rules that have been specified.

One of us had the privilege and the luck to follow undergraduate courses in game theory with Stef Tijs. There were courses in non-cooperative theory as well as in cooperative theory and both were fun. When that author had passed his final (oral) exam, he was still puzzled about the relationships between the models and the solution concepts that had been covered, and he asked Stef a practical question: when to use a cooperative model and when to use a non-cooperative one? That author does not recall the answer, but he now considers the question to be a nonsensical one: it all depends on what one wants to achieve and what is feasible to do. Frequently, it will not be possible to write down an explicit non-cooperative game, and even if this is possible, one should be aware that players may attempt to violate the rules that the analyst believes to apply. On the other hand, a cooperative model may be pitched at a too high level of abstraction and may contain too little detail to allow the theorist to come up with a precise prediction about the outcome. In a certain sense, the large variety of solution concepts that one finds in cooperative game theory is a natural consequence of the model that is used being very abstract It also follows from these considerations that cooperative and non-cooperative models are complements to each other, rather than competitors.

Our aim in this chapter is to demonstrate the complementarity between the two types of game theory models and to illustrate their usefulness for the analysis of actual markets. Section 3.2 provides a historical perspective and briefly discusses the views expressed in Von Neumann and Morgenstern (1953) and Nash (1953). Section 3.3 focuses on bargaining games, while Section 3.4 discusses oligopoly games and markets. Auctions are the topic of Section 3.5. Section 3.6 concludes.

3.2 Von Neumann, Morgenstern and Nash

As von Neumann and Morgenstern (1953) argue, there is not much point in forming a coalition in 2-person zero-sum games. In this case, both the cooperative and the non-cooperative theory predict the same outcome. Furthermore, in 2-person non-zero-sum games, there is only one coalition that can possibly form and it will form when it is attractive to form it and the rules of the game do not stand in the way. The remaining question then is how the players will divide the surplus, a question that we will return to in Section 3.3. The really interesting problems start to appear when there are at least three players. Von Neumann and Morgenstern (1953, Chapter V) argue that in this case the game cannot sensibly be analyzed without coalitions and side-payments, for, even if these are not explicitly allowed by the rules of the game, the players will try to form coalitions and make side payments outside of these formal rules.

To illustrate their claim, the founding fathers of game theory start from a simple non-cooperative game. Assume there are three players and each player can point to one of the others if he wants to form a coalition with him. In this case, the coalition $\{i, j\}$ forms if and only if i points to j and j points to i. The rules also stipulate that if $\{i, j\}$ forms, the third player, k, has to pay 1 money unit to each of i and j. Formally this game of coalition formation can, therefore, be represented by the normal form (non-cooperative) game in Figure 3.1.

	1	3
2	1, 1, -2	0, 0, 0
3	1, -2, 1	1, -2, 1
	1	

	1	3
2	1, 1, -2	-2, 1, 1
3	0, 0, 0	-2, 1, 1
	2	

Figure 3.1: *A non-cooperative game of coalition formation. Player 1 chooses a row, player 2 a column, and players 3 a matrix.*

The game in Figure 3.1 has several pure Nash equilibria; it also has a mixed Nash equilibrium in which each player chooses each of the others with equal probability. Von Neumann and Morgenstern start their analysis from a non-cooperative point of view, i.e. as if the above matrix tells the whole story:

> "Since each player makes his personal move in ignorance of those of the others, no collaboration of the players can be established during the course of play" (p. 223).

Nevertheless, von Neumann and Morgenstern argue that the whole point of the game is to form a coalition, and they conclude that, if players are prevented to do so within the game, they will attempt to do so outside. They realize that this raises the question of why such outside agreements will be kept, and they pose the crucial question what, if anything, *enforces* the "sanctity" of such agreements? They answer this question in the following way

> "There may be games which themselves—by virtue of the rules of the game (...) provide the mechanism for agreements and their enforcement. But we cannot base our considerations on this possibility since a game *need* not provide this mechanism; (...) Thus there seems no escape from the necessity of considering agreements concluded outside the game. If we do not allow for them, then it is hard to see what, if anything, will govern the conduct of a player in a simple majority game" (p. 223).

The reader may judge for himself whether, and in which circumstances, he considers this argument to be convincing. In any case, if one accepts the argument that a convincing theory cannot be formulated without auxiliary concepts such as "agreements" and "coalitions", then one is also naturally led to the conclusion that side-payments will form an integral part of the theory. This latter argument is easily seen by considering a minor modification of the game of Figure 3.1. Suppose that if the coalition $\{1,2\}$ would form the payoffs would be $(1+\varepsilon, 1-\varepsilon, -2)$ and that if $\{1,3\}$ would form, the payoffs would be $(1+\varepsilon, -2, 1-\varepsilon)$: what outcome of the game would result in this case? Von Neumann and Morgenstern argue that the advantage of player 1 is quite illusory: if player 1 would insist on getting $1+\varepsilon$ in the coalition $\{1,2\}$, then 2 would prefer to form the coalition with 3, and similarly with the roles of the weaker players reversed. Consequently, in order to prevent the coalition of the two "weaker" players from forming, player 1 will offer a side payment of ε to the one he is negotiating with. Consequently, von Neumann and Morgenstern conclude:

> "It seems that what a player can get in a definite coalition depends not only on what the rules of the game provide for that eventuality, but also on the other (competing) possibilities of coalitions for himself and for his partner. Since the rules of the game are absolute and inviolable, this means that under certain conditions *compensations* must be paid among coalition partners; i.e., that a player must have to pay a well-defined price to a prospective coalition partner. The amount of the compensations will depend on what other alternatives are open to each of the players" (p. 227).

Obviously, if one concludes that coalitions and side-payments have to be considered in the solution, then the natural next step is to see whether the solution can be determined by these aspects alone, and it is that problem that Von Neumann and Morgenstern then set out to solve in the remaining 400 pages of their book.

John Nash refused to accept that it was necessary to include elements outside the formal structure of the game to develop a convincing theory of games. His thesis (Nash, 1950a), of which the mathematical core was published a bit later (Nash, 1951) opens with

> "Von Neumann and Morgenstern have developed a very fruitful theory of two-person zero-sum games in their book *Theory of Games and Economic Behavior*. This book also contains a theory of n-person games of a type which we would call cooperative. This theory is based on an analysis of the interrelationships of the various coalitions which can be formed by the players of the game. Our theory, in contradistinction, is based on the *absence* of coalitions in that it is assumed that each participant acts independently, without collaboration or communication with any of the others. The notion of an *equilibrium point* is the basic ingredient in our theory."

Hence, Nash was the first to introduce the formal distinction between the two classes of games. After having given the formal definition of a non-cooperative game, Nash then defines the equilibrium notion, he proves that any finite game has at least one equilibrium, he derives properties of equilibria, he discusses issues of robustness and equilibrium selection and finally he discusses interpretational issues. Even though the thesis is short, it will be clear that it accomplishes a lot. In the remainder of this section, we give a brief sketch of the mathematical core of Nash's thesis, as it also allows us to introduce some notation.

A non-cooperative game is a tuple $< S_i, u_i >_{i \in I}$ where I is a nonempty set of players, S_i is the strategy set of player i and $u_i : S \to I\!R$ (where $S = X_{i \in I} S_i$) is the payoff function of player i. This formal structure had already been introduced by von Neumann and Morgenstern, who had also argued that, for finite S_i, it was natural to introduce mixed strategies. A mixed strategy σ_i of player i is a probability distribution on S_i. In what follows we write k to denote a generic pure strategy and we write σ_i^k for the probability that σ_i assigns to k. If $\sigma = (\sigma_1, \ldots, \sigma_I)$ is a combination of mixed strategies, we may write $u_i(\sigma)$ for player i's expected payoff when σ is played. Von Neumann and Morgenstern had proved the important result that for rational players it was sufficient to look at expected payoffs. In other words, it is assumed that payoffs are von Neumann Morgenstern utilities. Nash now defines an equilibrium point as a mixed strategy combination σ^* such that each player's mixed strategy σ_i^* maximizes his payoff if the strategies of the others (denoted by σ_{-i}^*) are held fixed, hence

$$u_i(\sigma^*) = \max_{\sigma_i} u_i(\sigma_i, \sigma_{-i}^*) \text{ for all } i \in I.$$

Nash's main result is that in finite games (i.e. I and all S_i are finite sets) at least one equilibrium exists. The proof is so elegant that it is worthwhile to give it here. For $i \in I$ and $k \in S_i$, write

$$U_i^k(\sigma) = \max(0, u_i(k, \sigma_{-i}) - u_i(\sigma))$$

and consider the map $f : \Sigma \to \Sigma$ defined (componentwise) by

$$f_i^k(\sigma) = (\sigma_i^k + U_i^k(\sigma))/(1 + \sum_{\ell \in S_i} U_i^\ell(\sigma)),$$

then f is a continuous map, that maps the convex set Σ (of all mixed strategy profiles) into itself, so that, by Brouwer's fixed point theorem, a fixed point σ^* exists. It is then easily seen that such a σ^* is an equilibrium point of the game.

The section "Motivation and Interpretation" from Nash's thesis was not included in the published version (Nash, 1951). In retrospect, this is to be regretted as it led to misunderstandings and delayed progress in game theory for some time. Nash provided two interpretations. The first "rationalistic interpretation" argues why equilibrium is relevant when the game is played by fully rational players, the second "mass action representation" argues that equilibrium might be obtained as a result of ignorant players learning to play the game over time when the game is repeated. We refer the reader to van Damme (1995) for further discussion on these interpretations; here we confine ourselves to the remark that the rationalistic interpretation, the view of a solution as a convincing theory of rationality, had already been proposed in von Neumann and Morgenstern, see Section 17.3 of their book. However, the founding fathers had not followed up their own suggestion. In addition, they had come to the conclusion that it was necessary to consider set-valued solution concepts. Again, Nash was not convinced by their arguments and he found it a weak spot in their theory.

3.3 Bargaining

In this section we illustrate the complementarity between game theory's two approaches for the special case of bargaining problems.

As referred to already at the end of the previous section, the theory that von Neumann and Morgenstern developed generally allows multiple outcomes. Consider the special case of a simple bargaining problem. Assume there is one seller who has one object for sale, who does not value this object himself, and that there is one buyer that attaches value 1 to it, with both players being risk neutral. For what price will the object be sold? Von Neumann and Morgenstern discuss this problem in Section 61 of their book where they come to the conclusion that "a satisfactory theory of this highly simplified model should leave the entire interval (i.e. in this case [0,1]) available for p" (p. 557).

The above is unsatisfactory to Nash. In Nash (1950b), he writes:

> "In *Theory of Games and Economic Behavior* a theory of n-person games is developed which includes as a special case the two-person bargaining problem. But the theory developed there makes no attempt to find a value for a given n-person game, that is, to determine what it is worth to each player to have the opportunity to engage in the game (...) It is our opinion that these n-person games should have values."

Nash then postulates that a value exists and he sets out to identify it. To do so, he uses the axiomatic method, that is

> "One states as axioms several properties that it would seem natural for the solution to have and then one discovers that the axioms actually determine the solution uniquely" (Nash, 1953, p. 129)

In his 1950b paper, Nash adopts the cooperative approach, hence, he assumes that the solution can be identified by using only information about what outcomes and coalitions are possible. Without loss of generality, let us normalize payoffs such that each player has payoff 0 if players do not cooperate and that cooperation pays, i.e., there is at least one feasible payoff vector u with $u_1, u_2 > 0$. In this case, the solution then should just depend on the set of payoffs that are possible when players do cooperate. Let us write $f(S)$ for the solution when the set of feasible payoff vectors is S. This set S will be convex, as players can randomize.

Obviously, such trivialities as $f(S) \in S$ and $f_i(S) \geq 0$ for $i = 1, 2$ should be satisfied. In addition, the solution should be independent of which utility function is used to represent the given players preferences and it should be symmetric ($u_1 = u_2$) when the game is symmetric. All these things are undebatable. It is quite remarkable that only one additional axiom is needed to uniquely determine the solution for each bargaining problem. This is the Axiom of Independence of Irrelevant Alternatives:

$$\text{If } S \subset T \text{ and } f(T) \in S, \text{ then } f(S) = f(T).$$

Again the proof of this major result is so elegant, that we cannot resist to give it. Define $g(S)$ as that point in S that maximizes $u_1 u_2$ in $S \cap I\!R^2_+$. Then by rescaling utilities we may assume $g(S) = (1, 1)$, and it follows that the line $u_2 + u_2 = 2$ is a supporting hyperplane for S at $(1, 1)$. (It separates the convex set S from the convex set $\{(u_1, u_2); u_1 u_2 \geq 1\}$.) Now let T be the set $\{(u_1, u_2) \in I\!R^2_+; u_1 + u_2 \leq 2\}$. Then, by symmetry, $f(T) = (1, 1)$, hence IIA implies $f(S) = (1, 1)$. We have, therefore, established that there is only one solution satisfying the IIA axiom: it is the point where the product of the players' utilities is maximized. As a corollary we obtain that, in the simple seller-buyer example that we started out with, the solution is a price of $\frac{1}{2}$.

One interpretation of the above solution is that it will result when players can bargain freely. Obviously, if the players would be severely restricted in their bargaining possibilities, then a different outcome may result. For example, in the above buyer-seller game, if the seller can make a take it or leave it offer, the buyer will be forced to pay a price of (almost) one. Similarly, if the buyer would have all the bargaining power, the price would be (close to) zero. The advantage of non-cooperative modelling is that it allows to analyze each specific bargaining procedure and to predict the outcome on the basis of detailed modelling of the rules; the drawback (or realism?) of that model is that the outcome may crucially depend on these details. The symmetry assumption in Nash's axiomatic model represents something like players having equal bargaining power and this is obviously violated in these take it or leave it games. It is not clear how such asymmetric games could be relevant for players that are otherwise completely symmetric. Nash (1953) contains important modelling advise for non-cooperative game theorists. He writes that in the non-cooperative approach

> "the cooperative game is reduced to an non-cooperative game. To do this, one makes the players' steps of negotiation in

> the cooperative game become moves in the non-cooperative model. Of course, one cannot represent all possible bargaining devices as moves in the non-cooperative game. The negotiation process must be formalized and restricted, but in such a way that each participant is still able to utilize all the essential strength of his position" (Nash, 1953, p. 129).

Nash also writes that the two approaches to solve a game are complementary and that each helps to justify and clarify the other. To complement his cooperative analysis, Nash studies the following simultaneous demand game: each player i demands a certain utility level u_i that he should get; if the demands are compatible, that is, if $(u_1, u_2) \in S$, then each player gets what he demanded, otherwise disagreement (with payoff 0) results. At first it seems that this non-cooperative game does not fulfill our aims, after all any Pareto optimal outcome of S corresponds to a Nash equilibrium of the game, and so does disagreement. Nash, however, argues that one of these equilibria is distinguished in the sense that it is the only one that is robust against small perturbations in the data. Of course, this unique robust equilibrium is then seen to correspond to the cooperative solution of the game. Specifically, Nash assumes that players are somewhat uncertain about what outcomes are feasible. Let $p(u)$ be the probability that u is feasible, with $p(u) = 1$ if $u \in S$ and p a continuous function that falls rapidly to zero outside of S. With uncertainty given by p, player i's payoff function is now given by $u_i p(u)$ and it is easily verified that any maximum of the map $u_1 u_2 p(u)$ is an equilibrium of this slightly perturbed game. Note that all these equilibria converge to the Nash solution (the maximum of $u_1 u_2$ on S) when p tends to the characteristic function of S and that, for nicely behaved p, the perturbed game will only have equilibria close to the Nash solution. Consequently, only the Nash solution constitutes a robust equilibrium of the original demand game.

The above coincidence certainly is not an isolated result, the Nash solution also arises in other natural non-cooperative bargaining models. As an example, we discuss Rubinstein's (1981) alternating offer bargaining game. Consider the simple seller buyer game that we started this section with and assume bargaining proceeds as follows, until agreement is reached or the game has come to an end. In odd numbered periods $(t = 1, 3, \ldots)$, the seller proposes a price to the buyer and the buyer responds by accepting or rejecting the offer; in even numbered periods $(t = 2, 4, \ldots)$, the roles of the players are reversed and the buyer has the initiative; after each rejection, the game stops with positive but small probability α. Rubinstein shows that this game has a unique (subgame perfect) equilibrium, and that, in equilibrium, agreement is reached immediately. Let p_s (resp. p_b) be the price proposed by the seller (resp. the buyer). The seller realizes that if the buyer rejects his first offer, the buyer's expected utility will be $(1-\alpha)(1-p_b)$, hence, the seller will not offer a higher utility, nor a lower. Consequently, in equilibrium we must have

$$1 - p_s = (1 - \alpha)(1 - p_b)$$

and, by a similar argument

$$p_b = (1 - \alpha)p_s.$$

It follows that the equilibrium prices are given by

$$p_s = 1/(2 - \alpha) \qquad p_b = (1 - \alpha)/(2 - \alpha)$$

and as α tends to zero (when the first mover advantage vanishes and the game becomes symmetric), we obtain the Nash bargaining solution.

We conclude this section with the observation that also in Von Neumann and Morgenstern (1953) both cooperative and non-cooperative approaches are mixed. In Section 3.2, we discussed the 3-player zero-sum game and the need to consider coalitions and side-payments. In Section 22.2 of the *Theory of Games and Economic Behavior*, the general such game is considered: if coalition $\{j, k\}$ forms, then player i has to pay a_i to this coalition ($i \neq j \neq k$). What coalition will form and how will it split the surplus? To answer this question, von Neumann and Morgenstern consider a demand game. They assume that each player i specifies a price p_i for his participation in each coalition. Obviously, if p_i is too large, j and k will prefer to cooperate together rather than to form a coalition with i. Given p_i, j cannot expect more than $a_k - p_i$ in $\{i, j\}$ while k cannot expect more than $a_j - p_i$ in $\{i, k\}$, hence i will price himself out of the market if

$$(a_k - p_i) + (a_j - p_i) < a_i.$$

Consequently, each player i cannot expect more than

$$p_i = (-a_i + a_j + a_k)/2.$$

If the game is essential and it pays to form a coalition, i.e. $a_1 + a_2 + a_3 > 0$, then the above system of three questions with three unknown

($i = 1, 2, 3$) has a unique solution. Each player i can reasonably demand p_i: we can predict how the coalition that will form will split the surplus, but all three possible coalitions are equally likely, hence, we cannot say which coalition will form.

3.4 Markets

In this section, we briefly discuss the application of game theory to oligopolistic markets. In line with the literature, most of the discussion will be based on non-cooperative models, but we will see that also here cooperative analysis plays its role.

In a non-cooperative *oligopoly game,* the players are firms, the strategy sets are compact and connected subsets of an Euclidean space, and the payoffs are the profits of the firms. As Nash's existence theorem only applies to finite games, a first question is whether equilibrium exists. Here we will confine ourselves to the specific case where the strategy set of player i, denoted X_i, is a closed and connected interval in $\mathbb{R}$. Hence, in essence we assume that each firm sells just one product, of which it either sets the price or the quantity. We speak of a *Cournot game* when the strategies are quantities, of a *Bertrand game* when the strategies are prices. Write X for the Cartesian product of all X_i. For player $i \in N$ his *best response correspondence* is the map $\mathcal{B}_i$ that assigns to each $x \in X$ the set of all $y_i \in X_i$ that maximize this player's payoff against x. Note that in the two-player case, $\mathcal{B}_i$ (viewed as a function of x_{-i}) will typically be decreasing in the case of a Cournot game and be increasing in the case of Bertrand. In the former case, we speak of strategic substitutes, in the latter of strategic complements. We write $\mathcal{B}$ for the vector of all $\mathcal{B}_i$. When for each player $i \in N$ the profit function is continuous on X and is quasi–concave in $x_i \in X_i$ for fixed $x_{-i} \in X_{-i}$, then the conditions of the Kakutani fixed point theorem are satisfied ($\mathcal{B}$ is an upper-hemi continuous map, for which all image sets are non-empty compact and convex), hence, the oligopoly game has a Nash equilibrium. When products are differentiated, these conditions will typically be satisfied, but with homogeneous products, they may be violated. For example, in the Bertrand case, without capacity constraints and with no possibility to ration demand, the firm with the lowest price will typically attract all demand, hence, demand functions and profit functions are discontinuous. Dasgupta and Maskin (1986) contains useful existence theorems for cases like these. (See also Furth, 1986.) Of course, the equilibrium is not necessarily unique.

The first formal analysis of an oligopolistic market was performed by Cournot, who analyzed a duopoly in which two firms sell a homogeneous (consumption) good to the consumers, see Cournot(1838). He writes

> "Let us now imagine two proprietors and two springs of which the qualities are identical, and which, on account of their similar positions, supply the same market in competition. In this case the price is necessarily the same for each proprietor. [...]; and *each of them independently* will seek to make this income as large as possible." (Cournot, 1838, cited from Daughety, 1988, p. 63)

In Cournot's model, a strategy of a firm is the quantity supplied to the market. Cournot argued that if firm i supplies q_i firm j will have an incentive to supply the quantity q_j that is the best response to q_i and he defined an equilibrium as a situation in which each of the duopolists is at a best response. Hence, the solution that Cournot proposed, the Cournot equilibrium, can be viewed as a Nash equilibrium. Nevertheless, Cournot's interpretation of the equilibrium seems to have been very different from the modern "rationalistic" interpretation of equilibrium. In retrospect, it seems to be more in line with the "mass action interpretation" of Nash. The following citations are revealing about the relationship between the works of Nash and Cournot:

> "After one hundred and fifty years the Cournot model remains the benchmark of price formation under oligopoly. Nash equilibrium has emerged as the central tool to analyze strategic interactions and this is a fundamental methodological contribution which goes back to Cournot's analysis." (Vives, 1989, p.511)

> "After the appearance of the Nash equilibrium, what we witness is the gradual injection of a certain ambiguity into Cournot's account in order to make it interpretable in terms of Nash. Following Nash, Cournot is reread and reinterpreted. This may have several different motivations, of which we here present concrete evidence of two. In one case, it is a way of anchoring, or stabilizing, the new and still floating idea of the Nash equilibrium. By showing that somebody in the past—and all the better if it is an eminent figure—seems

> to have had 'the same idea' in mind, the Nash equilibrium is given a history, it is legitimised, and the case for game theory is strengthened. In the other case, the motivation is to detract from the originality of Nash's idea, maintaining that 'it was always there', i.e., Nash has said nothing new." (Leonard, 1994, p. 505)

Bertrand (1883) criticized Cournot for taking quantities as strategic variables and he suggested to take prices instead. It matters a lot for the outcome what the strategic variables are. In a Cournot game, a player assumes that the opponent's quantity remains unchanged, hence, this corresponds to assuming that the opponent raises his price if I raise mine. Clearly such a situation is less competitive than one of Bertrand competition in which a firm assumes that the opponent maintains his price when it raises its own price. Consequently, prices are frequently lower in the Bertrand situation. In fact, when the firms produce identical products, marginal cost are constant and there are no capacity constraints, already with two firms Bertrand price competition results in the competitive price, that is, the price is equal to the marginal cost in this case.

This result, that in a Bertrand game with homogeneous products and constant marginal cost, the competitive price is already obtained with two firms is sometimes called the Bertrand paradox and it seems to have bothered many economists in the past. Edgeworth (1897) suggested that firms have capacity constraints and that such constraints might resolve the paradox; after all, with capacity constraints, the reaction of the opponent will be less aggressive, hence, the market less competitive. However, capacity constraints raise another puzzle. Suppose one firm sets the competitive price, but is not able to supply total demand at that price. After this firm has sold its full capacity, a 'residual' market remains and the other firm makes most profits when it charges the 'residual monopoly price' in this market. As Edgeworth observed, given the high price of the second firm, the first firm has an incentive to raise its price to just below this price. Obviously, at these higher prices, there is then a game of each firm trying to undercut the other, which is driving prices down again. As a consequence, a pure strategy equilibrium need not exist. We are led to *Edgeworth cycles*, see also Levitan and Shubik (1972). However, we note here that there always exists an equilibrium in mixed strategies: firms set prices randomly, according to some distribution function. It may be shown, see

Levitan and Shubik (1972), Kreps and Scheinkman (1983) and Osborne and Pitchik (1986), that for small capacities a Cournot type outcome results, i.e. supplies are sold against a market clearing price, while for sufficiently large capacities, the Bertrand outcome is the equilibrium, i.e. firms set the competitive price. For the remaining intermediate capacity levels, there is no equilibrium in pure strategies.

Kreps and Scheinkman (1983) also analyze the situation where firms can choose their capacity levels.They assume that in the first period firms choose their capacity levels k_1 and k_2, and that next, knowing these capacities, in the second period firms play the Bertrand Edgeworth price game. In this situation, high capacity levels are attractive as they allow to sell a lot, but they are likewise unattractive as they imply a very competitive market; in contrast, low levels imply high prices but low quantities. Kreps and Scheinkman (1983) show that, with efficient rationing in the second period, firms will choose the Cournot quantities in the first period and the corresponding market clearing prices in the second. Hence, the Cournot model can be viewed as a shortcut of the two-stage Bertrand-Edgeworth model. However, it turns out that the solution of the game depends on the rationing scheme, as Davidson and Deneckere (1986) have shown.

All the oligopoly games discussed thus far are games with imperfect information: players take their decisions simultaneously. Oligopoly games with perfect information, in which players take their decisions sequentially with they being informed about all the previous moves, are nowadays called Stackelberg games, after Stackelberg (1934). Moving sequentially is a way in which too intense competition might be avoided, for example, if players succeed in avoiding simultaneous price setting, prices will typically be higher. Von Stackelberg assumed that one of the players is the 'first mover', the *leader*, and the other is the *follower*. In Stackelberg's model, first 'the leader' decides and next, knowing what the leader has done, 'the follower' makes his decision, hence, we have a game with *perfect information*.We believe that Stackelberg meant 'leader' and 'follower' more as a behavior rule, rather than an exogenously imposed ordering of the moves, hence, in our view, he assumed asymmetries between different player types. Such an asymmetry results in a different outcome. The best a follower can do, is to play a best response against the action of the leader

$$x_F = B_F(x_L).$$

The leader knowing this, will therefore play

$$x_L = \arg\max_{x \in X_L} \pi_L(x; B_F(x)).$$

In a Cournot setting, this typically implies that the leader will produce more, and the follower will produce less than his Cournot quantity, hence, the follower is in a weaker position, and it pays to lead: there is a first-mover advantage. Bagwell (1995), however, has argued that this first-mover advantage is eliminated if the leader's quantity can only be observed with some noise. Specifically, he considers the situation where, if the leader choose x_L the follower observes x_L with probability $1 - \varepsilon$, while the follower sees a randomly drawn $\tilde{x}$ with the remaining positive probability ε, where $\tilde{x}$ has full support. As now the signal that the follower receives is completely uniformative, the follower will not condition on it, hence, it follows that in the unique pure equilibrium, the Cournot quantities are played. Hence, there is no longer a first mover advantage. Van Damme and Hurkens (1997) however show that there is always a mixed equilibrium, that there are good arguments for viewing this equilibrium as the solution of the game, and that this equilibrium converges to the Stackelberg equilibrium when the noise vanishes.

We note that, in this approach to the Stackelberg game with perfect information, leader and follower are determined exogenously. Now it is easy to see that, in Cournot type games, it is most advantageous to be the leader, while in Bertrand type games, the follower position is most advantageous. Hence, the question arises which player will take up which player role. There is a recent literature that addresses this question of endogenous leadership. In this literature, there are two-stage models in which players choose the role they want to play in a timing game. The trade-off is between moving early and enjoy the advantage of commitment, or moving late and having the possibility to best respond to the opponent. Obviously, when firms are 'identical' there will be no way to determine an endogenous leader, hence, these models assume some type of asymmetry: endogenous leaders may emerge from different capacities, different efficiency levels, different information, or product differentiation. In cases like these, one could argue that player i will become the leader when he profits more from it than player j does, hence, that player i will lead if

$$\pi_i^L - \pi_i^F > \pi_j^L - \pi_j^F,$$

or equivalently when

$$\pi_i^L + \pi_j^F > \pi_i^F + \pi_j^L,$$

in other words, that the leadership will be determined as if players had joint profits in mind. Based on such considerations, many papers come to the conclusion that the dominant or most efficient firm will become the leader, see Ono (1982), Deneckere and Kovenock (1992), Furth and Kovenock (1993), and van Cayseele and Furth (1996). To get some intuition for this result, let consider a simple asymmetric version of the 2-firm Bertrand game. Assume that the product is perfectly divisible, that the demand curve is given by $D(p) = 1$ for $p < 1$ and $D(p) = 0$ for $p > 1$, and that firm 2 has a capacity constraint of k. If firm 2 acts as a leader, firm 1 will undercut and firm 2's profit is zero. Firm 2's profit is also zero if price setting is simultaneous and in this case firm 1's profit is zero as well. If firm 1 commits to be leader, he will be undercut by firm 2, but given that firm 2 has a capacity constraint, firm 1 is not hurt that much by it. Firm 1 will simply commits to the monopoly price and profits will be $1 - k$ for firm 1 and k for firm 2. Hence, only in the case where firm 1 takes up the leadership position will profits be positive for each firm, and we may expect firm 1 to take up the leadership position.

Van Damme and Hurkens (1996, 1999) argue that the above profit calculation is not convincing and that the leadership position should result from individual risk considerations. Be that as it may, the interesting result that they derive is that these risk considerations do lead to exactly the above inequalities, hence, van Damme and Hurkens obtain that both in the price and in the quantity game, the efficient firm will lead. Note then, that the efficient firm obtains the most preferred position in the case of Cournot competition, but not in the case of Bertrand competition.

Above, we already briefly referred to the work of Edgeworth on Bertrand competition with capacity constraints. Edgeworth was also the one who introduced the Core as the concept that models unbridled competition. Shubik (1959) rediscovered this concept in the context of cooperative games, and the close relation between the Core of the cooperative exchange game and the competitive outcome was soon discovered. Hence, also here we see the close relation between cooperative and non-cooperative theory. In fact, it is perhaps most beautiful in the theory of matching, see Roth and Sotomayor (1990). In the remainder of this section, we illustrate this relationship for the most simple 3-person exchange game, a game that, incidentally, also was analyzed in von Neumann and Morgenstern (1953). The founding fathers indeed already mention the possibility of applying their theory in the context of an oligopoly. Specifically, in the Sections 62.1 and 62.4 of their book, they calculated their solution, the Stable Set, of a three-person non-constant sum game that arises in a situation with one buyer and two sellers. Shapley (1958) generalized their analysis to a game with n (≥ 1) buyers and m (≥ 1) sellers, see also Shapley and Shubik (1969). We will confine ourselves to the case with $m = 2$ and $n = 1$. Furthermore, for simplicity, we will assume that the sellers are identical, that they each have one single indivisible object for sale, that they do not value this object, and that the buyer is willing to pay 1 for it. Denoting the consumer by player 3, the situation can be represented by the (cooperative) 3-person characteristic function game given by $v(S) = 1$ if $3 \in S$ and $|S| \geq 2$; and $v(S) = 0$ otherwise. In this game, the Core consists of a single allocation (0,0,1), corresponding to the consumer buying from either producer for a price of 0, hence, the Core coincides with the competitive outcome, illustrating the well-known Core equivalence theorem.

When, in the mid 1970s, one of us took his first courses in game theory with Stef Tijs, he considered the solution prescribed by the Core in the above game to be very natural. As a consequence, he was bothered very much by the fact that the Shapley value of this game was not an element of the Core and that it predicted a positive expected utility for each of the sellers. (As is well-known, the Shapley value of this game is (1,1,4)/6). Why could the sellers expect a positive utility in this game? The answer is in fact quite simple: the sellers can form a cartel! Obviously, once the sellers realize that their profits will be competed away if they do not form a cartel, they will try to form one. Hence, in this game, coalitions arise quite naturally and, as a consequence, the Core actually provides a misleading picture. If the sellers succeed in forming a stable coalition, they transform the situation into a bilateral monopoly in which case the negotiated price will be $\frac{1}{2}$. By symmetry, each of the sellers will get $\frac{1}{4}$ in this case. But, anticipating this, the consumer will try to form a coalition with any of the sellers, if only to prevent these sellers from entering into a cartel agreement. As von Neumann and Morgenstern (1953) already realized, and as we discussed in Section 3.2, the game is really one in which players will rush to form a coalition and the price that the buyer will pay will depend on the ease with which various coalitions can form. But then the outcome will be determined by the coalition formation process, hence, following Nash's advise, non-cooperative modelling should focus on that process.

Let us here study one such process. Let us assume that the players bump into each other at random and that, if negotiations between two players are not successful (which, of course, will not happen in equilibrium), the match is dissolved and the process starts afresh. The remaining question is what price, p, the consumer will pay to the seller if a buyer-seller coalition is formed. (By symmetry, this price does not depend on which seller the buyer is matched with.) The outcome is determined by the players' outside options, i.e. by what players can expect if the negotiations break down. The next table provides the utilities players can expect depending on the first coalition that is formed

	Utility		
First Coalition	1	2	3
$\{1,3\}$	p	0	$1-p$
$\{2,3\}$	0	p	$1-p$
$\{1,2\}$	$\frac{1}{4}$	$\frac{1}{4}$	$\frac{1}{2}$

For the coalition $\{1,3\}$, the outside option of the seller is $\frac{1}{3}(p+\frac{1}{4})$, while the buyer's outside option is $\frac{2}{3}(1-p)+\frac{1}{6}$. (This follows since all three 2-person coalitions are equally likely to form in the next round.) The coalition loses $\frac{1}{3}(p+\frac{1}{4})$ if it does not come to an agreement, hence, it will split this surplus evenly. It follows that the price p must satisfy

$$p = \frac{1}{3}(p+\frac{1}{4}) + \frac{1}{6}(p+\frac{1}{4}).$$

Hence, $p=\frac{1}{4}$. Since all coalitions are equally likely, the expected payoff of a seller equals $\frac{1}{6}$, while the buyer's expected payoff equals $\frac{2}{3}$. The conclusion is that expected payoffs are equal to the Shapley value of the game. Furthermore, the outcome, naturally, lies outside of the Core. We refer that reader who thinks that we have skipped over too many details in the above derivation to Montero (2000), where all such details are filled in.

Of course, the exact price will depend on the details of the matching process and different processes may give rise to different prices, hence, different cooperative solution concepts. Viewed in this way, also von Neumann and Morgenstern's solution of this game appears quite natural. As they write (von Neumann and Morgenstern, 1953, pp. 572, 573), the solution consists of two branches, either the sellers compete (and then the buyer gets the surplus), a situation they call the classical solution, or the sellers form a coalition, and in this case, they will have to agree on a definite rule for how to split the surplus obtained; as different rules may be envisaged, multiple outcome may be a possibility.

3.5 Auctions

In this section, we illustrate the usefulness of game theory in the understanding of real life auctions. The section consists of three parts. First, we briefly discuss some auction theory. Next,we discuss an actual auction and provide a non-cooperative analysis to throw light on a policy issue. In the third part, we demonstrate that also in this non-cooperative domain, insights from cooperative game theory are very relevant.

Four basic auction forms are typically distinguished. The first type is the Dutch auction. If there is one object for sale, the auction proceeds by the seller starting the auction clock and continuously lowering the price until one of the bidders pushes the button, or shouts "mine"; that bidder then receives the item for the price at which he stopped the clock. The second basic auction form is the English (ascending) auction in which the auctioneer continuously increases the price until one bidder is left; this bidder then receives the item at the price where his final competitor dropped out. The two basic static auction forms are the sealed bid first price auction and the Vickrey auction (Vickrey, 1961). In the first price auction, bidders simultaneously and independently enter their bids, typically in sealed envelopes, and the object is awarded to the highest bidder who is required to pay his bid. In the Vickrey auction, players enter their bids in the same way, and the winner is again the one with the highest bid, however, the winner "only" pays the second highest bid.

As auctions are conducted by following explicit rules they can be represented as (non-cooperative) games. Milgrom and Weber (1982) have formulated a fairly general auction model. In this model, there are n bidders, that occupy symmetric positions. The game is one with incomplete information, each bidder i has a certain type θ_i that is known only to this bidder himself. In addition, there may be residual uncertainty, represented by θ_0, where 0 denotes the chance player. If $\theta = (\theta_0, \theta_1, \ldots, \theta_n)$ is the vector of types (including that of nature), then θ is called the state of the world, and θ is assumed to be drawn from a commonly known distribution F on a set Θ that is symmetric with respect to the last n arguments. (Symmetry thus means that F is invariant with respect to permutations of the bidders.) In addition to his type, each player i has a value function, $v_i(\theta)$, where again the assumption of symmetry is maintained, i.e. if θ_i and θ_j are interchanged, then v_i and v_j are interchanged as well. Under the additional assumption of affiliation (which roughly states that a higher value of θ_i makes a higher value of θ_j more likely), Milgrom and Weber derive a symmetric equilibrium for this model. For the Vickrey auction, the optimal bid is characterized by

$$b_i(\theta_i) = E(v_i | v_i = v^1 = v^2)$$

where v^k denotes the k-th largest component of the vector $v(\theta) = (v_1(\theta), \ldots, v_n(\theta))$. In words, in the Vickrey auction, the player bids the expected value of the object to him, conditional on his value being the highest, and this value also being equal to the second highest value. For the Dutch (first price) auction, the optimal bid is lower, and the formula will not be given here. (See Wilson, 1992.) We also note that, in addition to giving insights into actual auctions, game theory has also contributed to characterizing optimal auctions, where optimality either is defined with respect to seller revenue or with respect to some efficiency criterion (Myerson, 1981; Wilson, 1992; Klemperer, 1999).

In many cases, the seller will have more than one item for sale. In case the objects are identical (such as in the case of shares, or treasury bills), the generalizations of the model and the theory are relatively straightforward: only one price is relevant; players can indicate how much they demand at each possible price and the seller can adjust price (either upward, or downward, or in a static sealed bid format) to equate supply and demand. The issue is more complicated in case the objects are heterogenous. With m objects, the relevant price region would be $\mathbb{R}^m_+$ and, of course, one could imagine bidders expressing their demand $(d_i : \mathbb{R}^m_+ \rightarrow 2^{\{1,\ldots,m\}})$ for all possible price vectors, but this may get very complicated. Alternatively, each bidder expresses bids for collections of items, hence, if $S \subset \{1, \ldots, m\}$, then $B_i(S)$ is the maximum that i is willing to pay if he is awarded set S, where the auction rule would be completed by a winner determination rule. At present, there is active research on such combinatorial auctions. In connection with spectrum auctions in the US, game theorists designed the simultaneous multi round ascending auction, a generalization of the English auction. In this format, the objects are sold simultaneously in a sequence of rounds with at least one price increasing from one round to the next. In its most elementary form, each bidder can bid on all items and the auction continues to raise prices as long as at least one new bid is made; when the auction ends, the current highest bidders are awarded the objects at these respective prices. To speed up the auction, activity rules may be introduced that force the bidders to bid seriously already early on. We refer to Milgrom (2000) for more detailed description and analysis.

Having briefly gone over the theory, our aim in the remainder of this section is to show how game theory can contribute to better insight and to more rational discussion in several policy areas. Our examples are drawn from the Dutch policy context, and our first example relates to electricity. Electricity prices in the Netherlands are high, at least they are higher than in the neighboring Germany. As a result of the price difference, market parties are interested in exporting electricity from Germany into the Netherlands. Such imports into the Netherlands are limited by the limited capacity of the interconnectors at the border, which in turn implies that the price difference can persist. In 2000, it was decided to allocate this scarce capacity by means of an auction; on the website `www.tso-auction.org`, the interested reader can find the details about the auction rules and the auction outcomes. We discuss here a simplified (Cournot) model that focuses on some of the aspects involved.

As always in auction design, decisions have to be made about what is to be auctioned, whether the parties are to be treated symmetrically, and what the payment mechanism is going to be. Of course, these decisions have to be made to contribute optimally to the ultimate goal. In this specific case, the goal may be taken as to have an as low price for electricity in the Netherlands as possible. The simple point now is that adopting this goal implies that players cannot be treated symmetrically. The reason is that they are not in symmetric positions: some of them have electricity generating capacity in the Netherlands, while others do not, and members of the first group may have an incentive to block the interconnector in order to guarantee a higher price for the electricity that is produced domestically. To illustrate this possibility, we consider a simple example.

Suppose there is one domestic producer of electricity, who can produce at constant marginal cost c. Furthermore, assume that demand is linear, $D(p) = 1 - p$. If the domestic producer is shielded from competition, and is not regulated, he will produce the monopoly quantity q_m, found by solving:

$$\max_{q_m} q_m(1 - q_m) - cq_m.$$

Hence the quantity q_m, the price p_m and the profit π_m will be given by:

$$\begin{aligned} q_m &= (1-c)/2 \\ p_m &= (1+c)/2 \\ \pi_m &= (1-c)^2/4. \end{aligned}$$

Assume that the interconnector has capacity $k > 0$, and that in the neighboring country electricity is also produced at marginal cost c. In contrast to the home country, the foreign country is assumed to have a competitive market, so that the price in the foreign country $p_f = c$. As a result $p_f < p_m$ and there is interest in transporting electricity from the foreign to the home country. If all interconnector capacity would be available for competitors of the monopolist, the monopolist would instead solve the following problem:

$$\max_{q_m} q_m(1 - q - k) - cq,$$

hence, if he produces q, the total production is $q + k$, and the price $1 - q - k$. The quantity q_c that the monopolist produces in this competitive situation is:

$$q_c = (1 - k - c)/2,$$

while the resulting price p_c and the profit for the monopolist π_c are given by:

$$\begin{aligned} p_c &= (1 + c - k)/2 \\ \pi_c &= (1 - c - k)^2/4. \end{aligned}$$

The above calculations allow us to compute how much the capacity is worth for the competing (foreign) generators. If they acquire the capacity, they can produce electricity at price c and sell it at price p_c, thus making a margin $p_c - c = (1 - c - k)/2$ on k units, resulting in a profit of

$$\pi_f = k(1 - c - k)/2.$$

At the same time, the loss in profit for the monopolist is given by

$$\Delta\pi = \pi_m - \pi_c = k(1 - c - k/2)/2.$$

We see that

$$\pi_f < \Delta\pi,$$

so that the capacity is worth more to the monopolist. The intuition for this result is simple, and is already given in Gilbert and Newbery (1982): competition results in a lower price; this price is relevant for all units that one produces, hence, the more units that a player produces, the more he is hurt. It follows that, if the interconnector capacity would be sold in an ordinary auction, with all players being treated equally, then all the capacity would be bought by the home producer, who would then not use it. Consequently, a simple standard auction would not contribute to the goal of realizing a lower price in the home electricity market.

The above argument was taken somewhat into account by the designers of the interconnector auction, however it was not taken to its logical limit. In the actual auction rules, no distinction is being made between those players that do have generating capacity at home and those that do not: a uniform cap of 400 Mw of capacity is imposed on all players, (hence, the rule is that no player can have more than 400 Mw of interconnector capacity at its disposal, which corresponds with some 25 percent of all available capacity). This rule has obvious drawbacks. Most importantly, the price difference results because of the limited interconnector capacity that is available, hence, one would want to increase that capacity. As long as the price difference is positive, and sufficiently large, market parties will have an incentive to build extra interconnector capacity: the price margin will be larger than the investment cost. However, in such a situation, imposing a cap on the amount of capacity that one may hold, may actually deter the incentive to invest. Consequently, it would be better to have the cap only on players that do have generating capacity in the home country, and that profit from interconnector capacity being limited.

To prevent players with home generating capacity from buying, but not using interconnector capacity, the auction rules include "use it or lose it" clause. Clearly, such clauses are effective in ensuring that the capacity is used, however, they need not be effective in guaranteeing a lower price in the home electricity market. This can be easily seen in the explicit example that was calculated above. Suppose that a "use it or lose it" clause would be imposed on the monopolist, how would it change the value of the interconnector capacity for this monopolist? Note that the value is not changed for the foreign competitors, this is still π_f , as they will use the capacity in any way. The important insight now is that the clause also does not change the value for the monopolist: if the monopolist is forced to use k units at the interconnector, he will simply adjust by using k units less of his domestic production capacity. By behaving in this way, he will still produce q_m in total and obtain monopoly profits of π_m. Hence a "use it or lose it" clause has no effect, neither on the value of the interconnector for the incumbent, nor on the value for the entrants. Therefore, the value is larger for the incumbent, the incumbent will acquire the capacity and the price will remain unchanged, hence, the benefits of competition will not be realized.

This simple example has shown that the design that has been adopted can be improved: it would be better to impose capacity caps asymmetrically, and it should not be expected that "use it or lose it" clauses are very effective in lowering the price. Of course, the actual situation is much richer in detail than our model. However, the actual situation is also very complicated and one has to pick cherries to come to better grips with the overall situation. We hope it is clear that a simple model like the one that we have discussed in this section provides an appropriate starting point for coming to grips with a rather complicated situation.

Our second example relates to the high stakes telecommunications auctions that recently took place in Europe. During 2000, various European countries auctioned licenses for third generation mobile telephony (UMTS) services. Already a couple of years earlier, some of these countries had auctioned licenses for second generation (DCS-1800) services. In the remainder of this section, we briefly review some aspects of the Dutch auctions. For further detail, see Van Damme (1999, 2001, 2002a).

Van Damme (1999) describes the Dutch DCS-1800 auction and argues that, as a consequence of time pressure imposed on Dutch officials by the European Commission, that auction was badly designed. The main drawback was that the available spectrum was divided into very unequal lots: 2 large ones of 15 MHz each and 16 small ones of on average 2.5 MHz, which were sold simultaneously by using a variant of the multiround ascending auction that had been pioneered in the US. The rules stipulated that newcomers could bid on all lots, but that incumbents (at the time, KPN and Libertel) could bid only on the small lots. In this situation, new entrants had the choice between bidding on large lots, or trying to assemble a sufficient number of small lots so that enough spectrum would be obtained in total to create a viable national network. The latter strategy was risky. First of all, by bidding on the small lots one was competing with the incumbents. Secondly, one faced the risk of not obtaining enough spectrum. This is what is called in the literature "the exposure problem": if say 6 small lots were needed for a viable network, one had the risk of finding out that one could not obtain all six because of the intensity of competition, one might be left with three lots which would be essentially worthless. (At the time of auction, it was not clear whether such blocks could be resold, the auction rules stating that this was up to the Minister to decide.)

The structure of supply that was chosen had an interesting consequence. Most newcomers found it too risky to bid on the small lots, hence, bidding concentrated on the large lots and the price was driven up there. In the end, the winners of the large lots, Dutchtone and Telfort paid Dfl. 600 mln and Dfl. 545 mln, respectively for their licenses. Compared to the prices paid on the small lots, these prices are very high: van Damme (1999) calculates that, on the basis of prices paid for the small lots, these large lots were worth only Dfl. 246 mln, hence, less than half of what was paid. There was only one newcomer, Ben, who dared to take the risk of trying to assemble a national license from small lots and it was successful in doing so; it was rewarded by having to pay only a relatively small price for its license. It seems clear that if the available spectrum had been packaged in a different way, say 3 large lots of 15 MHz each and 10 small lots of an average 2.5 MHz each, the price difference would have been smaller, and the situation less attractive for the incumbents. Perhaps one might even argue that the design that was adopted in the Dutch DCS-1800 auction was very favorable for the incumbents.

In any case, the 1998 DCS-1800 auction led to a five player market, at least one player more than in most other European markets. This provides relevant background for the third generation (UMTS) auction that took place in the summer of 2000, and which was really favorable for the incumbents. At that time, the two "old" incumbents (KPN and Libertel) still had large market shares, with the market shares of the newer incumbents (Ben, Dutchtone and Libertel) being between 5 and 10 percent each. In this situation, it was decided to auction five 3G-licenses, two large ones (of 15 MHz each) and three smaller ones (of 10 MHz each). It is also relevant to know that the value of a license is larger for an incumbent than for a newcomer to the market, and this because of two reasons. First, an incumbent can use its existing network, hence, it will have lower cost in constructing the necessary infrastructure. Secondly, if an incumbent does not win a 3G-license, it will also risk to lose its 2G-customers. Finally, it is relevant to know that it was decided to use a simultaneous ascending auction.

The background provided in the previous paragraph makes clear why the Dutch 3G-auction was unfavorable to newcomers. First, the supply of licenses (2 large, 3 small) exactly matches the existing market structure (5 incumbents, of which 2 large ones). Secondly, an ascending auction was used, a format that allows incumbents to react to bids and thus to outbid new entrants. Thirdly, the value of a license being larger for an incumbent than for an entrant implies that an incumbent will also have an incentive to outbid a newcomer. In a situation like this, an entrant cannot expect to win a license, so why should it bother to participate in this auction? On the basis of these arguments, one should expect only the incumbents to participate and, hence, the revenues to remain small, see Maasland (2000).

The above arguments seem to have been well understood by the players in the market. Even though many potential entrants had expressed an interest to participate in the Dutch 3G-auction at first, all but one subsequently decided not to participate. In the end, only one newcomer, Versatel, participated in the auction. This participant had equally well understood that it could not win; in fact, it had started court cases (both at the European and the Dutch level) to argue that the auction rules were "unfair" and that it was impossible for a newcomer to win. If Versatel knew that it could not win a license in this auction, why did it then participate? A press release that Versatel posted on its website the day before the auction givens the answer to this question:

> "We would however not like that we end up with nothing whilst other players get their licenses for free. Versatel invites the incumbent mobile operators to immediately start negotiations for access to their existing 2G networks as well as entry to the 3G market either as a part owner of a license or as a mobile virtual network operator."

The press release that Versatel realizes, and want the competitors to realize, that it has power over the incumbents. By participating in the auction, Versatel drives up the price that the winners (the incumbents) will have to pay. (Viewed in this light, the court cases that Versatel had started signals to the incumbents that Versatel know that it cannot win, hence, that it must participate in the auction with another objective in mind.) On the other hand, by dropping out, Versatel does the incumbents a favour, since the auction will end as soon as Versatel does drop out. The press release signals that Versatel is willing to drop out, provided that the incumbents are willing to let Versatel share in the benefits that they obtain in this way. All in all then, Versatel appears to be following a smarter strategy than the newcomers that did not participate in the auction.

For the reader who has studied von Neumann and Morgenstern (1953), the above may all appear very familiar. Recall the basic three-player non-zero sum game from that book, with one seller, two buyers, one indivisible object, and one buyer attaching a higher value to this object than the other. Why would the weaker participate in the game, if he knows right from the start that he will not get the object anyway? The answer that the founding fathers give is that he has power over both other players: by being in the game, he forces the other buyer to pay a higher price and he benefits the seller; by stepping out he benefits the buyer, and by forming a coalition with one of these other players, he can exploit his power. This argument is also contained, and popularized, in Brandenburger and Nalebuff (1996), a book that also clearly demonstrates the value of combining cooperative and competitive analysis. If one knows that Nalebuff was an advisor to Versatel, then it is no longer that surprising that Versatel has used this strategy.

One would like to continue this story with a happy end for game theory, but unfortunately that is not possible in this situation. Even though Versatel's strategy was clever, it was not successful. Versatel stayed in the auction, but it did not succeed in reaching a sharing agreement with one of the incumbents, even though negotiations have been conducted with one of them. Perhaps, the other parties had not fully realized the cleverness of Versatel and, as Edgar Allen Poe already remarked, it pays to be one level smarter than your opponents, but not more. Eventually, Versatel dropped out and, in the end, only the Dutch government was the beneficiary of Versatel's strategy.

3.6 Conclusion

In this chapter, we have attempted to show that the cooperative and non-cooperative approaches to games are complementary, not only for bargaining games, as Nash had already argued and demonstrated, but also for market games. Specifically, we have demonstrated this for oligopoly games and for auctions. We have shown that each approach may give essential insights into the situation and that, by combining insights from both vantage points, a deeper understanding of the situation may be achieved.

The strength of the non-cooperative approach is that allows detailed modelling of actual institutions. Hence, many different institutional arrangements may be modelled and analysed, thus allowing an informed, rational debate about institutional reform. Indeed, the non-cooperative models show that outcomes can depend strongly on the rules of the game. The strength of this approach is at the same time its weakness:

why would players play by the rules of the game? Von Neumann and Morgenstern argued that, whenever it is advantageous to do so, players will always seek for possibilities to evade constraints, in particular, they will be motivated to form coalitions and make side-payments outside the formal rules. This insight is relevant for actual markets and even though competition laws attempt to avoid cartels and bribes, one should expect these laws to be not fully successful.

The cooperative approach aims to predict the outcome of the game on the basis of much less detailed information, it only takes account of the coalitions that can form and the payoffs that can be achieved. One lesson that the theory has taught us is that frequently this information is not enough to pin down the outcome. The multiplicity of cooperative solution concepts testifies to this. Hence, in many situations we may need a non-cooperative model to make progress. Such a non-cooperative model may also alert us to the fact that the efficiency assumption that frequently is routinely made in cooperative models may not be appropriate. On the other hand, when the cooperative approach is really successful, such as in the 2-person bargaining context, it is really powerful and beautiful.

We expect that that the tension between the two models will continue to be a powerful engine of innovation in the future.

References

Bagwell, K. (1995): "Commitment and observability in games," *Games and Economic Behavior*, 8, 271–280.

Bertrand, J. (1883): "Theorie mathématiques de la richesse sociale," *Journal des Savants*, 48, 499–508.

Brandenburger, A., and B. Nalebuff (1996): *Coopetion.* Currency/Doubleday.

Cayseele, P. van, and D. Furth (1996a): "Bertrand-Edgeworth duopoly with buyouts or first refusal contracts," *Games and Economic Behavior*, 16, 153–180.

Cayseele, P. van, and D. Furth(1996b): "Von Stackelberg equilibria for Bertrad-Edgeworth duopoly with buyouts," *Journal of Economic Studies*, 23, 96–109.

Cayseele, P. van, and D. Furth(2001): "Two is not too many for monopoly," *Journal of Economics*, 74, 231–258.

Cournot, A. (1838): *Recherches sur les Principes Mathématiques de la Théorie des Richesses.* Paris: L. Hachette.

Damme, E. van (1995): "On the contributions of John C. Harsanyi, John F. Nash and Reinhard Selten," *International Journal of Game Theory*, 24, 3–12.

Damme, E. van (1999): "The Dutch DCS-1800 auction," in: Patrone, Fioravante, I. García-Jurado & S. Tijs (eds.), *Game Practise: Contributions from Applied Game Theory.* Boston: Kluwer Academic Publishers, 53–73.

Damme, E. van (2001): "The Dutch UMTS-auction in retrospect," CPB Report 2001/2, 25–30.

Damme, E. van (2002a): "The European UMTS-auctions," *European Economic Review*, forthcoming.

Damme, E. van (2002b): "Strategic equilibrium," forthcoming in R.J. Aumann and S. Hart (eds.), *Handbook in Game Theory*, Vol. III., North Holland Publ. Company.

Damme, E. van, and S. Hurkens (1996): "Endogenous price leadership," *Discussion Paper* nr. 96115, CentER, Tilburg University.

Damme, E. van, and S. Hurkens (1997): "Games with imperfectly observable commitment," *Games and Economic Behavior*, 21, 282–308.

Damme, E. van, and S. Hurkens (1999): "Endogenous Stackelberg leadership," *Games and Economic Behavior*, 28, 105–129.

Dasgupta, P., and E. Maskin (1986): "The existence of equilibria in discontinuous games. I: Theory and II: Applications, *Review of Economic Studies*, 53, 1–26, and 27–41.

Daughety, A. (ed.) (1988): *Cournot Oligopoly.* Cambridge: Cambridge University Press.

Davidson, C., and R. Deneckere (1986): "Long-run competition in capacity, short-run competition in price, and the Cournot model," *Rand Journal of Economics*, 17, 404–415.

Deneckere, R., and D. Kovenock (1992): "Price leadership," *Review of Economic Studies*, 59, 143–162.

Edgeworth, F.Y. (1897): "The pure theory of monopoly," reprinted in William Baumol and Stephan Goldfeld *Precusors in Mathematical Economics: An Anthology*, London School of Economics, 1968.

Furth, D. (1986): "Stability and instability in oligopoly," *Journal of Economic Theory*, 40, 197–228.

Furth, D., and D. Kovenock(1993): "Price leadership in a duopoly with capacity constraints and product differentiation," *Journal of Economics*, 57, 1–35.

Gilbert, R., and D. Newbery (1982): "Preemptive patenting and the persistence of monopoly," *American Economic Review*, 72, 514–526.

Klemperer, P. (1999): "Auction theory: a guide to the literature," *Journal of Economic Surveys*, 13, 227–286.

Kreps, D., and J. Scheinkman (1983): "Quantity pre-commitment and Bertrand competition yields Cournot outcomes," *Bell Journal of Economics*, 14, 326–337.

Leonard, R. (1994): "Reading Cournot, reading Nash," *The Economic Journal*, 104, 492–511.

Levitan, R., and M. Shubik (1972): "Price duopoly and capacity constraints," *International Economic Review*, 13, 111–122.

Maasland, E. (2000): "Veilingmiljarden zijn een fictie," *Economisch Statistische Berichten*, 9 juni 2000, 479.

Milgrom, P. (2000): "Putting auction theory to work: the simultaneous ascending auction," *Journal of Political Economy*, 108, 245–272.

Milgrom, P., and R. Weber (1982): "A theory of auctions and competitive bidding," *Econometrica*, 50, 1089–1122.

Montero, M. (2000): *Endogeneous Coalition Formation and Bargaining.* PhD thesis, CentER, Tilburg University.

Myerson, R. (1981): "Optimal auction design," *Mathematics of Operations Research*, 6, 58–73.

Nash, J.F. (1950a): *Non-Cooperative Games.* Ph.D. Dissertation, Princeton University.

Nash, J.F. (1950b): "The bargaining problem," *Econometrica*, 18, 155–162.

Nash, J.F. (1951): "Non-cooperative games," *Annals of Mathematics*, 54, 286–295.

Nash, J.F. (1953): "Two-person cooperative games," *Econometrica*, 21, 128–140.

von Neumann, J., and O. Morgenstern (1953): *Theory of Games and Economic Behavior.* Princeton, NJ: Princeton University Press (First edition, 1944).

Ono, Y. (1982): "Price leadership: a theoretical analysis," *Economica*, 49, 11–20.

Osborne, M., and C. Pitchick (1986): "Price competition in a capacity-constrained duopoly," *Journal of Economic Theory*, 38, 238–260.

Roth, A., and M. Sotomayor (1990): *Two-sided Matching: A study in Game Theoretic Modelling and Analysis.* Cambridge, Mass.: Cambridge University Press.

Rubinstein, A. (1982): "Perfect equilibrium in a bargaining model," *Econometrica*, 50, 97–109.

Shapley, L. (1958): "The solution of a symmetric market game," in: A.W, Tucker and R.D. Luce (eds.): *Contributions to the Theory of Games* IV, Annals of Mathematics Studies 40, Princeton, Princeton University Press, 145–162.

Shapley, L., and M. Shubik (1969): "On market games," *Journal of Economic Theory*, 1, 9–25.

Shubik, M. (1959): "Edgeworth market games," in A.W. Tucker and R.D. Luce (eds.) *Contributions to the Theory of Games* IV, Annals of Mathematics Studies 40, Princeton, NJ: Princeton University Press.

von Stackelberg, H. (1934): *Marktform und Gleichgewicht.* Berlin: Julius Springer.

Vickrey, W. (1961): "Counterspeculation, auctions and competitive sealed tenders," *Journal of Finance*, 16, 8–37.

Vives, X. (1989): "Cournot and the oligopoly problem," *European Economic Review*, 33, 503–514.

Wilson, R. (1992): "Strategic analysis of auctions," in R.J. Aumann and S. Hart (eds.), *Handbook in Game Theory*, Vol. I., North Holland Publ. Company, 227–279.

Chapter 4

On the Number of Extreme Points of the Core of a Transferable Utility Game

BY JEAN DERKS AND JEROEN KUIPERS

4.1 Introduction

Stability of an allocation among a group of players is normally considered to refer to the property that there is no incentive among subgroups or coalitions of players to deviate from the given allocation and choose the alternative of cooperation. In a transferable utility game the stable allocations are exactly the elements of the upper core. These allocations always exist but may not be feasible in the sense that the total payoff exceeds the total earnings of the grand coalition. The core of a game is the set of feasible allocations within the upper core. It is a (possibly empty) face of the upper core.

The core is perhaps the best known solution concept within Cooperative Game Theory. The first contributions within this context are found in Gillies (1953). It is generally believed that the core and core-like structured sets have at most $n!$ extreme points. This is indeed the case and the main contribution of this note is to provide a proof.

With core-like structured sets we denote those sets that can appear as a core of a game. Examples are the so-called core covers, which are generalizations of the core, and are introduced mainly in order to bypass

P. Borm and H. Peters (eds.), Chapters in Game Theory, 83–97.

the dissatisfactory property of the core that it may be empty. The first results in this direction are found in Tijs (1981) and Tijs and Lipperts (1982). Other examples of core-structures are the anti-core, the least core and the Selectope. Vasilev (1981) and, recently, Derks, Haller and Peters (2000) are contributions dealing with the core structure of the Selectope.

Although our first concern is the core, the main results and concepts deal with the upper core. The upper core of a game can be described as the feasible region of a suitably chosen linear program, where the matrix is 0,1-valued and the constraint vector coefficients are the values of the coalitions. Actually, we are dealing with polyhedra of the type $P_{A,b} = \{x \in \mathbb{R}^n | \, Ax \geq b\}$, where A is an $m \times n$ integer valued matrix and $b \in \mathbb{R}^m$. In the literature there is a comprehensive study on the upper bound on the number of extreme points of such polyhedra. It is well-known that $x \in P_{A,b}$ is an extreme point of $P_{A,b}$ if and only if there is a set of n linearly independent vectors among the rows a of A for which the equality $a\,x = b_a$ (with b_a the coefficient of b associated with row a) holds. Hence, a trivial upper bound for the number of extreme points of $P_{A,b}$ is $\binom{m}{n}$. McMullen (1970) showed that this is an overestimate and he proved that the polyhedron $P_{A,b}$ has at most

$$f(m,n) = \binom{m-\lfloor\frac{n+1}{2}\rfloor}{m-n} + \binom{m-\lfloor\frac{n+2}{2}\rfloor}{m-n}$$

extreme points. Furthermore, Gale (1963) constructed examples of polyhedra having precisely $f(m,n)$ extreme points, so that McMullen's bound cannot be further improved for arbitrary matrices A (see also Chvátal, 1983, for these results).

Our main result is that for polyhedra $P_{A,b}$, where A is an $m \times n$-matrix with 0,1-valued coefficients, an upper bound of $n!$ different extreme points exists. This is an improvement of McMullen's upper bound in case A contains all possible 0,1-valued row vectors: $f(2^n, n) \geq 2(\frac{n}{2})^n \geq n!$, and by the Stirling approximation of $n!$, being approximately $\sqrt{(\pi n)}(\frac{n}{e})^n$, we observe that the McMullen upper bound exponentially exceeds the value $n!$.

The proof of our main result is based on a polar version of an argument, stated by Imre Bárány (see Ziegler, 1995, p. 25) in the context of the related search for upper bounds for the number of facets of a 0/1-polytope, which is defined to be the convex hull of a set of elements with 0,1-valued coefficients. This argument shows that $n! + 2n$ is an upper bound of the number of facets of any n-dimensional 0/1-polytope. The polar translation of a 0/1-polytope is a polyhedron of type $-P_{A,-\mathbf{1}}$, with A an $m \times n$-matrix with 0,1-valued coefficients, and $\mathbf{1}$ the m-vector $(1, \ldots, 1)$, so that for these polyhedra Bárány's upper bound directly implies a maximum number of $n! + 2n$ extreme points. With a bit more effort we will obtain a better upper bound for the larger class of polyhedra where the constraint vector b may admit arbitrary values.

Let us define the polytope Q_A as the convex hull of the origin 0 and all row vectors of the matrix A. In Section 4.2 we shall prove that the number of extreme points of $P_{A,b}$ is bounded by $n!$ times the n-dimensional volume of the polytope Q_A. Since Q_A is contained in the unit hypercube, described by the restrictions $0 \leq x_i \leq 1$ for all $i \in \{1, \ldots, n\}$ if A is a 0,1-valued matrix, Q_A has a volume of at most 1, and it follows that the polyhedron $P_{A,b}$ has at most $n!$ extreme points.

In Section 4.3 we formally introduce the cooperative game model, and state that the core, being a face of a polyhedron of type $P_{A,b}$, with A a (0,1)-matrix, has at most $n!$ different extreme core points. Strict convexity implies that the core actually has $n!$ different extreme points, but we show that there are more games with this property. We further discuss an intuitive and direct approach for listing $n!$ extreme points of the core (possibly with duplicates) but we show that this approach may fail to list all extreme points, thus showing that it cannot be used for establishing a maximum on the number of extreme points.

Section 4.4 describes some properties that are induced by having $n!$ extreme core points. These are the large core property, a kind of strict exactness, and a non-degeneracy property. We supply an example that these properties are not sufficient for obtaining $n!$ extreme core points.

In Section 4.5 we conclude the paper with a summary.

4.2 Main Results

Let $X \subseteq \mathbb{R}^n$ and $x \in X$. The vector x is said to be an *interior* point of X if there exists $\epsilon > 0$ such that for all $d \in \mathbb{R}^n$ with $||d|| \leq 1$ and all δ with $0 \leq \delta \leq \epsilon$ we have $x + \delta d \in X$. Here $||d||$ denotes the Euclidian norm of the vector d. The set of all interior points of X is called the *interior* of X.

For any two vectors $x, y \in \mathbb{R}^n$ their inner product is denoted by $x\,y$. We shall denote the righthand side associated with row a of matrix A

by b_a.

For $x \in P_{A,b}$ let I_x denote the set of rows a for which $a\,x = b_a$. Further, let $Q_x \subseteq Q_A$ denote the convex hull of 0 and the vectors of I_x. It is intuitively quite clear that $Q_x \cap Q_y$ has an empty interior for any two distinct extreme points $x, y \in P_{A,b}$. For the sake of completeness we provide a proof here.

Lemma 4.1 *Let x and y be two distinct extreme points of $P_{A,b}$. Then $Q_x \cap Q_y$ has an empty interior.*

Proof. Let x and y be two distinct extreme points of $P_{A,b}$ and suppose that $Q_x \cap Q_y$ has a non-empty interior. Choose z in the interior of $Q_x \cap Q_y$. So, z lies also in the interior of Q_x and therefore, it can be written as a convex combination of the extreme points of the convex set Q_x with all coefficients strictly positive, i.e. $z = \sum_{a \in I_x} \lambda_a a$ with $\lambda_a > 0$. We have that $y\,\hat{a} > b_{\hat{a}}$ for at least one $\hat{a} \in I_x$. Since $\lambda_{\hat{a}} > 0$, it follows that $zy = \sum_{a \in I_x} \lambda_a\, ay > \sum_{a \in I_x} \lambda_a ax = zx$. Analogously one proves that $zx > zy$, a contradiction. □

As a consequence of Lemma 4.1, the n-dimensional volume of the union of polytopes Q_x is simply the sum of their volumes. In the following we shall provide a lower bound on the volume of Q_x. This gives us then an upper bound on the number of polytopes Q_x that can be contained in Q_A, or equivalently, it gives us an upper bound on the number of extreme points of $P_{A,b}$.

Let us denote the volume of an n-dimensional body $X \subseteq \mathbb{R}^n$ by $\mathcal{V}(X)$. The following theorem is well-known in linear algebra. For a proof we refer to Birkhoff and MacLane (1963).

Theorem 4.2 *Let $X \subseteq \mathbb{R}^n$ be a $\mathcal{V}$-measurable set, and let A be a square matrix of dimension n. Then $\mathcal{V}(\{Ax \mid x \in X\}) = |\det(A)|\,\mathcal{V}(X)$, where $\det(A)$ denotes the determinant of the matrix A.*

Now we are in a position to provide a lower bound on $\mathcal{V}(Q_x)$.

Lemma 4.3 *Let A be an $m \times n$ integer valued matrix and let $b \in \mathbb{R}^n$. Furthermore, let x be an extreme point of $P_{A,b}$. Then $\mathcal{V}(Q_x) \geq 1/n!$.*

Proof. Since x is an extreme point of $P_{A,b}$, I_x contains a set of n independent vectors, say $x^1, x^2, \ldots, x^n$. According to Theorem 4.2 the volume of the convex hull of the points $x^1, x^2, \ldots, x^n$ and 0 equals

$|\det(x^1,\ldots,x^n)|\mathcal{V}(\{\lambda \in \mathbb{R}^n_+ : \sum_{i=1}^n \lambda_i \leq 1\})$, where $\det(x^1,\ldots,x^n)$ denotes the determinant of the matrix with columns $x^1,\ldots,x^n$. All entries of this matrix are integer, so its determinant is also integer. The independent nature of the columns in the matrix ensures that the determinant is unequal to 0 and therefore, $|\det(x^1,\ldots,x^n)| \geq 1$. Consequently, the volume of the convex hull of the points $x^1, x^2, \ldots, x^n$ and 0 is at least the volume of $\{\lambda \in \mathbb{R}^n_+ : \sum_{i=1}^n \lambda_i \leq 1\})$, which equals $1/n!$. Since this convex hull is contained in Q_x, also Q_x has a volume of at least $1/n!$. □

Observe that the lower bound of $1/n!$ can be achieved by $\mathcal{V}(Q_x)$ if and only if there is a set of n independent vectors $x^1, x^2, \ldots, x^n$ in Q_x with $|\det(x^1,\ldots,x^n)| = 1$, and there are no elements of Q_x outside the convex hull of $x^1, x^2, \ldots, x^n$ and 0.

Theorem 4.4 *Let A be an $m \times n$ integer valued matrix and let $b \in \mathbb{R}^n$. Then $P_{A,b}$ has at most $n!\mathcal{V}(Q_A)$ extreme points.*

Proof. Let E denote the set of extreme points of $P_{A,b}$. Clearly, $Q_x \subseteq Q_A$ for all $x \in E$. Hence, $\cup_{x \in E} Q_x \subseteq Q_A$ and

$$\mathcal{V}(Q_A) \geq \mathcal{V}(\cup_{x\in E} Q_x).$$

According to Lemma 4.1, the intersection of any two polytopes Q_x and Q_y with $x, y \in E$ $(x \neq y)$ has an empty interior, and therefore

$$\mathcal{V}(\cup_{x\in E} Q_x) = \sum_{x\in E} \mathcal{V}(Q_x).$$

Furthermore, each polytope Q_x has a volume of at least $1/n!$. Hence,

$$\sum_{x\in E} \mathcal{V}(Q_x) \geq |E|/n!.$$

Combining these results the theorem follows. □

Corollary 4.5 *For any $m \times n$ (0,1)-matrix A and m-vector b the polyhedron $P_{A,b}$ has at most $n!$ extreme points.*

The maximum of $n!$ can only be achieved if every $(0, 1)$-vector except the null vector is a row of A. Clearly, if A has less rows, then $\mathcal{V}(Q_A)$ is strictly less than 1, and hence the bound in Theorem 4.4 is strictly less than $n!$. If not every $(0, 1)$-vector is a row of the matrix A we therefore obtain a stronger bound.

The maximum of $n!$ extreme points of $P_{A,b}$ can actually be achieved, with A chosen 'maximal' as indicated. Examples are given e.g. in Edmonds (1970), and in Shapley (1971) in the context of transferable utility games.

To obtain $n!$ different extreme points in $P_{A,b}$ the sets Q_x, being 0/1-polytopes, should all have volume $1/n!$ for each extreme point x of $P_{A,b}$, and this is only possible if Q_x is a simplex, i.e., the convex hull of a set of $n+1$ affine independent vectors (see the observation following the proof of Lemma 3). Further, the union of these simplices should coincide with the unit hypercube. This gives rise to a simplicial subdivision of the unit hypercube, also called a triangulation. The main issue in the literature on triangulations is the minimal number of simplices needed to form a triangulation (Mara, 1976; Hughes, 1994). The so-called standard triangulation is the subdivision of the hypercube in simplices of the form $Q^\pi = \{x \in I\!R^n : x_{\pi(1)} \geq x_{\pi(2)} \geq \dots x_{\pi(n)} \geq 0, \sum_{i=1}^n x_i = 1\}$, with π running over all permutations on $\{1, 2, \dots, n\}$. See Freudenthal (1942) for an early reference (Todd, 1976, 29–30). The standard triangulation pops up in many situations. The next section will provide examples.

4.3 The Core of a Transferable Utility Game

An n-person transferable utility game (or *game* for short), with $n \in I\!N$, is a real valued map on the set of subsets of the *player set* $N = \{1, 2, \dots, n\}$, the empty set excluded. A non-empty subset S of N is referred to as a *coalition*, and its *value* $v(S)$ in the game v is interpreted as the net gain of the cooperation of the players in S.

A game v is said to be *additive* if each coalition value is obtained by summing up the one-person coalition values: $v(S) = \sum_{i \in S} v(\{i\})$ for all coalitions S. Given an *allocation* $x \in I\!R^N$ the corresponding additive game is the game, also denoted by x, with coalition values $x(S) = \sum_{i \in S} x_i$. An allocation x is called *stable* in the game v if the corresponding additive game majorizes v: $x(S) \geq v(S)$ for all coalitions S (or $x \geq v$ for short).

The *upper core* of a game v, denoted UCore(v), is the set of stable allocations. Its elements are interpreted as those payoffs to the players that are preferred to playing the game. However, not all stable allocations are feasible in the sense that they can be afforded by the players. Here, we assume that an allocation x is *feasible* in the game v if $v(N) \geq \sum_{i \in N} x_i$ holds. The *core* of a game v, denoted Core(v), is defined as the set of stable and feasible allocations of v.

Consider a fixed sequence $S_1, S_2, \ldots, S_{2^n-1}$ of the coalitions in N. Let A denote the $(2^n - 1) \times n$-dimensional matrix with j,i-coefficient equal to 1 if player i is a member of coalition S_j, and 0 otherwise. For a game v the upper core obviously equals P_{A,b^v}, with b^v the $(2^n - 1)$-dimensional constraint vector with j-coefficient $v(S_j)$. For a stable allocation x the set Q_x is the convex hull of the zero vector and the indicator functions, the rows of A, corresponding to coalitions S for which equality $x(S) = v(S)$ holds. We will refer to these coalitions as being *tight* .

Feasibility of a stable allocation of course imply that the grand coalition N has to be tight. Therefore, the core of v is equal to the face of the upper core P_{A,b^v} of v, determined by the constraint corresponding to coalition N.

With the help of Corollary 4.5 we conclude that

Corollary 4.6 *The core of an n-person cooperative game has at most $n!$ extreme points.*

We will first show that there is a large class of n-person games for which the number of different core points equals the maximum possible number of $n!$. For this we need the following. A game $v : 2^N \to I\!R$ is called *convex* if

$$v(S) + v(T) \leq v(S \cup T) + v(S \cap T) \quad \text{for all coalitions } S, T \tag{4.1}$$

(with the convention that the game value of the empty set equals 0). The game is called *strictly convex* if the convexity inequalities hold, and none of them with equality whenever $S \not\subseteq T$ or $T \not\subseteq S$.

It is well known that the extreme points of the core of a convex game are among the so called marginal contribution vectors (see Shapley, 1971; and Ichiishi, 1981, for the converse statement). For a permutation π on the player set N the *marginal contribution* allocation $m(v, \pi)$ in the game v is defined by

$$m_i(v, \pi) = v(P_\pi(i)) - v(P_\pi(i) \backslash \{i\}),\ i \in N,$$

with $P_\pi(i)$ denoting the predecessors j of player i in π: $\pi(j) \leq \pi(i)$. The allocation $m(v, \pi)$ is the final outcome of the procedure where the players enter a room one by one in the order given by π and each player obtains the value of the coalition of players in the room minus what already has been allocated.

Some of the $n!$ marginal contribution allocation vectors may coincide, but if the game is strictly convex then all these allocations are different. To show this, first observe that $m(v,\pi)(P_\pi(i)) = v(P_\pi(i))$ for all permutations π and players i, and games v not necessarily (strictly) convex. Now let v be a strictly convex game, π a permutation of N, and S a coalition unequal to any predecessor set of π. We will prove that $m(v,\pi)(S) > v(S)$. Let σ be a permutation such that the players in S are the first and such that if $\pi(i) < \pi(j)$ for $i, j \in S$ then also $\sigma(i) < \sigma(j)$. A permutation with this property is constructed, for example, by interchanging the positions (in π) of any player $j \notin S$ with a player $i \in S$ with $\pi(j) < \pi(i)$ until there are no players left with this property. For each $i \in S$ we have $P_\sigma(i) \subseteq P_\pi(i)$ so that by applying (4.1), with $S = P_\sigma(i)$ and $T = P_\pi(i)\backslash\{i\}$, we obtain

$$\begin{aligned} m_i(v,\sigma) &= v(P_\sigma(i)) - v(P_\sigma(i)\backslash\{i\}) \\ &\leq v(P_\pi(i)) - v(P_\pi(i)\backslash\{i\}) = m_i(v,\pi). \end{aligned} \tag{4.2}$$

Therefore, $m(v,\pi)(S) \geq m(v,\sigma)(S)$, and since S is a predecessor set of σ it follows that $m(v,\pi)(S) \geq v(S)$ (thus proving that the marginal contribution allocations are elements of the upper core). There is an $i \in S$ such that $P_\sigma(i)$ is a proper subset of $P_\pi(i)$, and for this player strict inequality holds in (4.2) because of strict convexity of the game. Therefore, $m(v,\pi)(S) > v(S)$ for all coalitions S except the predecessor sets of π. From this, one easily derives that there are no two marginal contribution allocations equal to each other. Consequently, the core has precisely $n!$ extreme points (Shapley, 1971). This shows that the bound in Corollary 4.6 is sharp.

Let us call a collection of coalitions of $\{1, 2, \ldots, n\}$ *regular* if the indicator functions span $\mathbb{R}^n$. It is evident that a stable allocation of a game is extreme in the upper core if and only if its tight coalitions form a regular collection, and a stable allocation is an extreme core point if and only if N is tight, and the set of its tight coalitions is regular.

Observe that we actually proved that the tight coalitions of the marginal contribution allocation $m(v,\pi)$, with v strictly convex, are precisely its predecessor sets, so that the tight coalitions form a regular collection. The corresponding set $Q_{m(v,\pi)}$, the convex hull of the zero vector and the indicator functions of the predecessor sets of π, is easily seen to equal $Q^\pi = \{x \in \mathbb{R}^n : x_{\pi(1)} \geq x_{\pi(2)} \geq \ldots x_{\pi(n)} \geq 0, \sum_{i=1}^n x_i = 1\}$, a typical simplex of the standard triangulation of the unit hypercube. On the other hand, if the tight coalitions of a stable allocation give rise to a simplex of the form Q^π then the allocation has to be a marginal contribution allocation. Therefore, the strict convex games are exactly the games that give rise to the standard triangulation.

The strictly convex games are not the only games with core having the maximum number of extreme points as the following example shows.

Consider the non-convex (symmetric) 4-player game v' with values 0,7,12, and 22 for coalitions with number of players respectively, 1, 2, 3, and 4:

$\lvert S\rvert$:	1	2	3	4
$v'(S)$:	0	7	12	22

Consider the allocations $(2,5,5,10)$ and $(0,7,7,8)$. Obviously, both belong to the core of v', with tight coalition sets respectively $\{N,\{1,2\},\{1,3\},\ \{1,2,3\}\}$ and $\{N,\{1\},\{1,2\},\{1,3\}\}$. The two collections are regular, so that we may conclude that the two allocations are extreme in the core. Because of the symmetry among the players in the game any of the 12 allocations with coefficients 2,5,5,10, and the 12 allocations with coefficients 0,7,7,8 are extreme core points. Therefore the game has at least 24 extreme core points. There are no other since 24 is the maximum number: $4! = 24$.

There is an intuitive approach for obtaining the extreme core points. First, take any ordering of the players. Then, take the first player and maximize its payoff among the core allocations. Thereafter, take the next player, and maximize his payoff among the core allocations where the first player gets his maximal payoff. Continue in this way until the last player. Following this way we obtain an extreme point of the core. Since there are $n!$ different orderings of the players we obtain $n!$ extreme points (possibly, there may be duplicates).

The above example, however, shows that we may not obtain all extreme points in this way. Observe that if we maximize the payoff to a player among the core allocations we obtain the value 10, and therefore, we will never end up in an extreme core allocation with coefficients 0,7,7,8. Analogously, if we minimize the payoff, instead of maximize, we will not terminate in a core allocation with coefficients 2,5,5,10.

4.4 Strict Exact Games

It is not only of mathematical interest to provide necessary and sufficient conditions for games having the maximum number of extreme core points. One may, for example, argue that the extreme points of the core are precisely the outcomes of a game where the players choose their actions in an extreme social way. The number of extreme core points may as such serve as a measure for social complexity (whatever these terms may indicate in an appropriate context). Also, procedures or protocols that construct or give rise to core allocations may endure a complexity that is dependent on the number of extreme core points, especially when, depending on the settings, any core point may occur as outcome.

It is therefore of interest to deduce properties that are implied by the fact that the number of extreme core points is maximal. First, one can easily verify that the number of tight coalitions in an extreme (upper) core allocation should not exceed the dimension of the allocation space. This means that the indicator functions of the tight coalitions are linearly independent. Collections of coalitions with this property are called *non-degenerate*, and a game is called *non-degenerate* if the collections of tight coalitions is non-degenerate for each extreme upper core allocation.

To obtain the maximum number of extreme core points in an n-person game v, the upper core should not have extreme points outside the face corresponding to the grand coalition. This is equivalent to the upper core being equal to the core and all the points lying above the core: $\text{UCore}(v) = \text{Core}(v) + \mathbb{R}^n_+$. If this holds we say that v has a *large core*. A game v has a large core if and only if for each stable allocation x there is a core allocation x' such that $x' \leq x$.

For the upper core having the maximum amount of different extreme points it is essential that in its description as a polyhedral set $P_{A,b}$ no rows of A can be deleted (see the remark following Corollary 4.5). This hints to the condition that for each coalition there is a corresponding face in the upper core, which has to be of maximal dimension $n-1$, a facet. In other words, for each coalition S there is a stable allocation for which S is the only tight coalition. If this is the case then the game is called *strict upper exact*. Without going into details, it is not hard to prove that strict upper exactness is equivalent to the property that each subgame has a core of maximal dimension.

Proposition 4.7 *If the core of a game has the maximum of $n!$ different extreme core points then the core has to be large and the game has to be non-degenerate and strict upper exact.*

The next example shows that the converse does not hold. Consider the 5-person game v'' defined by

$\|S\|$:	1	2	3	4	5
$v''(S)$:	0	1	8	11	23

We will show that the extreme stable allocations in the upper core of v'' are the following points:
(1) the 20 allocations with coefficients 0,4,4,4,11,
(2) the 20 allocations with coefficients 2,3,3,3,12, and
(3) the 60 allocations with coefficients 0,1,7,7,8.
It is left to the reader to check the stability property of these allocations, 100 in total (and less than the maximum possible of 5!=120). Also, with the help of these allocations one easily derives that the game is strict upper exact.

The tight coalitions of the stable allocation $(0,4,4,4,11)$ are the player set $N = \{1,2,3,4,5\}$, $\{1\}$, $\{1,2,3\}$, $\{1,2,4\}$, $\{1,3,4\}$. The independence of the corresponding indicator functions follows from determining the determinant value of the matrix consisting of these indicator functions, say in the given order. The value equals -2, so that we may conclude that $(0,4,4,4,11)$ is extreme in the upper core, and due to the symmetry among the players in the game the other 19 allocations with the same coefficients are also extreme in the upper core. Further, the computed determinant value implies that the volume of $Q_{(0,4,4,4,11)}$, the convex hull of the zero vector and the indicator functions of the tight coalitions, equals 2/5!, so that the 20 allocations of type (1) consume $20 \cdot 2/5!$ of the available volume of the unit hypercube, implying that the upper core can have at most $20 + 120 - 40 = 100$ extreme points.

The other two types of allocations can be derived in the same way. The tight coalitions of the stable allocation $(2,3,3,3,12)$ are N, $\{1,2,3\}$, $\{1,2,4\}$, $\{1,3,4\}$, $\{1,2,3,4\}$, and form a regular collection, implying extremality in the upper core for $(2,3,3,3,12)$ and the other 19 allocations with the same coefficients.

Finally, the tight coalitions of the stable allocation $(0,1,7,7,8)$ form the regular collection $\{N, \{1\}, \{1,2\}, \{1,2,3\}, \{1,2,4\}\}$, implying extremality in the upper core of $(0,1,7,7,8)$ and the other 59 allocations with the same coefficients.

This shows that the mentioned allocations are the extreme upper core points. All allocations are feasible, implying that the game has a large core. Further, all collections of tight coalitions are non-degenerate, showing that the game is non-degenerate.

A game is called *strict exact* if for each coalition S a *core* allocation exist for which S and N are the only tight coalitions. Strict exactness implies strict upper exactness. To see this, let v be strict exact, and let S be an arbitrary coalition. A core allocation x exists for which S and N are the only tight coalitions. Then the sum of x and the indicator function of the complement of S, $x + e_{N\setminus S}$, is a stable allocation with S as the only tight coalition (this argument captures also the case $S = N$).

One easily derives the strict exactness of the game v'' in the previous example. This is not coincidental as the following result shows.

Proposition 4.8 *If a game is non-degenerate and strict upper exact, and has a large core, then it is strict exact.*

Proof. Let v be non-degenerate, strict upper exact, and let its core be large. For an arbitrary coalition S take a stable allocation x for which S is the only tight coalition. There is a core allocation x'' such that $x' \leq x$. Obviously, S and N are tight for x'. Since v is non-degenerate the collection of tight coalitions of x' has to be non-degenerate, and we may therefore assume that an $y \in I\!R^n$ exists such that $y(S) = y(N) = 0$ and $y(T) > 0$ for the other tight coalitions T of x'. For $\epsilon > 0$ sufficiently small the allocation $x'' = x' + \epsilon y$ belongs to the core of v. Its tight coalitions are S and N, thus showing that v is strict exact. □

We cannot leave out the non-degenerate property or the large core condition. This can be derived from the following two symmetric games w' and w'' on the player set $N = \{1, 2, 3\}$: $w'(S) = w''(S) = 0$ for coalitions S with 1 player, $w'(S) = w''(S) = 4$ if S consists of 2 players, and $w'(N) = 7$, and $w''(N) = 8$. It is left to the reader to check that both games are strict upper exact but not strict exact, w' is non-degenerate but does not have a large core, and w'' has a large core but fails to be non-degenerate.

Combining the previous two propositions we conclude that:

Corollary 4.9 *A game is strict exact if its core consists of the maximum of $n!$ extreme points.*

4.5 Concluding Remarks

Summarizing the contents of the paper, we proved that polyhedral sets of the form $P_{A,b} = \{x \in I\!R^n | \, Ax \geq b\}$ have at most $n!$ times the volume of the convex hull of the zero vector and the rows of the matrix A. We applied this result on 0,1-valued matrices and obtained the upper bound

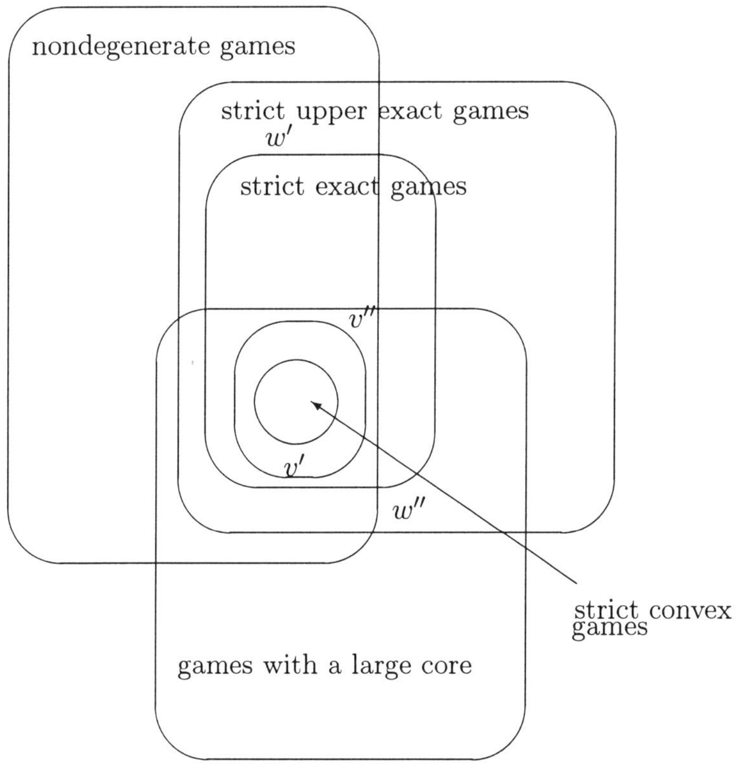

Figure 4.1: *The oval containing v' denotes the relative position of the games with maximum number of extreme core points.*

of $n!$ for the number of extreme points of the upper core and the core of a game. The maximum number is attained by the strictly convex games but other games may have this property as well. These games have to be strict upper exact, must have a large core and fulfill a kind of non-degeneracy. We showed that not all games with these properties have $n!$ different extreme points. See also Figure 4.1.

Future research is concentrated on the dependence relations of the mentioned properties and on the impact of the non-degenerate condition which seems to involve combinatorial techniques for obtaining and analyzing the triangulations of the unit hypercube.

References

Birkhoff, G., and S. Maclane (1963): *A Survey of Modern Algebra*. New York: MacMillan.

Chvátal, V. (1983): *Linear Programming*. New York: Freeman.

Derks J., H. Haller and H. Peters (2000): "The selectope for cooperative games," *International Journal of Game Theory*, 29, 23–38.

Edmonds, J. (1970): "Submodular functions, matroids, and certain polyhedra," in: Richard Guy et al., (eds.), *Combinatorial Structures and their Applications*. Gordon and Breach, 69–87.

Freudenthal, H. (1942): "Simplizialzerlegungen von Beschränkter Flachheit," *Annals of Mathematics*, 43, 580–582.

Gale, D. (1963): "Neighborly and cyclic polytopes," in: V. Klee (ed.), *Convexity, Proceedings of Symposia in Pure Mathematics*, 7, American Mathematical Society, 225–232.

Gillies (1953): *Some theorems on n-person games*. Dissertation, Department of Mathematics, Princeton University.

Hughes, R.B. (1994): "Lower bounds on cube simplexity," *Discrete Mathematics*, 133, 123–138.

Ichiishi, T. (1981): "Super-modularity: application to convex games and to the greedy algorithm for LP," *Journal of Economic Theory*, 25, 283–286.

Kuipers, J. (1994): *Combinatorial Methods in Cooperative Game Theory*. Ph.D. thesis, Universiteit Maastricht, The Netherlands.

Mara, P.S. (1976): "Triangulations for the cube," *Journal of Combinatorial Theory*, Ser. A, 20, 170–177.

McMullen, P. (1970): "The maximum number of faces of a convex polytope," *Mathematika*, 17, 179–184.

Schmeidler, D. (1972): "Cores of exact games," *Journal of Mathematical Analysis and Applications*, 40, 214–225.

Shapley, L.S. (1971): "Cores of convex games," *International Journal of Game Theory*, 1, 11–26.

Tijs, S.H. (1981): "Bounds for the core and the τ-value," in: O. Moeschlin and D. Pallaschke (eds.), *Game Theory and Mathematical Economics*. Amsterdam: North-Holland Publishing Company, 23–132.

Tijs, S.H., and F.A.S. Lipperts (1982): "The Hypercube and the core cover of n-Person cooperative games," *Cahiers du Centre d'Etudes de Recherche Opérationelle*, 24, 27–37.

Todd, M.J. (1976): *The Computation of Fixed Points and Applications.* Lecture notes in Economics and Mathematical Systems, 124, Springer-Verlag.

Vasilev, V.A. (1981): "On a class of imputations in cooperative games," *Soviet Math. Dokl.*, 23, 53–57.

Ziegler, G.M. (1995): *Lectures on Polytopes.* Graduate Texts in Mathematics 152, New York: Springer.

Chapter 5

Consistency and Potentials in Cooperative TU-Games: Sobolev's Reduced Game Revived

BY THEO DRIESSEN

5.1 Introduction

In physics a vector field u is said to be conservative if there exists a continuously differentiable function U called potential the gradient of which agrees with the vector field (notation: $\nabla U = u$). There exist several characterizations of conservative vector fields (e.g., $\nabla_j u_i = \nabla_i u_j$, or every contour integral with respect to the vector field u is zero). Surprisingly, the successful treatment of the potential in physics turned out to be reproducible, in the late eighties, in the mathematical field called cooperative game theory. Informally, a solution concept ψ on the universal game space $\mathcal{G}$ is said to possess a potential representation if it is the discrete gradient of a real-valued function P on $\mathcal{G}$ called potential (notation: $\nabla P = \psi$). In other words, if possible, each component of the game-theoretic solution may be interpreted as the incremental return with respect to the potential function. In their innovative paper, Hart and Mas-Colell (1989) showed that the well-known game-theoretic solution called Shapley value is the unique solution that has a potential

P. Borm and H. Peters (eds.), Chapters in Game Theory, 99–120.

representation and meets the efficiency principle as well. In the second stage (the nineties) of the potential research into the solution part of cooperative game theory, various researchers contributed different, but equivalent characterizations of (not necessarily efficient) solutions that admit a potential (cf. Ortmann, 1989; Calvo and Santos, 1997; Sánchez, 1997). Almost all of these characterizations of solutions, stated in terms of the potential approach applied in cooperative game theory, resemble similar ones stated in physical terminology. For instance, the characterization $\nabla_j u_i = \nabla_i u_j$ of a conservative vector field u is analogous to its discrete version $\nabla_j \psi_i = \nabla_i \psi_j$ with respect to a game-theoretic solution ψ, commonly known as the law of preservation of discrete differences (cf. Ortmann, 1998), also called the balanced contributions principle (cf. Calvo and Santos, 1997; Myerson, 1989; Sánchez, 1997).

One characterization with no counterpart in physics states that a game-theoretic solution possesses a potential representation if and only if the solution for any game equals the Shapley value of another game induced by both the initial game and the relevant solution concept (cf. Calvo and Santos, 1997). Our main goal is to exploit this particular characterization whenever one deals with the so-called reduced game property for solutions, also called consistency property. Informally, the key notions of a reduced game and consistency can be elucidated as follows. A cooperative game is always described by a finite player set as well as a real-valued "characteristic function" on the collection of subsets of the player set. A so-called reduced game is deducible from a given cooperative game by removing one or more players on the understanding that the removed players will be paid according to a specific principle (e.g. a proposed payoff vector). The remaining players form the player set of the reduced game; the characteristic function of which is composed of the original characteristic function, the proposed payoff vector, and/or the solution in question. The consistency property for the solution states that if all the players are supposed to be paid according to a payoff vector in the solution set of the original game, then the players of the reduced game can achieve the corresponding payoff in the solution set of the reduced game. In other words, there is no inconsistency in what the players of the reduced game can achieve, in either the original game or the reduced game.

Generally speaking, the consistency property is a very powerful and widely used tool to axiomatize game-theoretic solutions (cf. the surveys on consistency in Driessen, 1991, and Maschler, 1992). In the early seventies Sobolev (1973) established the consistency property for the well-known Shapley value with respect to an appropriately chosen reduced game. With Sobolev's result at hand, we are in a position to establish, under certain circumstances, the consistency property for a solution that has a potential representation. For that purpose the consistency property is formulated with respect to a strongly adapted version of the reduced game used by Sobolev. Section 5.2 is devoted to the whole treatment of the relevant consistency property. The proof of this specific consistency property (see Theorem 5.6) is based on the particular characterization of solutions that admit a potential. In summary, this chapter solves the open problem concerning a suitably chosen consistency property for a wide class of game-theoretic solutions, inclusive of the Shapley value. In addition, for any solution that admits a potential representation, we provide an axiomatization in terms of the new consistency property, together with some kind of standardness for two-person games (see Theorem 5.8).

Our modified reduced game differs from Sobolev's reduced game only in that any game is replaced by its image under a bijective mapping on the universal game space (induced by the solution in question). The particular bijective mapping, induced by the Shapley value, equals the identity. To be exact, Sobolev's explicit description of the reduced game refers to the initial game itself, whereas our similar, but implicit definition of the modified reduced game is formulated in terms of the image of both the modified reduced game and the initial game (see Theorem 5.6).

In the general framework concerning an arbitrary solution that admits a potential, there is no way to acquire more information about the associated bijective mapping and consequently, the implicit definition of the modified reduced game can not be explored any further to strengthen the consistency property for this solution. For a certain type of solutions called semivalues (cf. Dubey *et al.*, 1981) however, the associated bijective mapping and its inverse are computable and hence, under these particular circumstances, one gains an insight into the modified reduced game itself. Section 5.3 is devoted to a thorough study of these semivalues and, in the setting of the consistency property for these semivalues, we provide various elegant interpretations of the modified reduced game (see Theorems 5.11 and 5.12).

5.2 Consistency Property for Solutions that Admit a Potential

A *cooperative game* with transferable utility (TU) is a pair $\langle N, v\rangle$, where N is a nonempty, finite set and $v : 2^N \to I\!R$ is a *characteristic function*, defined on the power set of N, satisfying $v(\emptyset) := 0$. An element of N (notation: $i \in N$) and a nonempty subset S of N (notation: $S \subseteq N$ or $S \in 2^N$ with $S \neq \emptyset$) is called a *player* and *coalition* respectively, and the associated real number $v(S)$ is called the *worth* of coalition S. The size (cardinality) of coalition S is denoted by $|S|$ or, if no ambiguity is possible, by s. Particularly, n denotes the size of the player set N. Given a (transferable utility) game $\langle N, v\rangle$ and a coalition S, we write $\langle S, v\rangle$ for the *subgame* obtained by restricting v to subsets of S only (i.e., to 2^S). Let $\mathcal{G}$ denote the set of all cooperative games with an arbitrary player set, whereas $\mathcal{G}^N$ denotes the (vector) space of all games with reference to a player set N which is fixed beforehand.

Concerning the solution theory for cooperative TU-games, the chapter is devoted to single-valued solution concepts. Formally, a *solution* ψ on $\mathcal{G}$ (or on a particular subclass of $\mathcal{G}$) associates a single payoff vector $\psi(N, v) = (\psi_i(N, v))_{i\in N} \in I\!R^N$ with every TU-game $\langle N, v\rangle$. The so-called *value* $\psi_i(N, v)$ of player i in the game $\langle N, v\rangle$ represents an assessment by i of his gains from participating in the game. Until further notice, no constraints are imposed upon a solution ψ on $\mathcal{G}$. In the next definition (cf. Calvo and Santos, 1997; Dragan, 1996; Hart and Mas-Colell, 1989; Ortmann, 1998; Sanchez, 1997) we present two key notions (out of four).

Definition 5.1 Let ψ be a solution on $\mathcal{G}$.

(i) We say the solution ψ admits a *potential* if there exists a function $P_\psi : \mathcal{G} \to I\!R$ satisfying

$$P_\psi(N, v) - P_\psi(N\backslash\{i\}, v) = \psi_i(N, v) \quad \text{for all } \langle N, v\rangle \in \mathcal{G},\ i \in N. \tag{5.1}$$

(ii) The mapping $F_\psi : \mathcal{G} \to \mathcal{G}$ associates with any game $\langle N, v\rangle$ its *solution game* $\langle N, F_\psi^v\rangle$; the characteristic function of which is defined to be

$$F_\psi^v(S) := \sum_{i\in S} \psi_i(S, v) \qquad \text{for all } S \subseteq N,\ S \neq \emptyset. \tag{5.2}$$

◁

In words, the potential function P_ψ represents a scalar evaluation for cooperative TU-games, of which any player's marginal contribution agrees with the player's value according to the relevant solution ψ (notation: $\nabla P_\psi = \psi$). If the potential exists, it is uniquely determined up to an additive constant by the recursive formula
$n \cdot P_\psi(N,v) - \sum_{i \in N} P_\psi(N \backslash \{i\}, v) = F_\psi^v(N)$. Usually, it is tacitly assumed that the potential is zero-normalized (i.e., $P_\psi(\emptyset, v) := 0$). In fact, it is well-known that the potential function P_ψ (if it exists) is given by

$$P_\psi(N,v) = \sum_{\substack{S \subseteq N, \\ S \neq \emptyset}} \frac{F_\psi^v(S)}{s \cdot \binom{n}{s}} \qquad \text{for all } \langle N,v \rangle \in \mathcal{G}, \text{ where } P_\psi(\emptyset, v) := 0.$$

By (5.2), the worth $F_\psi^v(S)$ of coalition S in the solution game $\langle N, F_\psi^v \rangle$ represents the overall gains (according to the solution ψ) to the members of S from participating in the induced subgame $\langle S, v \rangle$ (on the understanding that players outside S are not supposed to cooperate). Generally speaking, the solution game differs from the initial game. Notice that both games are the same if and only if the solution ψ meets the *efficiency* principle, i.e.,

$$\sum_{i \in N} \psi_i(N,v) = v(N) \qquad \text{for all } \langle N,v \rangle \in \mathcal{G}.$$

The core topic involves the so-called consistency treatment for solutions that admit a potential. For that purpose, we need to recall one basic theorem from Calvo and Santos (1997); the main result of which is referring to the well-known Shapley value. With the help of Sobolev's (1973) pioneer work in the early seventies on the consistency property for the Shapley value, we are able to prove, under certain circumstances, a similar consistency property for (not necessarily efficient) solutions that admit a potential.

Definition 5.2 The *Shapley value* $Sh_i(N,v)$ of player $i \in N$ in the n-person game $\langle N,v \rangle$ is defined as follows (cf. Shapley, 1953):

$$Sh_i(N,v) := \sum_{S \subseteq N \backslash \{i\}} \frac{1}{n \cdot \binom{n-1}{s}} \cdot \Big[v(S \cup \{i\}) - v(S) \Big] \quad \text{for all } \langle N,v \rangle \in \mathcal{G}, \text{ all } i \in N. \tag{5.3}$$

◁

Theorem 5.3 *Consider the setting of Definitions 5.1 and 5.2.*

(i) (Cf. Calvo and Santos, 1997, Theorem, page 178.) Let ψ be a solution on $\mathcal{G}$. Then ψ admits a potential if and only if $\psi(N,v) = Sh(N, F_\psi^v)$ for all $\langle N,v\rangle \in \mathcal{G}$. In words, any solution that admits a potential equals the Shapley value of the associated solution game.

(ii) (Cf. Hart and Mas-Colell, 1989, Theorem A, page 591.) The Shapley value is the unique solution ψ on $\mathcal{G}$ that admits a potential and is efficient as well.

Definition 5.4 (Cf. Sobolev, 1973; Driessen, 1991.)

(i) With an n-person game $\langle N,v\rangle$, a player $i \in N$, and his payoff $x_i \in \mathbb{R}$ (provided $n \geq 2$), there is associated the *reduced game* $\langle N\backslash\{i\}, v_{N\backslash\{i\}}^{x_i}\rangle$ with player set $N\backslash\{i\}$ defined by

$$v_{N\backslash\{i\}}^{x_i}(S) := \frac{s}{n-1}\cdot\left[v(S\cup\{i\}) - x_i\right] + \frac{n-1-s}{n-1}\cdot v(S) \quad \text{for all } S \subseteq N\backslash\{i\}. \tag{5.4}$$

(ii) A solution ψ on $\mathcal{G}$ is said to be *consistent* with respect to this reduced game if it holds

$$\psi_j(N\backslash\{i\}, v_{N\backslash\{i\}}^{\psi_i(N,v)}) = \psi_j(N,v) \quad \text{for all } \langle N,v\rangle \in \mathcal{G},\ i \in N,\ j \in N\backslash\{i\}. \tag{5.5}$$

In words, there is no inconsistency in what each of the players in the reduced game will get according to the solution ψ, in either the reduced game or the initial game. ◁

Theorem 5.5 *(Cf. Sobolev, 1973; Driessen, 1991.)*
The Shapley value Sh on $\mathcal{G}$ is consistent with respect to the reduced game of the form (5.4).

Now we are in a position to state and prove a similar consistency property for solutions that admit a potential. Actually, for a given solution ψ, the appropriately chosen reduced game resembles Sobolev's reduced game (5.4), but they differ in that the initial game v is replaced by the associated solution game F_ψ^v. In summary, it turns out that the cornerstone of the consistency approach to (not necessarily efficient) solutions is the solution game instead of the game itself. Consequently, we have to define the modified reduced game implicitly by means of its associated solution game, on the understanding that a one-to-one correspondence (bijection) between games and solution games is supposed to be available.[1]

Theorem 5.6 *Let ψ be a solution on $\mathcal{G}$ that admits a potential. Suppose that the induced mapping $F_\psi : \mathcal{G} \to \mathcal{G}$, as given by (5.2), is a bijection.*

With an n-person game $\langle N, v\rangle$, a player $i \in N$, and his payoff $y_i \in \mathbb{R}$ (provided $n \geq 2$), there is associated the modified reduced game *$\langle N\backslash\{i\}, \tilde{v}^{y_i}_{N\backslash\{i\}}\rangle$ with player set $N\backslash\{i\}$ which is defined implicitly by its associated solution game $\langle N\backslash\{i\}, F_\psi^{(\tilde{v}^{y_i}_{N\backslash\{i\}})}\rangle$; the characteristic function of which is defined to be*

$$F_\psi^{(\tilde{v}^{y_i}_{N\backslash\{i\}})}(S) \;:=\; \frac{s}{n-1}\cdot\left[F_\psi^v(S\cup\{i\}) - y_i\right] + \frac{n-1-s}{n-1}\cdot F_\psi^v(S) \quad \text{for all } S \subseteq N\backslash\{i\}. \tag{5.6}$$

Then the solution ψ on $\mathcal{G}$ is consistent with respect to this modified reduced game, i.e.,

$$\psi_j(N\backslash\{i\}, \tilde{v}^{\psi_i(N,v)}_{N\backslash\{i\}}) \;=\; \psi_j(N,v) \quad \text{for all } \langle N,v\rangle \in \mathcal{G},\ i \in N,\ j \in N\backslash\{i\}. \tag{5.7}$$

That is, there is no inconsistency in what each of the players in the reduced game will get according to the solution ψ, in either the reduced game or the initial game.

Proof. Fix both the n-person game $\langle N, v\rangle$ and a player $i \in N$ (where $n \geq 2$). Write w instead of F_ψ^v. Since ψ admits a potential, it holds, by Theorem 5.3(i), $\psi(N,v) = Sh(N, F_\psi^v) = Sh(N,w)$. The essential part of the proof concerns the claim that the solution game (5.6) technique applied to the modified reduced game agrees with Sobolev's reduced

[1] In Dragan (1996, Definition 11, page 459), the solution game plays an identically prominent role in defining the reduced game; the characteristic function of which is, however, from a different type since it deals with the reduced game in the sense of Hart and Mas-Colell (1989). Our model deals with the reduced game in the sense of Sobolev.

game (5.4) technique applied to the initial solution game. Formally, we claim the following:

$$\langle N\backslash\{i\}, F_{\psi}^{(\tilde{v}_{N\backslash\{i\}}^{\psi_i(N,v)})}\rangle = \langle N\backslash\{i\}, w_{N\backslash\{i\}}^{Sh_i(N,w)}\rangle. \tag{5.8}$$

Indeed, from both types of reduced games, we deduce that, for all $S \subseteq N\backslash\{i\}$, it holds

$$\begin{aligned} F_{\psi}^{(\tilde{v}_{N\backslash\{i\}}^{\psi_i(N,v)})}(S) &\overset{(5.6)}{=} \frac{s}{n-1}\Big[F_{\psi}^{v}(S\cup\{i\}) - \psi_i(N,v)\Big] + \frac{n-1-s}{n-1}F_{\psi}^{v}(S) \\ &= \frac{s}{n-1}\Big[w(S\cup\{i\}) - Sh_i(N,w)\Big] + \frac{n-1-s}{n-1}w(S) \\ &\overset{(5.4)}{=} w_{N\backslash\{i\}}^{Sh_i(N,w)}(S). \end{aligned}$$

This proves (5.8). From this we deduce that the following chain of four equalities holds:

$$\begin{aligned} \psi_j(N\backslash\{i\}, \tilde{v}_{N\backslash\{i\}}^{\psi_i(N,v)}) &= Sh_j(N\backslash\{i\}, F_{\psi}^{(\tilde{v}_{N\backslash\{i\}}^{\psi_i(N,v)})}) \\ &\overset{(5.8)}{=} Sh_j(N\backslash\{i\}, w_{N\backslash\{i\}}^{Sh_i(N,w)}) \\ &\overset{(5.5)}{=} Sh_j(N,w) \\ &= \psi_j(N,v) \qquad \text{for all } j \in N\backslash\{i\}. \end{aligned}$$

where the first and last equality are due to Theorem 5.3(i) and the third equality is due to Theorem 5.5 concerning the consistency property (5.5) for the Shapley value Sh. This completes the full proof of the consistency property for the solution ψ. □

Definition 5.7 Let ψ and Φ be two solutions on $\mathcal{G}$. We say the solution Φ is *F_ψ-standard for two-person games* if, for every two-person game $\langle\{i,j\},v\rangle$ and every player $k \in \{i,j\}$, it holds that

$$\Phi_k(\{i,j\},v) = F_{\psi}^{v}(\{k\}) + \tfrac{1}{2}\cdot\Big[F_{\psi}^{v}(\{i,j\}) - F_{\psi}^{v}(\{i\}) - F_{\psi}^{v}(\{j\})\Big]$$

◁

Clearly, by (5.1)–(5.2), a solution ψ that admits a (zero-normalized) potential, satisfies the F_ψ–standardness for two-person games. We conclude this section with the next axiomatization.

Theorem 5.8 *Let ψ be a solution on $\mathcal{G}$ that admits a potential. Suppose that the induced mapping $F_\psi : \mathcal{G} \to \mathcal{G}$, as given by (5.2), is a bijection. Then ψ is the unique solution Φ on $\mathcal{G}$ that satisfies the following two properties:*

(i) Consistency with respect to the modified reduced game implicitly defined through its associated solution game (5.6) (with reference to the given solution ψ).

(ii) F_ψ–standardness for two-person games.

Proof. We show the uniqueness part of Theorem 5.8. Besides the given solution ψ, suppose that a solution Φ on $\mathcal{G}$ satisfies the consistency property and the F_ψ–standardness for two-person games. We prove by induction on the size n of the player set N that $\Phi(N,v) = \psi(N,v)$ for every game $\langle N,v\rangle$. The case $n = 2$ holds trivially because of the F_ψ–standardness for two-person games applied to both solutions. From now on fix an n-person game $\langle N,v\rangle$ with $n \geq 3$. Due to the induction hypothesis, it holds that $\Phi(M,u) = \psi(M,u)$ for every game $\langle M,u\rangle$ with $2 \leq |M| \leq n-1$. Note that, for all $i \in N$, all $x_i \in I\!R$, $y_i \in I\!R$, it follows immediately from (5.6) that

$$F_\psi^{(\tilde{v}^{y_i}_{N\backslash\{i\}})}(S) = F_\psi^{(\tilde{v}^{x_i}_{N\backslash\{i\}})}(S) + s \cdot \frac{x_i - y_i}{n-1} \qquad \text{for all } S \subseteq N\backslash\{i\}.$$

In other words, the two solution games $\langle N\backslash\{i\}, F_\psi^{(\tilde{v}^{y_i}_{N\backslash\{i\}})}\rangle$ and $\langle N\backslash\{i\}, F_\psi^{(\tilde{v}^{x_i}_{N\backslash\{i\}})}\rangle$ are strategically equivalent (with reference to the translation vector $\frac{x_i-y_i}{n-1} \cdot (1,1,\ldots,1) \in I\!R^{N\backslash\{i\}}$) and thus, the covariance property for the Shapley value Sh applies in the sense that it holds

$$\begin{aligned} Sh_j(N\backslash\{i\}, F_\psi^{(\tilde{v}^{y_i}_{N\backslash\{i\}})}) \;&=\; Sh_j(N\backslash\{i\}, F_\psi^{(\tilde{v}^{x_i}_{N\backslash\{i\}})}) + \frac{x_i - y_i}{n-1} \\ &\quad \text{for all } j \in N\backslash\{i\}. \end{aligned} \tag{5.9}$$

For all $i \in N$ and all $j \in N\backslash\{i\}$, we obtain the following chain of equalities:

$$\begin{aligned} \Phi_j(N,v) &= \Phi_j(N\backslash\{i\}, \tilde{v}^{\Phi_i(N,v)}_{N\backslash\{i\}}) && \text{by consistency for } \Phi \\ &= \psi_j(N\backslash\{i\}, \tilde{v}^{\Phi_i(N,v)}_{N\backslash\{i\}}) && \text{by induction hypothesis} \\ &= Sh_j(N\backslash\{i\}, F_\psi^{(\tilde{v}^{\Phi_i(N,v)}_{N\backslash\{i\}})}) && \text{by Theorem 5.3(i)} \end{aligned}$$

$$\stackrel{(5.9)}{=} \quad Sh_j(N\backslash\{i\}, F_\psi^{(\tilde{v}_{N\backslash\{i\}}^{\psi_i(N,v)})}) + \frac{\psi_i(N,v) - \Phi_i(N,v)}{n-1}$$

by covariance for Sh

$$= \quad \psi_j(N\backslash\{i\}, \tilde{v}_{N\backslash\{i\}}^{\psi_i(N,v)}) + \frac{\psi_i(N,v) - \Phi_i(N,v)}{n-1}$$

by Theorem 5.3(i)

$$= \quad \psi_j(N,v) + \frac{\psi_i(N,v) - \Phi_i(N,v)}{n-1}$$

by consistency for ψ.

We conclude that

$$\Phi_j(N,v) - \psi_j(N,v) = \frac{\psi_i(N,v) - \Phi_i(N,v)}{n-1} \text{ for all } i \in N,\ j \in N\backslash\{i\}.$$

By interchanging the roles of players i and j, the latter result yields

$$\Phi_i(N,v) - \psi_i(N,v) = \frac{\Phi_i(N,v) - \psi_i(N,v)}{(n-1)^2} \text{ for all } i \in N.$$

Since $n \geq 3$, we arrive at the conclusion that $\Phi_i(N,v) - \psi_i(N,v) = 0$ for all $i \in N$. Thus, $\Phi(N,v) = \psi(N,v)$ for every game $\langle N, v\rangle$ as was to be shown. □

5.3 Consistency Property for Pseudovalues: a Detailed Exposition

In this section we aim to clarify that, if we deal with a particular type of solutions called pseudovalues, then various elegant interpretations arise in the study of the modified reduced game as given by (5.6). Besides the various appealing interpretations in some kind of terminology, we claim that the implicit definition of the modified reduced game can be transformed into an explicit one, although the resulting explicit description becomes rather laborious.

In Dubey *et al.* (1981) a *semivalue* on $\mathcal{G}^N$ is defined to be a function $\psi : \mathcal{G}^N \to \mathbb{R}^N$ which satisfies the linearity, symmetry, monotonicity, and projection axioms. It was shown (Theorem 1, page 123) that every semivalue can be expressed by the following formula which will be used as our starting point (but we omit certain non-negativity constraints). Throughout this section, lower-case letters s, t, n, and so on, are supposed to be non-negative integers because they are meant to refer to sizes of coalitions. For the sake of notation, let $\mathcal{P} = \{p_s^n\}$ represent an arbitrary collection of real numbers called *weights*, meant to be read as $\mathcal{P} := \{p_s^n \in \mathbb{R} \mid n \in \{1, 2, \ldots\}, s \in \{0, 1, \ldots, n-1\}\}$.

Definition 5.9 We say a solution ψ on $\mathcal{G}$ is a *pseudovalue* on $\mathcal{G}$ if there exists a collection of weights $\mathcal{P} = \{p_s^n\}$ such that the following two conditions hold:

(i)

$$\psi_i(N, v) = \sum_{S \subseteq N \setminus \{i\}} p_s^n \cdot \Big[v(S \cup \{i\}) - v(S) \Big] \quad \text{for all } \langle N, v \rangle \in \mathcal{G}, \text{ all } i \in N; \tag{5.10}$$

(ii) the collection of weights $\mathcal{P} = \{p_s^n\}$ possesses the *upwards triangle property*, i.e.,

$$p_s^{n+1} + p_{s+1}^{n+1} = p_s^n \qquad \text{for all } n \geq 1 \text{ and all } 0 \leq s \leq n-1. \tag{5.11}$$

In words, in the setting of populations with a variable size, the "weight" of the formation of a coalition of size s in an n-person population equals the sum of the "weights" of the two events which may arise by enlarging the population with one person (namely, two coalitions of consecutive sizes s and $s+1$ respectively in an $(n+1)$-person population).

◁

For reasons that will be explained later on, no further constraints are imposed upon the weights (e.g., they are not necessarily non-negative). A pseudovalue with reference to non-negative weights is known as a semivalue (Dubey *et al.*, 1981). It is straightforward to check that any pseudovalue ψ admits a potential (due to the upwards triangle property for $\mathcal{P} = \{p_s^n\}$), where the potential function $P_\psi : \mathcal{G} \to \mathbb{R}$ is given by

$$P_\psi(N, v) = \sum_{\substack{S \subseteq N, \\ S \neq \emptyset}} p_{s-1}^n \cdot v(S) \qquad \text{for all } \langle N, v \rangle \in \mathcal{G}.$$

To start with, we determine an explicit formula for the associated solution game. As an adjunct, we become engaged with induced collections of weights verifying the upwards triangle property.

Proposition 5.10 *Let $\mathcal{P} = \{p_s^n\}$ be an arbitrary collection of weights.*

(i) If the collection $\mathcal{P} = \{p_s^n\}$ possesses the upwards triangle property, so does the induced collection $\tilde{\mathcal{P}} = \{\tilde{p}_s^n\}$ defined by

$$\tilde{p}_s^n := (s+1) \cdot p_s^n - (n-1-s) \cdot p_{s+1}^n \text{ for all } n \geq 1,\ 0 \leq s \leq n-1. \quad (5.12)$$

(ii) For every $n \geq 1$, given that $\tilde{p}_s^n = (s+1) \cdot p_s^n - (n-1-s) \cdot p_{s+1}^n$ for all $0 \leq s \leq n-1$, then the weights p_s^n can be re-discovered as follows:

$$p_s^n = \frac{1}{n \cdot \binom{n-1}{s}} \cdot \sum_{t=s}^{n-1} \binom{n}{t+1} \cdot \tilde{p}_t^n \text{ for all } 0 \leq s \leq n-1. \quad (5.13)$$

(iii) Suppose (5.13) holds for all $n \geq 1$. If the collection $\tilde{\mathcal{P}} = \{\tilde{p}_s^n\}$ possesses the upwards triangle property, so does the induced collection $\mathcal{P} = \{p_s^n\}$.

(iv) The special case $p_s^n = \frac{1}{n \cdot \binom{n-1}{s}}$ yields $\tilde{p}_{n-1}^n = 1$ and $\tilde{p}_s^n = 0$ for all $n \geq 1$, all $0 \leq s \leq n-2$.

For expositional convenience, the computational, but straightforward proof of Proposition 5.10 is postponed until Section 5.5. By (5.12)–(5.13), there exists a natural one-to-one correspondence between collections of weights that satisfy the upwards triangle property. Particularly, any pseudovalue ψ induces another pseudovalue $\tilde{\psi}$, the weights of which are given by (5.12) (and vice versa, by (5.13)). For instance, by part (iv), the Shapley value induces the pseudovalue $\tilde{\psi}$ that agrees with the marginal contribution principle in the sense that $\tilde{\psi}_i(N, v) = v(N) - v(N \backslash \{i\})$ for every game $\langle N, v \rangle$ and all $i \in N$. Another well-known pseudovalue, called Banzhaf value, corresponds to the uniform weights $p_s^n = \frac{1}{2^{n-1}}$ for all $0 \leq s \leq n-1$, while the induced pseudovalue is associated with the weights $\tilde{p}_s^n = \frac{2 \cdot s + 2 - n}{2^{n-1}}$ for all $0 \leq s \leq n-1$. Note that the smallest weights $\tilde{p}_s^n$ are negative. Because of this observation, we do not want to exclude pseudovalues associated with not necessarily non-negative weights. Throughout the remainder of this section, the induced pseudovalue turns out to be of particular interest in order to provide an appealing explicit and implicit interpretation of the modified reduced game.

In the second stage we claim two preliminary results each of which is of interest on its own. Firstly, by (5.15), we state that the mapping F_ψ induced by an initial pseudovalue ψ may be interpreted as the potential function of the induced pseudovalue $\tilde{\psi}$ (in the sense that $P_{\tilde{\psi}}(N,v) := F_\psi^v(N)$ for every game $\langle N,v\rangle$). Secondly, by (5.16), in comparing the two solution games associated with the modified reduced game and the initial game respectively, the increase (decrease) to the worth of any coalition turns out to be coalitionally-size-proportional to the increase (decrease) to the payoff of the removed player, taking into account his initial payoff and his payoff according to the induced pseudovalue $\tilde{\psi}$ (with respect to the subgame the player set of which consists of the partnership between the coalition involved and the removed player).

In the third and final stage we claim, by (5.19), that a specifically chosen weighted sum of the latter increases (decreases) to the payoff of the removed player represents the increase (decrease) to the worth of any coalition, in comparing the modified reduced game and the initial game respectively. The recursively computable coefficients used in the relevant weighted sum are identical to those which appear in the explicit determination of the inverse of the bijective mapping F_ψ associated with the pseudovalue ψ. This mapping turns out to be bijective under very mild conditions imposed upon the underlying collection of weights $\mathcal{P}$ that prescribe the pseudovalue ψ.

Theorem 5.11 *Let ψ be a pseudovalue on $\mathcal{G}$ of the form (5.10) associated with the collection of weights $\mathcal{P} = \{p_s^n\}$. Let $F_\psi : \mathcal{G} \to \mathcal{G}$ be the induced mapping as given by (5.2). Further, let $\tilde{\psi}$ be the induced pseudovalue on $\mathcal{G}$ associated with the induced collection of weights $\tilde{\mathcal{P}} = \{\tilde{p}_s^n\}$ as given by (5.12). Then the following holds:*
(i)

$$F_\psi^v(T) = \sum_{S\subseteq T}\left[s\cdot p_{s-1}^t - (t-s)\cdot p_s^t\right]\cdot v(S) \text{ for all } \langle N,v\rangle \in \mathcal{G},\ T\subseteq N. \tag{5.14}$$

(ii)

$$F_\psi^v(T) - F_\psi^v(T\backslash\{i\}) = \tilde{\psi}_i(T,v) \text{ for all } \langle N,v\rangle \in \mathcal{G},\ T\subseteq N,\ i\in T. \tag{5.15}$$

(iii)

$$F_\psi^{(\tilde{v}^{y_i}_{N\backslash\{i\}})}(S) - F_\psi^v(S) = \frac{s}{n-1}\cdot\left[\tilde{\psi}_i(S\cup\{i\},v) - y_i\right] \tag{5.16}$$

for all $\langle N,v\rangle \in \mathcal{G}$, all $i\in N$, all $S\subseteq N\backslash\{i\}$, and all $y_i \in \mathbb{R}$ (provided $n\geq 2$).

Proof. (i) Let $\langle N, v\rangle \in \mathcal{G}$ and $T \subseteq N$, $T \neq \emptyset$. By assumption of a pseudovalue ψ of the form (5.10) applied to the subgame $\langle T, v\rangle$, and by some straightforward combinatorial computations, we obtain

$$F_\psi^v(T) \overset{(5.2)}{=} \sum_{i\in T} \psi_i(T,v) \overset{(5.10)}{=} \sum_{i\in T} \sum_{S\subseteq T\backslash\{i\}} p_s^t \cdot \Big[v(S\cup\{i\}) - v(S)\Big]$$

$$= \sum_{S\subseteq T} \Big[s\cdot p_{s-1}^t - (t-s)\cdot p_s^t\Big]\cdot v(S) \overset{(5.12)}{=} \sum_{S\subseteq T} \tilde{p}_{s-1}^t \cdot v(S).$$

(ii) Let $\langle N, v\rangle \in \mathcal{G}$, $T \subseteq N$ and $i \in T$. By (5.14), player i's incremental return with respect to the coalition T in the associated solution game $\langle N, F_\psi^v\rangle$ is determined as follows:

$$F_\psi^v(T) - F_\psi^v(T\backslash\{i\}) \overset{(5.14)}{=} \sum_{S\subseteq T} \tilde{p}_{s-1}^t v(S) - \sum_{S\subseteq T\backslash\{i\}} \tilde{p}_{s-1}^{t-1} v(S)$$

$$= \sum_{\substack{S\subseteq T,\\ S\ni i}} \tilde{p}_{s-1}^t v(S) + \sum_{S\subseteq T\backslash\{i\}} \Big[\tilde{p}_{s-1}^t - \tilde{p}_{s-1}^{t-1}\Big] v(S)$$

$$= \sum_{S\subseteq T\backslash\{i\}} \tilde{p}_s^t v(S\cup\{i\}) - \sum_{S\subseteq T\backslash\{i\}} \tilde{p}_s^t v(S)$$

$$= \sum_{S\subseteq T\backslash\{i\}} \tilde{p}_s^t \Big[v(S\cup\{i\}) - v(S)\Big] \overset{(5.10)}{=} \tilde{\psi}_i(T,v)$$

where the third equality is due to the upwards triangle property for $\tilde{\mathcal{P}}$ (see Proposition 5.10(i)).
(iii) Let $\langle N, v\rangle \in \mathcal{G}$, $i \in N$, and $y_i \in \mathbb{R}$ (provided $n \geq 2$). From the implicit definition of the modified reduced game as given by (5.6), and (5.15) applied to $T = S\cup\{i\}$ respectively, we derive the following:

$$F_\psi^{(\tilde{v}_{N\backslash\{i\}}^{y_i})}(S) - F_\psi^v(S) \overset{(5.6)}{=} \frac{s}{n-1}\cdot\Big[F_\psi^v(S\cup\{i\}) - F_\psi^v(S) - y_i\Big]$$

$$\overset{(5.15)}{=} \frac{s}{n-1}\cdot\Big[\tilde{\psi}_i(S\cup\{i\}, v) - y_i\Big]$$

for all $S \subseteq N\backslash\{i\}$.

□

Theorem 5.12 *Let ψ be a pseudovalue on $\mathcal{G}$ of the form (5.10) associated with the collection of weights $\mathcal{P} = \{p_s^n\}$ satisfying $p_{n-1}^n > 0$*

for all $n \geq 1$. *Let* $F_\psi : \mathcal{G} \to \mathcal{G}$ *be the induced mapping as given by (5.2). Further, let* $\tilde{\mathcal{P}} = \{\tilde{p}_s^n\}$ *be the induced collection of weights as given by (5.12). For every* $t \geq 1$, *let the induced collection of constants* $\{q_k^t(s) \in I\!R \mid k \in \{1,2,\ldots,t\}, s \in \{1,2,\ldots,t+1-k\}\}$ *be defined recursively by*

$$q_1^t(s) := \tilde{p}_{s-1}^t \qquad \text{for all } 1 \leq s \leq t \quad \text{and}$$

$$q_{k+1}^t(s) := \binom{t-s}{k} \cdot \frac{\tilde{p}_{s-1}^{t-k}}{\tilde{p}_{t-1-k}^{t-k}} \cdot q_k^t(t-k) - q_k^t(s)$$
$$\text{for all } 1 \leq k \leq t-1, \text{ all } 1 \leq s \leq t-k. \tag{5.17}$$

Then the following holds:

(i) Given that $F_\psi^v(T) = \sum_{S \subseteq T} \tilde{p}_{s-1}^t \cdot v(S)$ *for every game* $\langle N, v\rangle$ *and all* $T \subseteq N$ *(see (5.14)), the data* $v(T)$, $T \subseteq N$, *of any game* $\langle N, v\rangle$ *can be re-discovered as follows:*

$$\tilde{p}_{t-1}^t \cdot v(T) = \sum_{\substack{S \subseteq T,\\ S \neq \emptyset}} (-1)^{t-s} \cdot \frac{q_{t-s}^t(s)}{\tilde{p}_{s-1}^s} \cdot F_\psi^v(S)$$
$$\text{for all } T \subseteq N, \quad \text{where } q_0^t(t) := \tilde{p}_{t-1}^t \tag{5.18}$$

(ii) Let $\tilde{\psi}$ *be the induced pseudovalue on* $\mathcal{G}$ *associated with* $\tilde{\mathcal{P}} = \{\tilde{p}_s^n\}$. *Then it holds*

$$\tilde{p}_{t-1}^t \cdot \left[\tilde{v}_{N\setminus\{i\}}^{y_i}(T) - v(T)\right]$$

$$= \sum_{\substack{S \subseteq T,\\ S \neq \emptyset}} (-1)^{t-s} \cdot \frac{q_{t-s}^t(s)}{\tilde{p}_{s-1}^s} \cdot \frac{s}{n-1} \cdot \left[\tilde{\psi}_i(S \cup \{i\}, v) - y_i\right] \tag{5.19}$$

for all $\langle N, v\rangle \in \mathcal{G}$, *all* $i \in N$, *all* $T \subseteq N\setminus\{i\}$, $T \neq \emptyset$, *and all* $y_i \in I\!R$ *(provided* $n \geq 2$*).*

(iii) For the special case $p_s^n = \frac{1}{n \cdot \binom{n-1}{s}}$, *then (5.19) reduces to Sobolev's reduced game (5.4).*

The rather technical proof of Theorem 5.12 will be postponed until Section 5.5.

Remark 5.13 To conclude with, we specify the explicit determination for the worth of one- and two-person coalitions in the $(n-1)$-person modified reduced game (5.6), without regard to the number n of players in the initial game.

Let ψ be a pseudovalue on $\mathcal{G}$ of the form (5.10) associated with the collection of weights $\mathcal{P} = \{p_s^n\}$ satisfying $p_{n-1}^n > 0$ for all $n \geq 1$. Let $\tilde{\psi}$ be the induced pseudovalue on $\mathcal{G}$ associated with the induced collection of weights $\tilde{\mathcal{P}} = \{\tilde{p}_s^n\}$ as given by (5.12).

Consider an arbitrary n-person game $\langle N, v\rangle$ and let $i \in N$, $y_i \in \mathbb{R}$. By applying (5.19) to one- and two-person coalitions $T = \{j\}$ and $T = \{j,k\}$ respectively, and (5.10) to the pseudovalue $\tilde{\psi}$, we obtain that the worth of one- and two-person coalitions in the $(n-1)$-person modified reduced game $\langle N\backslash\{i\}, \tilde{v}_{N\backslash\{i\}}^{y_i}\rangle$ of the form (5.6) is determined as follows (recall that, by (5.17), $q_1^2(1) = \tilde{p}_0^2$):

$$\tilde{v}_{N\backslash\{i\}}^{y_i}(\{j\})$$

$$= v(\{j\}) + \frac{1}{(n-1)\cdot \tilde{p}_0^1}\cdot\Big[\tilde{p}_1^2\cdot[v(\{i,j\}) - v(\{j\})] + \tilde{p}_0^2\cdot v(\{i\}) - y_i\Big];$$

$$\tilde{v}_{N\backslash\{i\}}^{y_i}(\{j,k\})$$

$$= v(\{j,k\}) + \frac{2\cdot\tilde{p}_2^3}{(n-1)\cdot\tilde{p}_1^2}\cdot\Big[v(\{i,j,k\}) - v(\{j,k\})\Big]$$

$$+\frac{[2\cdot\tilde{p}_1^3\cdot\tilde{p}_0^1 - \tilde{p}_1^2\cdot\tilde{p}_0^2]}{(n-1)\cdot\tilde{p}_1^2\cdot\tilde{p}_0^1}\cdot\Big[v(\{i,j\}) - v(\{j\}) + v(\{i,k\}) - v(\{k\})\Big]$$

$$+\frac{2\cdot[\tilde{p}_0^3\cdot\tilde{p}_0^1 - \tilde{p}_0^2\cdot\tilde{p}_0^2]}{(n-1)\cdot\tilde{p}_1^2\cdot\tilde{p}_0^1}\cdot v(\{i\}) + \frac{2\cdot[\tilde{p}_0^2 - \tilde{p}_0^1]}{(n-1)\cdot\tilde{p}_1^2\cdot\tilde{p}_0^1}\cdot y_i.$$

In the framework of three-person games, we obtained a complete description of the two-person modified reduced game and by tedious, but straightforward calculations, one may verify that the consistency property $\psi_j(N\backslash\{i\}, \tilde{v}_{N\backslash\{i\}}^{\psi_i(N,v)}) = \psi_j(N,v)$ holds true for the pseudovalue ψ with respect to three-person games. One useful tool concerns the upwards triangle property for $\tilde{\mathcal{P}}$. $\triangleleft$

Remark 5.14 The relationship (5.15) is also useful to provide, in the framework of pseudovalues, an alternative proof of the fundamental equivalence theorem between any pseudovalue ψ and the Shapley value, that is $\psi(N,v) = Sh(N, F^v_\psi)$ for every game $\langle N,v\rangle$. Let us outline this alternative proof that differs from the proofs of Calvo and Santos (1997) and Sánchez (1997) of the equivalence theorem applied to solutions that admit a potential.

Let $\langle N,v\rangle \in \mathcal{G}$. Recall that, by straightforward combinatorial computations, the solution game $\langle N, F^v_\psi\rangle$ is determined by (5.14) and in turn, the incremental returns of any player in the solution game are determined by (5.15), i.e.,

$$F^v_\psi(T\cup\{i\}) - F^v_\psi(T) \overset{(5.15)}{=} \tilde{\psi}_i(T\cup\{i\}, v) \tag{5.20}$$

$$\overset{(5.10)}{=} \sum_{R\subseteq T} \tilde{p}^{t+1}_r \cdot \Big[v(R\cup\{i\}) - v(R)\Big] \tag{5.21}$$

for all $i \in N$ and all $T \subseteq N\backslash\{i\}$. From this and some additional combinatorial computations, we deduce that, for all $i \in N$, the following chain of equalities holds:

$$\begin{aligned}
Sh_i(N, F^v_\psi) &\overset{(5.3)}{=} \sum_{T\subseteq N\backslash\{i\}} \frac{1}{n\cdot\binom{n-1}{t}} \cdot \Big[F^v_\psi(T\cup\{i\}) - F^v_\psi(T)\Big] \\
&= \sum_{T\subseteq N\backslash\{i\}} \frac{1}{n\cdot\binom{n-1}{t}} \cdot \sum_{R\subseteq T} \tilde{p}^{t+1}_r \cdot \Big[v(R\cup\{i\}) - v(R)\Big] \\
&= \sum_{S\subseteq N\backslash\{i\}} \Big[\sum_{t=s}^{n-1} \binom{n-1-s}{t-s} \cdot \frac{1}{n\cdot\binom{n-1}{t}} \cdot \tilde{p}^{t+1}_s\Big] \\
&\quad \cdot \Big[v(S\cup\{i\}) - v(S)\Big] \\
&= \sum_{S\subseteq N\backslash\{i\}} p^n_s \cdot \Big[v(S\cup\{i\}) - v(S)\Big] \overset{(5.10)}{=} \psi_i(N,v).
\end{aligned}$$

For the sake of the last equality but one, we need to establish the following claim:

$$\sum_{t=s}^{n-1} \binom{n-1-s}{t-s} \cdot \frac{\tilde{p}^{t+1}_s}{n\cdot\binom{n-1}{t}} = p^n_s$$

or equivalently,

$$\sum_{t=s}^{n-1} \binom{t}{s} \cdot \tilde{p}^{t+1}_s = (n-s) \cdot \binom{n}{s} \cdot p^n_s \tag{5.22}$$

for all $0 \leq s \leq n-1$. The proof of the claim (5.22) proceeds by induction on the size n, $n \geq s+1$, where $s \geq 0$ is fixed. Recall (5.12) and the upwards triangle property of $\mathcal{P} = \{p_s^n\}$. The inductive proof of (5.22) is left to the reader. $\triangleleft$

5.4 Concluding remarks

Definition 5.1 deals with the existence of the so-called *additive potential representation* in the sense that each component of the game-theoretic solution may be interpreted as the incremental return with respect to the potential function. In Ortmann (2000) the multiplicative potential approach to the solution theory for cooperative games is based on the quotient instead of the difference.

Definition 5.15 (Cf. Ortmann, 2000.) Let ψ be a solution on the set $\mathcal{G}_+$ of positive cooperative games. We say the solution ψ admits a *multiplicative potential* if there exists a function $P_\psi : \mathcal{G}_+ \to \mathbb{R}_+$ satisfying $P_\psi(\emptyset, v) := 1$ and

$$\frac{P_\psi(N,v)}{P_\psi(N\backslash\{i\},v)} = \psi_i(N,v) \qquad \text{for all } \langle N,v\rangle \in \mathcal{G}_+ \text{ and all } i \in N. \quad (5.23)$$

$\triangleleft$

As noted in Ortmann (2000), there exists a unique solution on $\mathcal{G}_+$ that admits a multiplicative potential and is efficient as well. This unique solution, however, can not be represented in an explicit manner, opposite to the explicit formula (5.3) for the Shapley value in the framework of efficient solutions that admit an additive potential. In addition to the pioneer work by Ortmann (2000), a more detailed theory about solutions that admit a multiplicative potential is presented in Driessen and Calvo (2001). It is still an outstanding problem to study the various types of consistency properties for these solutions that admit a multiplicative potential.

5.5 Two technical proofs

Proof of Proposition 5.10.
(i) Let $n \geq 1$. By (5.12) and the upwards triangle property (5.11) for

$\mathcal{P} = \{p_s^n\}$, it holds

$$\begin{aligned}
\tilde{p}_s^{n+1} + \tilde{p}_{s+1}^{n+1} &\overset{(5.12)}{=} \Big[(s+1)\cdot p_s^{n+1} - (n-s)\cdot p_{s+1}^{n+1}\Big] \\
&+ \Big[(s+2)\cdot p_{s+1}^{n+1} - (n-1-s)\cdot p_{s+2}^{n+1}\Big] \\
&= (s+1)\Big[p_s^{n+1} + p_{s+1}^{n+1}\Big] - (n-1-s)\Big[p_{s+1}^{n+1} + p_{s+2}^{n+1}\Big] \\
&\overset{(5.11)}{=} (s+1)\cdot p_s^n - (n-1-s)\cdot p_{s+1}^n \\
&\overset{(5.12)}{=} \tilde{p}_s^n \text{ for all } 0 \le s \le n-1.
\end{aligned}$$

(ii) Fix $n \geq 1$. The proof of (5.13) proceeds by backwards induction on the size s, $0 \leq s \leq n-1$. For $s = n-1$, (5.13) holds because of $\tilde{p}_{n-1}^n = n \cdot p_{n-1}^n$. For $1 \leq s \leq n-1$, we deduce from (5.12) and the induction hypothesis applied to s that it holds

$$\begin{aligned}
p_{s-1}^n &\overset{(5.12)}{=} \frac{\tilde{p}_{s-1}^n}{s} + \frac{n-s}{s}\cdot p_s^n \\
&= \frac{\tilde{p}_{s-1}^n}{s} + \frac{n-s}{s}\cdot\frac{1}{n\cdot\binom{n-1}{s}}\cdot\sum_{t=s}^{n-1}\binom{n}{t+1}\cdot\tilde{p}_t^n \\
&= \frac{1}{n\cdot\binom{n-1}{s-1}}\cdot\sum_{t=s-1}^{n-1}\binom{n}{t+1}\cdot\tilde{p}_t^n.
\end{aligned}$$

(iii) For every $n \geq 1$ and $0 \leq s \leq n-1$, write $\beta_s^n := \frac{1}{n\cdot\binom{n-1}{s}}$. Let $n \geq 1$ and $0 \leq s \leq n-1$. On the one hand, we deduce from the assumption (5.13) that it holds

$$\begin{aligned}
p_s^{n+1} + p_{s+1}^{n+1} &\overset{(5.13)}{=} \beta_s^{n+1}\sum_{t=s}^{n}\binom{n+1}{t+1}\tilde{p}_t^{n+1} + \beta_{s+1}^{n+1}\sum_{t=s+1}^{n}\binom{n+1}{t+1}\tilde{p}_t^{n+1} \\
&= \beta_s^{n+1}\binom{n+1}{s+1}\tilde{p}_s^{n+1} + \Big[\beta_s^{n+1} + \beta_{s+1}^{n+1}\Big]\sum_{t=s+1}^{n}\binom{n+1}{t+1}\tilde{p}_t^{n+1} \\
&= \frac{\tilde{p}_s^{n+1}}{s+1} + \beta_s^n\sum_{t=s+1}^{n}\binom{n+1}{t+1}\tilde{p}_t^{n+1}.
\end{aligned}$$

On the other hand, we deduce from the upwards triangle property for $\tilde{\mathcal{P}} = \{\tilde{p}_s^n\}$ that it holds

$$\begin{aligned} p_s^n &\overset{(5.13)}{=} \beta_s^n \cdot \sum_{t=s}^{n-1} \binom{n}{t+1} \cdot \tilde{p}_t^n \\ &= \beta_s^n \cdot \sum_{t=s}^{n-1} \binom{n}{t+1} \cdot \left[\tilde{p}_t^{n+1} + \tilde{p}_{t+1}^{n+1}\right] \\ &= \beta_s^n \cdot \binom{n}{s+1} \cdot \tilde{p}_s^{n+1} + \beta_s^n \cdot \sum_{t=s+1}^{n-1} \left[\binom{n}{t+1} + \binom{n}{t}\right] \cdot \tilde{p}_t^{n+1} + \beta_s^n \cdot \tilde{p}_n^{n+1} \\ &= \frac{\tilde{p}_s^{n+1}}{s+1} + \beta_s^n \cdot \sum_{t=s+1}^{n} \binom{n+1}{t+1} \cdot \tilde{p}_t^{n+1}. \end{aligned}$$

Since both computational methods yield the very same outcome, we conclude that $p_s^{n+1} + p_{s+1}^{n+1} = p_s^n$. Finally, the statement in part (iv) is a direct consequence of (5.12). □

Proof of Theorem 5.12. Let $\langle N, v\rangle \in \mathcal{G}$.
(i) For every $T \subseteq N$, $T \neq \emptyset$, it suffices to prove the next equality:

$$\sum_{S\subseteq T,\ S\neq T, S\neq\emptyset} \tilde{p}_{s-1}^t v(S) = \sum_{S\subseteq T,\ S\neq T, S\neq\emptyset} (-1)^{t+1-s} \frac{q_{t-s}^t(s)}{\tilde{p}_{s-1}^s} F_\psi^v(S). \tag{5.24}$$

Fix $S \subseteq T \subseteq N$ with $S \neq T$ and $S \neq \emptyset$. We aim to determine the coefficient of the term $v(S)$ in the sum given by the right hand of (5.24). The term $v(S)$ occurs in any expression $F_\psi^v(R)$ as long as $S \subseteq R$, provided that $R \subseteq T$, $R \neq T$. Thus, we need only to consider those coalitions R satisfying $S \subseteq R \subseteq T$ with $R \neq T$, and each such coalition R, say of size r, $s \leq r \leq t-1$, induces the term

$$(-1)^{t+1-r} \cdot \frac{q_{t-r}^t(r)}{\tilde{p}_{r-1}^r} \cdot \tilde{p}_{s-1}^r \cdot v(S).$$

Notice that, for any size r, $s \leq r \leq t-1$, there exists $\binom{t-s}{t-r}$ coalitions R of size r satisfying $S \subseteq R \subseteq T$, $R \neq T$. Hence, for every fixed $S \subseteq T$, $\emptyset \neq S \neq T$, the coefficient of the term $v(S)$ in the sum given by the right hand of (5.24) is determined by the next sum:

$$\sum_{r=s}^{t-1} \binom{t-s}{t-r} \cdot (-1)^{t+1-r} \cdot q_{t-r}^t(r) \cdot \frac{\tilde{p}_{s-1}^r}{\tilde{p}_{r-1}^r}.$$

By construction based on (5.17), for all $1 \leq s \leq t-1$, it holds

$$\begin{aligned}\sum_{r=s}^{t-1} \tbinom{t-s}{t-r}(-1)^{t+1-r} q_{t-r}^t(r)\frac{\tilde{p}_{s-1}^r}{\tilde{p}_{r-1}^r} &= \sum_{k=1}^{t-s} \tbinom{t-s}{k}(-1)^{k+1} q_k^t(t-k)\frac{\tilde{p}_{s-1}^{t-k}}{\tilde{p}_{t-1-k}^{t-k}} \\ &\overset{(5.17)}{=} \sum_{k=1}^{t-s}(-1)^{k+1}\left[q_{k+1}^t(s)+q_k^t(s)\right] \\ &= q_1^t(s)+(-1)^{t+1-s}q_{t+1-s}^t(s) \\ &\overset{(5.17)}{=} \tilde{p}_{s-1}^t+0=\tilde{p}_{s-1}^t.\end{aligned}$$

This proves (5.24).
(ii) (5.19) is a direct consequence of both (5.16) and (5.18) applied to the initial game and the reduced game as well.
(iii) By (5.12), $p_s^n = \frac{1}{n\cdot\binom{n-1}{s}}$ implies $\tilde{p}_{n-1}^n = 1$ and $\tilde{p}_s^n = 0$ for all $n \geq 1$, all $0 \leq s \leq n-2$. By (5.17), $q_1^t(s) = 0$ whenever $1 \leq s \leq t-1$ and thus, $q_k^t(s) = 0$ whenever $k \geq 2$. Therefore, (5.19) reduces to the next equality:

$$\begin{aligned}\tilde{v}_{N\setminus\{i\}}^{y_i}(T)-v(T) &\overset{(5.19)}{=} \frac{t}{n-1}\cdot\left[\tilde{\psi}_i(T\cup\{i\},v)-y_i\right] \\ &\overset{(5.10)}{=} \frac{t}{n-1}\cdot\left[v(T\cup\{i\})-v(T)-y_i\right].\end{aligned}$$

Obviously, the relevant equality agrees with Sobolev's reduced game (5.4). □

References

Calvo, E., and J.C. Santos (1997): "Potentials in cooperative TU-games," *Mathematical Social Sciences*, 34, 175–190.

Dragan, I. (1996): "New mathematical properties of the Banzhaf value," *European Journal of Operational Research*, 95, 451–463.

Driessen, T.S.H. (1988): *Cooperative Games, Solutions, and Applications.* Dordrecht: Kluwer Academic Publishers.

Driessen, T.S.H., (1991): "A survey of consistency properties in cooperative game theory," *SIAM Review*, 33, 43–59.

Driessen, T.S.H., and E. Calvo (2001): "A multiplicative potential approach to solutions for cooperative TU-games," Memorandum No. 1570,

Faculty of Mathematical Sciences, University of Twente, Enschede, The Netherlands.

Dubey, P., A. Neyman, and R.J. Weber (1981): "Value theory without efficiency," *Mathematics of Operations Research*, 6, 122–128.

Hart, S., and A. Mas-Colell (1989): "Potential, value, and consistency," *Econometrica*, 57, 589–614.

Maschler, M. (1992): "The bargaining set, kernel, and nucleolus," in: Aumann, R.J., and S. Hart (eds.), *Handbook of Game Theory with Economic Applications*, Volume 1. Amsterdam: Elsevier Science Publishers, 591–667.

Myerson, R. (1980): "Conference structures and fair allocation rules," *International Journal of Game Theory*, 9, 169–182.

Ortmann, K.M. (1998): "Conservation of energy in value theory," *Mathematical Methods of Operations Research*, 47, 423–450.

Ortmann, K.M. (2000): "The proportional value for positive cooperative games," *Mathematical Methods of Operations Research*, 51, 235–248.

Sánchez S., F. (1997): "Balanced contributions in the solution of cooperative games," *Games and Economic Behavior*, 20, 161–168.

Shapley, L.S. (1953): "A value for n-person games," *Annals of Mathematics Studies*, 28, 307–317.

Sobolev, A.I. (1973): "The functional equations that give the payoffs of the players in an n-person game," in: Vilkas. E. (ed.), *Advances in Game Theory*. Vilnius: Izdat. "Mintis", 151–153.

Chapter 6

On the Set of Equilibria of a Bimatrix Game: a Survey

BY MATHIJS JANSEN, PETER JURG, AND DRIES VERMEULEN

6.1 Introduction

Any survey on this topic should start with the celebrated results obtained by Nash. First of all he showed that every non-cooperative game in normal form has an equilibrium in mixed strategies (cf. Nash, 1950). He also established the well-known characterization of the equilibrium condition stating that a strategy profile is an equilibrium if and only if each player only puts positive weight on those pure strategies that are pure best responses to the strategies currently played by the other players (cf. Nash, 1951).

In the special case of matrix games the existence of equilibria was already established by von Neumann and Morgenstern (1944). Their results though show more than just that. They show for example that the collection of equilibria is a polytope. Furthermore they explain how one can use linear programming techniques to actually compute such an equilibrium.

Once the existence of equilibria was also established for bimatrix games, several authors, e.g. Vorobev, Kuhn, Mangasarian, Mills and Winkels, tried to develop methods based on linear programming to compute equilibria for bimatrix games. Later on authors like Winkels and Jansen also generalized the structure result and showed that the set of

P. Borm and H. Peters (eds.), Chapters in Game Theory, 121–142.

equilibria of a bimatrix game can be written as the union of a finite number of polytopes. Such a representation of the set of equilibria is called a *decomposition* in this survey.

SEVEN PROOFS During the last few decades several different decompositions have been given. We will discuss seven of them and briefly comment on the differences and similarities between these decompositions. The first three can be seen as variations on the same line of reasoning. In this approach, first the extreme points of the polytopes involved in the decomposition of the equilibrium set are characterized. Subsequently an analysis is given of exactly how groups of extreme points generate one such polytope of the decomposition. We will first discuss these three methods.

(i) In the approach by Vorobev (1958) and Kuhn (1961) (as it is described in this survey) first a description is given of the collection $\mathbb{E}_1$ of strategies of player 1 that can be combined to an extreme equilibrium. Then it is shown that

(1) this collection is finite

(2) the Cartesian product of the convex hull of a subset of $\mathbb{E}_1$ with all strategies of player 2 that combine to an equilibrium with any one strategy of the subset in question is a polytope, and

(3) any equilibrium is an element of such a product set.

(ii) Winkels (1979) basically uses the same steps in his proofs. The improvement over the proof of Vorobev and Kuhn is that the definition of the set $\mathbb{E}_1$ is a bit different. This difference has the advantage that the proofs become shorter and more transparent.

(iii) Mangasarian's (1964) proof is based on a more symmetric treatment of the players. He looks at Cartesian products of subsets of $\mathbb{E}_1$ with subsets of $\mathbb{E}_2$ and shows that, whenever such a product is included in the equilibrium set, so will the convex hull of this product. Moreover, any one equilibrium is an element of the convex hull of at least one such a product.

The latter four proofs take what can be called a dual approach. Based on the characterization of the notion of an equilibrium in terms of carriers and best responses the defining systems of linear inequalities are given directly. Subsequently it is shown that any solution of such a system is an equilibrium and that any equilibrium is indeed a solution of at least one of the systems generated by the approach in question.

(iv) The proof in Jansen (1981) is based on the two observations that any convex subset of the equilibrium set is contained in a maximal convex subset of the equilibrium set and that any maximal convex subset is a polytope. Thus, since each equilibrium by itself constitutes a convex subset of the equilibrium set, we again get the result that the equilibrium set is the union of polytopes. The fact that these polytopes are finite in number follows from the characterization of maximal convex subsets of the equilibrium set in terms of the carriers and best responses of the equilibria in such a subset.

(v) Undoubtedly the shortest proof is by Quintas (1989). He shows how to associate with each subset of the collection of pure strategies of player 1 and each subset of the collection of pure strategies of player 2 a polytope of equilibria. Since each equilibrium is evidently contained in such a polytope we easily get the result of Vorobev.

(vi) The approach of Jurg and Jansen (cf. Jurg, 1993) looks very much like the proof by Quintas. However, their approach yields a straightforward correspondence between the subsets of pure strategies used to generate the polytopes of the decomposition and faces of the equilibrium set.

(vii) The approach of Vermeulen and Jansen (1994) can be seen as a geometrical variation on the same theme. Its advantage though is that it can easily be adjusted to a proof of the same result for the collection of perfect and proper equilibria.

NEW ASPECTS Although this chapter is intended to be a survey, we would like to point out that we also used modern insights to get shorter or more transparent proofs of the original results. Further we used an idea of Winkels in order to show how the Mangasarian approach can be used to obtain the decomposition result. Finally we prove that the two decompositions of Vorobev and Winkels are in fact identical by showing that their (different) definitions of extreme strategies coincide.

Notation $\Delta_t = \{p \in \mathbb{R}^t | \, p \geq 0, \sum_{i=1}^t p_i = 1\}$. The unit vectors in $\mathbb{R}^t$ are denoted by $e_1, e_2, \ldots, e_t$. For $x, y \in \mathbb{R}^t$, we write $x \cdot y = \sum_{i=1}^t x_i y_i$. For $S \subset \mathbb{R}^t$ we denote by $\text{conv}(S)$ the convex hull of S and by $\text{cl}(S)$ the closure of S. For a convex set $C \subset \mathbb{R}^t$ we denote by $\text{relint}(C)$ the relative interior of C and by $\text{ext}(C)$ the set of extreme points of C. For a finite set T, the collection of non-empty subsets of T is denoted by 2^T.

6.2 Bimatrix Games and Equilibria

An $m \times n$-*bimatrix game* is played by two players, player 1 and player 2. Player 1 has a finite set $M = \{1, \ldots, m\}$ and player 2 has a finite set $N = \{1, \ldots, n\}$ of pure strategies. The payoff matrices $[a_{ij}]_{i \in M, j \in N}$ of player 1 and $[b_{ij}]_{i \in M, j \in N}$ of player 2 are denoted by A and B respectively. This game is denoted by (A, B).

Now the game (A, B) is played as follows. Players 1 and 2 choose, independent of each other, a strategy $p \in \Delta_m$ and $q \in \Delta_n$, respectively. Here p_i (q_j) can be seen as the probability that player 1 (2) chooses his i-th row (j-th column). The (expected) payoff for player 1 is pAq and the expected payoff to player 2 is pBq.

A strategy pair $(p^*, q^*) \in \Delta_m \times \Delta_n$ is an *equilibrium* for the game (A, B) if

$$p^* A q^* \geq p A q^* \text{ for all } p \in \Delta_m$$

and

$$p^* B q^* \geq p^* B q \text{ for all } q \in \Delta_n.$$

The set of all equilibria for the game (A, B) is denoted by $E(A, B)$. By a theorem of Nash (1950) this set is non-empty for all bimatrix games.

6.3 Some Observations by Nash

In a survey about equilibria it is inevitable to start with a description of concepts and results that can be found in John Nash' seminal paper 'Non-cooperative games' of 1951. Even in this first paper on the existence of equilibria Nash evidently realized that the key to a polyhedral description of the equilibrium set lies in the characterization of equilibria in terms of what are nowadays called carriers and best responses.

Since it is indeed the key to *all* known polyhedral descriptions of the Nash equilibrium set we will first have a look at his characterization of equilibria. It can be found at the bottom of page 287 of Nash's paper, but we will use the more modern terminology of Heuer and Millham (1976). Following them, we introduce for a strategy $p \in \Delta_m$ the *carrier* $C(p) = \{i \mid p_i > 0\}$ and the set $PB_2(p) = \{j \mid pBe_j \geq pBe_k \text{ for all } k\}$ of *pure best replies* of player 2 to p. For a strategy $q \in \Delta_n$ the sets $C(q)$ and $PB_1(q)$ are defined in the same way.

Lemma 6.1 *Let (A, B) be a bimatrix game and let (p, q) be a strategy pair. Then $(p, q) \in E(A, B)$ if and only if $C(p) \subset PB_1(q)$ and $C(q) \subset PB_2(p)$.*

Proof. (a) If $(p, q) \in E(A, B)$, then

$$pAq = \sum_{i \in C(p)} p_i e_i Aq \leq \sum_{i \in C(p)} p_i \max_k e_k Aq = \max_k e_k Aq \leq pAq.$$

So $p_i > 0$ implies that $e_i Aq = \max_k e_k Aq$, that is: $C(p) \subset PB_1(q)$. Similarly, one shows that $C(q) \subset PB_2(p)$.

(b) If $C(p) \subset PB_1(q)$ and $C(q) \subset PB_2(p)$, then for all i

$$pAq = \sum_{\ell \in C(p)} p_\ell e_\ell Aq = \sum_{\ell \in C(p)} p_\ell \max_k e_k Aq = \max_k e_k Aq \geq e_i Aq.$$

Similarly, $pBq \geq pBe_j$ for all j. Hence, $(p, q) \in E(A, B)$. □

Next we consider the concepts of interchangeability and sub-solutions introduced by Nash.

A subset S of the set of equilibria of a bimatrix game satisfies the *interchangeability* condition if for any pair $(p, q), (p', q') \in S$ we also have that $(p, q'), (p', q) \in S$.

If a subset S of the set of equilibria has the interchangeability property, then $S = S' \times S''$, where

$$S' = \{p \mid (p, q) \in S \text{ for some } q\}$$

and

$$S'' = \{q \mid (p, q) \in S \text{ for some } p\}$$

are called the *factor sets* of S. Since, obviously, a set $S = P \times Q$ of equilibria has the interchangeability property, sets of this form are precisely the sets with the interchangeability property. In Heuer and Millham (1976) sets of equilibria of the form $P \times Q$ were called *Nash sets*.

Nash used the term *sub-solution* for a Nash set that is not properly contained in another Nash set. In this chapter we prefer the term *maximal Nash set*, a term that was introduced by Heuer and Millham as well.

Nash gave a proof of the following result. Because it will be generalized in 6.15, the proof is left to the reader for now.

Lemma 6.2 *For a bimatrix game, a maximal Nash set is the product of two convex, compact sets.*

Finally, Nash proved that for a bimatrix game with only one maximal Nash set —he called such a game *solvable*—the set of equilibria is the product of two polytopes.

6.4 The Approach of Vorobev and Kuhn

In this section we will describe a result of Vorobev (1958) and its improved version of Kuhn (1961). Their method can be seen as a "one-sided" approach to the decomposition of the Nash equilibrium set into a finite number of (bounded) polyhedral sets.

They place themselves in the position of player 1. First they analyze which strategies of player 1 occur as extreme elements of certain polytopes of strategies that, combined with a finite number of strategies of player 2, are equilibria. This set of extreme elements is indicated by $\mathbb{E}_1$. Then they show that, for any subset P of $\mathbb{E}_1$, the collection $L(P)$ of strategies of player 2 that combine to an equilibrium for any element in P, is polyhedral. Finally they show that $\text{conv}(P) \times L(P)$, a polytope, is a subset of the Nash equilibrium set. Hence, since any equilibrium is indeed also an element of such a polytope, we get that the Nash equilibrium set is a, necessarily finite, union of polytopes.

In order to verify these claims, let (A, B) be an $m \times n$-bimatrix game. For $p \in \Delta_m$ and $q \in \Delta_n$ we introduce the sets

$$L(p) = \{q' \in \Delta_n | \ (p, q') \in E(A, B)\}$$

and

$$K(q) = \{p' \in \Delta_m | \ (p', q) \in E(A, B)\}.$$

Since

$$K(q) = \Delta_m \cap \bigcap_{i=1}^{m} \{p \in \mathbb{R}^m | \ pAq \geq e_i Aq\} \cap \bigcap_{j=1}^{n} \{p \in \mathbb{R}^m | \ pBq \geq pBe_j\}$$

is the intersection of the bounded polyhedral set Δ_m and a finite number of halfspaces, $K(q)$ is a bounded polyhedral set. So $K(q)$ is a polytope. Similarly, $L(p)$ is a polytope.

For a set P of strategies of player 1 and a set Q of strategies of player 2, Vorobev introduces the sets

$$L(P) = \bigcap_{p \in P} L(p) \quad \text{and} \quad K(Q) = \bigcap_{q \in Q} K(q).$$

Obviously these sets are convex and compact.

Vorobev calls a strategy p of player 1 *extreme* if $p \in \text{ext}(K(Q))$ for some finite set Q of strategies of player 2. Let $\mathbb{E}_1$ denote the set of extreme strategies of player 1.

In order to prove that $\mathbb{E}_1$ is a finite set, Kuhn introduces the sets

$$P_B = \{(p, \beta) | \; pBe_j \leq \beta \text{ for all } j\}$$

and

$$Q_A = \{(q, \alpha) | \; e_i Aq \leq \alpha \text{ for all } i\}.$$

In words one could say that the set P_B is the collection of pairs (p, β) for which p is a strategy of player 1 and β is an upper bound on the payoffs player 2 can obtain given that player 1 plays p.

Since the sets P_B and Q_A are obviously polyhedral we can easily see that they only have a finite number of extreme points. Thus, the following lemma implies the finiteness of $\mathbb{E}_1$.

Lemma 6.3 *If $p \in \text{ext}(K(Q))$ for some finite set Q of strategies of player 2, then $(p, pBq) \in \text{ext}(P_B)$ for all $q \in Q$.*

Proof. Let $q \in Q$. Suppose that $(p, pBq) = \frac{1}{2}(p', \beta') + \frac{1}{2}(p'', \beta'')$, where $(p', \beta'), (p'', \beta'') \in P_B$. We have to prove that $(p', \beta') = (p'', \beta'')$. First we will show that $p', p'' \in K(Q)$. So let $q' \in Q$.

Since, for all j,

$$pBe_j = \tfrac{1}{2}p'Be_j + \tfrac{1}{2}p''Be_j \leq \tfrac{1}{2}\beta' + \tfrac{1}{2}\beta'' = pBq,$$

we have for $j \in PB_2(p)$ that $p'Be_j = \beta'$. This implies that $\beta' = \max_k p'Be_k$. So $j \in PB_2(p')$. Furthermore, since $(p, q') \in E(A, B)$,

$$C(q') \subset PB_2(p) \subset PB_2(p'). \tag{1}$$

Since $p = \frac{1}{2}p' + \frac{1}{2}p''$,

$$C(p') \subset C(p) \subset PB_1(q'). \tag{2}$$

By (1) and (2), $(p', q') \in E(A, B)$. Similarly, $(p'', q') \in E(A, B)$. Hence, $p', p'' \in K(q')$.

Because $p \in \text{ext}(K(Q))$, this leads to the equality $p' = p''$. Since, for a $j \in PB_2(p)$, $\beta' = p'Be_j = p''Be_j = \beta''$, this proves that $(p, pBq) \in \text{ext}(P_B)$. □

In a similar way one shows that, for a finite set P of strategies of player 1, the set $L(P)$ has a finite number of extreme points. Therefore the following theorem implies that the set of equilibria of a bimatrix game is the union of a finite number of polytopes.

Theorem 6.4 *For any bimatrix game* (A, B)

$$E(A, B) = \bigcup_{P \subset \mathbb{E}_1} \text{conv}(P) \times L(P).$$

Proof. (a) Let P be a non-empty subset of $\mathbb{E}_1$ such that $L(P) \neq \phi$. Suppose that $p \in \text{conv}(P)$ and $q \in L(P)$. Then $P \subset K(q)$ and the convexity of $K(q)$ implies that $p \in \text{conv}(P) \subset K(q)$. So $(p, q) \in E(A, B)$.

(b) Suppose that (p^*, q^*) is an element of $E(A, B)$. Then, by definition, the set $P = \text{ext}(K(q^*))$ is a subset of $\mathbb{E}_1$. Since $p^* \in K(q^*)$ and $K(q^*)$ is a polytope, $p^* \in K(q^*) = \text{conv}(\text{ext}(K(q^*)) = \text{conv}(P)$. Clearly, $(p, q^*) \in E(A, B)$ for all $p \in P$, that is: $q^* \in L(P)$. So $(p^*, q^*) \in \text{conv}(P) \times L(P)$. □

Since, as we already observed, $\mathbb{E}_1$ is a finite set, the above theorem immediately implies that the equilibrium set is the finite union of maximal Nash sets.

Another observation we would like to make at this point is that the previous approach also yields a way to index maximal Nash sets. This works as follows.

Lemma 6.5 *Let P be a set of strategies of player 1 and Q be a set of strategies of player 2. Then $P \times Q$ is a Nash set if and only if P is a subset of $K(Q)$ and Q is a subset of $L(P)$. It is a maximal Nash set if and only if P equals $K(Q)$ and Q equals $L(P)$.*

6.5 The Approach of Mangasarian and Winkels

In his proof Mangasarian (1964) also employs the polyhedral sets P_B and Q_A as they were introduced by Kuhn (1961). However, compared with the previous approach, Mangasarian's method of proof is based on a more symmetric treatment of the players.

Mangasarian proved that each equilibrium of a bimatrix game can be constructed by means of the finite set of extreme points of the two polyhedral sets P_B and Q_A corresponding to the game. In this section we will describe Mangasarian's ideas. Furthermore we will incorporate the concept of a Nash pair due to Winkels (1979) to show that for any bimatrix game the set of equilibria is the finite union of polytopes. The exposition of the proof we present here is a slightly streamlined version of the original proof by Winkels.

Mangasarian's approach is based on the following result, also proved by Mills (1960): a pair (p, q) of strategies is an equilibrium of a bimatrix game (A, B) if and only if there exist scalars α and β such that

$$\begin{array}{lcl} (q,\alpha) & \in & Q_A \\ (p,\beta) & \in & P_B \\ pAq + pBq & = & \alpha + \beta. \end{array}$$

Mangasarian calls a quartet (p, pBq, q, pAq) *extreme* if $(p, pBq) \in \text{ext}(P_B)$ and $(q, pAq) \in \text{ext}(Q_A)$. Obviously, in this case, (p, q) is an equilibrium. In order to prove that all equilibria can be found with the help of the finite number of extreme quartets, we need the following lemma due to Winkels (1979).

Lemma 6.6 *Let (p^*, q^*) be an equilibrium of a bimatrix game (A, B), and let (p^*, p^*Bq^*) be a strict convex combination of pairs (p^1, β_1), ..., (p^r, β_r) in P_B. Then, for all i, (p^i, q^*) is an equilibrium of the game (A, B) and $\beta_i = p^iBq^*$.*

Proof. Suppose that $(p^*, p^*Bq^*) = \sum_k \lambda_k(p^k, \beta_k)$ where $\sum_k \lambda_k = 1$ and $\lambda_k > 0$ for all k.

Consider a strategy p^i. Then

$$C(p^i) \subset C(p^*) \subset PB_1(q^*) \tag{1}$$

and if $q_j^* > 0$, then

$$p^*Bq^* = p^*Be_j = \sum_k \lambda_k p^k Be_j \leq \sum_k \lambda_k \beta_k = p^*Bq^*.$$

Hence, $p^iBe_j = \beta_i$. This implies that $\beta_i = \max_k p^iBe_k$ so that

$$C(q^*) \subset PB_2(p^i). \tag{2}$$

In view of (1) and (2), $(p^i, q^*) \in E(A, B)$ and $\beta_i = p^iBq^*$. □

The following result is due to Mangasarian (1964).

Theorem 6.7 *Let (p^*, q^*) be an equilibrium of a bimatrix game (A, B). Then the quartet $(p^*, p^*Bq^*, q^*, p^*Aq^*)$ is a convex combination of extreme quartets.*

Proof. (a) First we will show that (p^*, p^*Bq^*) is a convex combination of extreme points of P_B.

Consider the linear function ℓ on P_B defined by

$$\ell(p, \beta) = p(A + B)q^* - p^*Aq^* - \beta.$$

Then $\ell(p, \beta) \leq 0$ for any pair $(p, \beta) \in P_B$ and (p^*, p^*Bq^*) is an element of the compact, convex set

$$M = \{(p, \beta) \in P_B | \ \ell(p, \beta) = 0\}.$$

So, the theorem of Krein-Milman states that (p^*, p^*Bq^*) is a convex combination of elements of the set $\text{ext}(M)$. Since ℓ is a linear function, $\text{ext}(M) \subset \text{ext}(P_B)$. So (p^*, p^*Bq^*) is a convex combination of extreme points of P_B.

(b) According to part (a), we can write (p^*, p^*Bq^*) as a strict convex combination of pairs $(p^1, \beta_1), \ldots, (p^r, \beta_r)$ in $\text{ext}(P_B)$. By Lemma 6.6, for all i, $(p^i, q^*) \in E(A, B)$ and $\beta_i = p^iBq^*$.

Similarly, we can write (q^*, p^*Aq^*) as a strict convex combination of pairs $(q^1, \alpha_1), \ldots, (q^s, \alpha_s)$ in $\text{ext}(Q_A)$ such that, for all j, $(p^*, q^j) \in E(A, B)$ and $\alpha_j = p^*Aq^j$.

The inclusion $C(q^j) \subset C(q^*)$ implies that $\beta_i = p^iBq^* = p^iBq^j$. Similarly, $\alpha_j = p^*Aq^j = p^iAq^j$. So, for all i and j, $(p^i, p^iBq^j, q^j, p^iAq^j)$ is an extreme quartet. Since $(p^*, p^*Bq^*, q^*, p^*Aq^*)$ is a convex combination of the quartets $(p^i, p^iBq^j, q^j, p^iAq^j)$, the proof is complete. □

Following Winkels we call a strategy p of player 1 *extreme* if there exists a strategy q of player 2 such that (p, pBq, q, pAq) is an extreme quartet. Extreme strategies for player 2 are defined in a similar way. Let $\mathbb{E}_i^*$ denote the (finite) set of extreme strategies of player i. Note that we will show in Lemma 6.17 that the extreme strategies in the sense of Winkels coincide with the extreme strategies as introduced by Vorobev.

We call a pair (P, Q) with $P \subset \mathbb{E}_1^*$ and $Q \subset \mathbb{E}_2^*$ a *Nash pair* for the game (A, B) if $P \times Q$ is a Nash set.

Lemma 6.8 *If $P \times Q$ is a Nash set for a bimatrix game (A, B), then* $\text{conv}(P) \times \text{conv}(Q)$ *is a Nash set too.*

Proof. If $(p, q) \in \text{conv}(P) \times \text{conv}(Q)$, then p and q are a convex combination of strategies $p_1, \ldots, p_s \in P$ and $q_1, \ldots, q_t \in Q$, respectively. Since $(p_i, q_j) \in P \times Q \subset E(A, B)$ for all j, $q_j \in L(p_i)$ for all j. By the convexity of $L(p_i)$, $q \in L(p_i)$. Hence $p_i \in K(q)$ for all i, which leads to $p \in K(q)$, that is: $(p, q) \in E(A, B)$. □

Since $\mathbb{E}_1^* \times \mathbb{E}_2^*$ is a finite set, the number of Nash pairs is finite too. Furthermore, for every Nash pair (P, Q), $\text{conv}(P) \times \text{conv}(Q) \subset E(A, B)$ and, by Theorem 6.7, each equilibrium is contained in a set $\text{conv}(P) \times \text{conv}(Q)$, where (P, Q) is a Nash pair. This proves that the set of equilibria of a bimatrix game is the finite union of polytopes.

Theorem 6.9 *For any bimatrix game (A, B)*

$$E(A, B) = \bigcup_{(P,Q) \text{ is a Nash pair}} \text{conv}(P) \times \text{conv}(Q).$$

Note that, due to the definition of a Nash pair, not all Nash sets used in this decomposition are necessarily maximal. Thus, some of them may be redundant.

6.6 The Approach of Winkels

In this section we will describe the result of Vorobev and Kuhn again. This time we will follow the ideas developed by Winkels (1979) by using his definition of an extreme strategy of a player. Winkels came to his definition by combining the ideas of Mangasarian and Kuhn.

Lemma 6.10 *If P is a set of strategies of player* 1, *then*

(a) $P \subset K(L(P))$ *if* $L(P) \neq \phi$

(b) $L(\text{conv}(P)) = L(P)$.

Proof. We will give a proof of part (b) only. Because $P \subset \text{conv}(P)$, $L(P) \supset L(\text{conv}(P))$.

Now suppose that $q \in L(P)$ and that p is a convex combination of strategies $p^i \in P$. Then the convexity of $K(q)$ implies that $p \in K(q)$. That is: (p, q) is an equilibrium. This proves that $q \in L(\text{conv}(P))$. □

Theorem 6.11 stated below is Winkels' version of Vorobev's result. In fact, by Lemma 11, this theorem is identical to (Vorobev's) Theorem 6.4.

Theorem 6.11 *For any bimatrix game* (A, B)

$$E(A,B) = \bigcup_{P \subset \mathbb{E}_1^*} \text{conv}(P) \times L(P).$$

Proof. (a) Let P be a non-empty subset of $\mathbb{E}_1^*$ such that $L(P) \neq \phi$. According to Lemma 6.10(a), $P \subset K(L(P))$, whereas the convexity of the right-hand set implies that $\text{conv}(P) \subset K(L(P))$. In combination with Lemma 6.10(b) and Lemma 6.5 this inclusion proves that

$$\text{conv}(P) \times L(P) \subset E(A,B).$$

(b) In order to prove the converse inclusion, assume that $(p, q) \in E(A, B)$. According to Theorem 6.7, the quartet (p, pBq, q, pAq) is a strict convex combination of extreme quartets, say $(p^1, \beta_1, q^1, \alpha_1), \ldots, (p^r, \beta_r, q^r, \alpha_r)$. Now let $P = \{p^1, \ldots, p^r\}$. Then $P \subset \mathbb{E}_1^*$ and $p \in \text{conv}(P)$. By Lemma 6.6, $q \in L(p^i)$ for all i which implies that $q \in L(P)$. Hence, (p, q) is an element of $\text{conv}(P) \times L(P)$, and the proof is complete. □

In order to prove that the sets described in the foregoing theorem are in fact polytopes, Winkels introduces for a subset P of $\mathbb{E}_1^*$ the finite set

$$L^{\text{ext}}(P) = \{q \in \mathbb{E}_2^* | \ (p,q) \in E(A,B) \text{ for all } p \in P\}$$

and he concludes that $L(P)$ is a polytope on the basis of the following result.

Lemma 6.12 *If P is a subset of $\mathbb{E}_1^*$, then* $L(P) = \text{conv}(L^{\text{ext}}(P))$.

Proof. Since $L^{\text{ext}}(P) \subset L(P)$ and $L(P)$ is convex, $\text{conv}(L^{\text{ext}}(P)) \subset L(P)$.

Now let $q \in L(P)$ and $\tilde{p} \in P$. As in the proof of Theorem 6.11 one shows that a set $Q \subset \mathbb{E}_2^*$ exists such that $(\tilde{p}, q) \in K(Q) \times \text{conv}(Q)$.

Since $(p, q) \in E(A, B)$ for all $p \in P$, Lemma 6.6 implies that, for any $q^j \in Q$, $(p, q^j) \in E(A, B)$ for all $p \in P$. Therefore $q^j \in L^{\text{ext}}(P)$ for all $q^j \in Q$. Since $q \in \text{conv}(Q)$, $q \in \text{conv}(L^{\text{ext}}(P))$. □

6.7 The Approach of Jansen

In the approaches described in the foregoing two sections, extreme strategies were the central issue. In the work of Jansen (1981) though the starting point was the notion of a maximal Nash set. In fact the source of inspiration for the research of Jansen was Heuer and Millham (1976), where several properties of (the intersection of) these maximal Nash sets were obtained.

Lemma 6.8 states in fact that any Nash set is contained in a convex Nash set. As a consequence of this result, a maximal Nash set is a convex set. Before we can show that the maximal Nash sets are in fact the maximal convex sets, we first need a lemma.

Lemma 6.13 *Any convex subset C of the set of equilibria of a bimatrix game (A, B) is contained in a (convex) Nash set.*

Proof. Assume that $(p, q), (p', q') \in C$. We will show that (p, q') and (p', q) are equilibria.

Consider, for $t \in (0, 1)$, the strategies $p(t) = tp + (1 - t)p'$ and $q(t) = tq + (1 - t)q'$. Since $(p(t), q(t)) \in C \subset E(A, B)$, for t close to 1

$$C(p) \cup C(p') = C(p(t)) \subset PB_1(q(t)) \subset PB_1(q).$$

So, $C(p') \subset PB_1(q)$. Similarly, for t close to 0

$$C(q) \cup C(q') = C(q(t)) \subset PB_2(p(t)) \subset PB_2(p')$$

and therefore $C(q) \subset PB_2(p')$. Hence, $(p', q) \in E(A, B)$. Similarly, $(p, q') \in E(A, B)$. This proves that

$$\{p | \text{ there is a } q \text{ with } (p, q) \in C\} \times \{q | \text{ there is a } p \text{ with } (p, q) \in C\}$$

is a (convex) Nash set containing C. □

Theorem 6.14 *Let C be a convex subset of the set of equilibria of a bimatrix game (A, B). Then C is a maximal convex subset if and only if C is a maximal Nash set.*

Proof. (a) Suppose that C is a maximal convex subset of $E(A, B)$. Then according to Lemma 6.13, C is contained in and hence equal to a convex Nash set. In view of Lemma 6.8, this Nash set must be maximal.

(b) Let C be a maximal Nash set and suppose that C is contained in the convex set T. According to Lemma 6.13, T is contained in a Nash set, say T'. So, by the maximality of C, this is possible only if $C = T = T'$. Hence, C is a maximal convex set. □

If (p, q) is an equilibrium of a bimatrix game (A, B), then $\{(p, q)\}$ is a convex subset of $E(A, B)$. Hence we can find, applying Zorn's Lemma, a maximal convex subset of $E(A, B)$ containing (p, q). In view of Theorem 6.14, each equilibrium of the game (A, B) is contained in a maximal Nash set and $E(A, B)$ is the union of such sets. In order to show that the number of maximal Nash sets is finite, we need the following lemma.

Lemma 6.15 *Let $S = S_1 \times S_2$ be a maximal Nash set for a bimatrix game (A, B). Further, let $(\mathring{p}, \mathring{q}) \in \operatorname{relint}(S)$ and let (p, q) be a strategy pair. Then*

(a) $S = K(\mathring{q}) \times L(\mathring{p})$

(b) $(p, q) \in S$ *if and only if* $C(p) \subset C(\mathring{p}), C(q) \subset C(\mathring{q})$, $PB_1(q) \supset PB_1(\mathring{q})$ *and* $PB_2(p) \supset PB_2(\mathring{p})$.

Proof. (a) Obviously, $S \subset S_1 \times L(\mathring{p})$. In order to show that $S_1 \times L(\mathring{p})$ is a Nash set, suppose that $(p, q) \in S_1 \times L(\mathring{p})$. Since $\mathring{p} \in \operatorname{relint}(S_1)$, there exists a $t > 1$ such that $p(t) = (1 - t)p + t\mathring{p} \in S_1$. Then $C(p) \subset C(\mathring{p})$. Since $(p(t), \mathring{q}), (p, \mathring{q}) \in S$, for $j \in PB_2(\mathring{p})$

$$\begin{aligned} \mathring{p}B\mathring{q} = \mathring{p}Be_j &= t^{-1}p(t)Be_j + (1 - t^{-1})pBe_j \\ &\leq t^{-1}p(t)B\mathring{q} + (1 - t^{-1})pB\mathring{q} = \mathring{p}B\mathring{q}. \end{aligned}$$

This proves that $pBe_j = pB\mathring{q}$ and, since $(p, \mathring{q}) \in E(A, B)$, that $j \in PB_2(p)$.

Thus we may conclude that $PB_2(\mathring{p}) \subset PB_2(p)$. This implies, in combination with the fact that $C(p) \subset C(\mathring{p})$ and $(\mathring{p}, q) \in E(A, B)$, that

$$C(p) \subset C(\mathring{p}) \subset PB_1(q) \quad \text{and} \quad C(q) \subset PB_2(\mathring{p}) \subset PB_2(p).$$

So $(p, q) \in E(A, B)$, which proves that $S_1 \times L(\mathring{p})$ is a Nash set containing S. Since S is maximal, $S_2 = L(\mathring{p})$. In a similar manner one shows that $S_1 = K(\mathring{q})$.

(b) In part (a) it has been proved that the four inclusions mentioned in the theorem hold for a $(p, q) \in S$. If, on the other hand, the four inclusions hold, then it follows that $(p, \mathring{q}), (\mathring{p}, q) \in E(A, B)$. This implies that $p \in K(\mathring{q})$ and $q \in L(\mathring{p})$, that is: $(p, q) \in K(\mathring{q}) \times L(\mathring{p}) = S$. □

By Lemma 6.15 a maximal Nash set is completely determined by the quartet $(C(\mathring{p}), PB_2(\mathring{p}), C(\mathring{q}), PB_1(\mathring{q}))$, where $(\mathring{p}, \mathring{q})$ is some equilibrium in its relative interior. Since there is only a finite number of such quartets, we obtain the following result of Jansen (1981).

Theorem 6.16 *The set of equilibria of a bimatrix game is a (not necessarily disjoint) union of a finite number of maximal Nash sets.*

Finally we will show that the extreme strategies as introduced by Winkels coincide with the extreme strategies in the sense of Vorobev.

Lemma 6.17 *For a strategy p of player 1 the following statements are equivalent:*

(1) *there exist a strategy q of player 2 and a maximal Nash set S such that $(p, q) \in \text{ext}(S)$*

(2) $p \in \mathbb{E}_1$

(3) $p \in \mathbb{E}_1^*$.

Proof. We will prove the implications (1) $\Longrightarrow$ (2) $\Longrightarrow$ (3) $\Longrightarrow$ (1).

(a) Suppose that $(p, q) \in \text{ext}(S)$ for some strategy q of player 2 and some maximal Nash set S. By Lemma 6.15, $p \in \text{ext}(K(\mathring{q}))$ and $q \in \text{ext}(L(\mathring{p}))$, where $(\mathring{p}, \mathring{q}) \in \text{relint}(S)$. Hence, $p \in \mathbb{E}_1$.

(b) Suppose that $p \in \mathbb{E}_1$. Let $q \in \mathbb{E}_2$. Then finite sets P and Q of strategies of player 1 and 2 exist such that $p \in \text{ext}(K(Q))$ and $q \in \text{ext}(L(P))$. In view of Lemma 6.3, this implies that $(p, \beta) \in \text{ext}(P_B)$ for some β and $(q, \alpha) \in \text{ext}(Q_A)$ for some α. Since $(p, q) \in E(A, B)$, $\beta = pBq$ and $\alpha = pAq$. So (p, pBq, q, pAq) is an extreme quartet, that is: $(p, q) \in \mathbb{E}_1^* \times \mathbb{E}_2^*$. So $p \in \mathbb{E}_1^*$.

(c) Suppose that $p \in \mathbb{E}_1^*$. By definition (of $\mathbb{E}_1^*$) there is a strategy q in $\mathbb{E}_2^*$ such that $(p, q) \in E(A, B)$. Then $(p, q) \in S$ for some maximal Nash set S. If $(p, q) \notin \text{ext}(S)$, then there exist $(p', q'), (p'', q'') \in S$ such that $(p, q) = \frac{1}{2}(p', q') + \frac{1}{2}(p'', q'')$ and $p' \neq p''$ or $q' \neq q''$. Let $p' \neq p''$. Then $(p', q), (p'', q) \in S$ so that $(p', p'Bq)$ and $(p'', p''Bq)$ are elements of P_B. Since $(p, pBq) = \frac{1}{2}(p', p'Bq) + \frac{1}{2}(p'', p''Bq)$, this contradicts the fact that $(p, pBq) \in \text{ext}(P_B)$. □

A similar results holds for strategies of player 2.

6.8 The Approach of Quintas

A very short and straightforward proof is the following one by Quintas (1989). With each set I of pure strategies of player 1 and set J of pure strategies of player 2 he associates the collection of strategy pairs (p,q) such that the carrier of p is contained in I, all pure strategies in J are best responses to p, the carrier of q is contained in J and all pure strategies in I are best responses to q. It is straightforward that such a collection is a polytope, that there is only a finite number of them, and that each equilibrium is contained in such a polytope.

More formally, for an $m \times n$ bimatrix game (A,B) and a pair $(I,J) \in 2^M \times 2^N$, Quintas introduces the subset

$$\begin{aligned} H(I,J) &= \{(p,q) \in \Delta_m \times \Delta_n |\ C(p) \subset I \subset PB_1(q) \\ &\quad \text{and } C(q) \subset J \subset PB_2(p)\} \end{aligned}$$

of $E(A,B)$. Because this set is bounded and determined by finitely many inequalities, it is a polytope.

If, for an equilibrium (p,q), we take $I = PB_1(q)$ and $J = PB_2(p)$, then obviously $(p,q) \in H(I,J)$. So

$$E(A,B) = \bigcup_{(I,J)\in 2^M \times 2^N} H(I,J).$$

One can show that for a pair $(I,J) \in 2^M \times 2^N$ the polytope $H(I,J)$ is a face of a maximal Nash set. However, generally there is not a nice relation between elements of $2^M \times 2^N$ and faces of maximal Nash sets as can be seen by considering the game

$$(A,B) = \begin{bmatrix} (2,1) & (1,0) & (1,1) \\ (2,0) & (1,1) & (0,0) \end{bmatrix}.$$

Although $((\frac{1}{2},\frac{1}{2}),e_1)$ is an extreme equilibrium of this game (and hence a face of some maximal Nash set), there is no pair $(I,J) \in 2^{\{1,2\}} \times 2^{\{1,2,3\}}$ such that $H(I,J) = \{((\frac{1}{2},\frac{1}{2}),e_1)\}$.

Moreover for this game $H(\{1,2\},\{1,2\}) = H(\{1,2\},\{1,2,3\})$. In the next section we will describe an approach not suffering from this drawback.

6.9 The Approach of Jurg and Jansen

In this section we describe the approach of Jurg and Jansen (cf. Jurg, 1993) who adapted the method of Quintas by replacing the pairs he dealt with by quartets consisting of the two carriers and the two sets of pure best replies of a strategy pair. Their approach reveals more of the structure of the set of equilibria and in particular of maximal Nash sets.

By Lemma 6.1 a strategy pair (p, q) is an equilibrium of an $m \times n$-bimatrix game (A, B) if and only if the (equilibrium) inclusions $C(p) \subset PB_1(q)$ and $C(q) \subset PB_2(p)$ are satisfied. To check this relation we need the quartet

$$(C(p), PB_2(p), C(q), PB_1(q)).$$

If (p, q) is an equilibrium of (A, B), then this quartet is called the *characteristic quartet* of (p, q). The set of all characteristic quartets for the bimatrix game (A, B) is denoted by $Char(A, B)$. Clearly, as a subset of $2^M \times 2^N \times 2^N \times 2^M$, this set is finite and it partitions the set of equilibria.

For a quartet $(I, J, K, L) \in Char(A, B)$ the set $F(I, J, K, L)$ is the collection of pairs of strategies (p, q) for which

$$C(p) \subset I, PB_2(p) \supset J, C(q) \subset K \text{ and } PB_1(q) \supset L\},$$

and it is called the *characteristic set* corresponding to this quartet. If (p, q) is an element of $F(I, J, K, L)$, then

$$C(p) \subset \quad I \subset L \subset PB_1(q)$$

and

$$C(q) \subset \quad K \subset J \subset PB_2(p),$$

which implies that (p, q) satisfies the equilibrium inclusions. Hence, $F(I, J, K, L)$ is a subset of $E(A, B)$. Clearly an equilibrium (p, q) is contained in the characteristic set corresponding to the characteristic quartet of (p, q), so we have

$$E(A, B) = \bigcup_{(I,J,K,L) \in Char(A,B)} F(I, J, K, L).$$

Since there are only finitely many different characteristic quartets, there are also finitely many different characteristic sets. Again, each characteristic set is bounded and described by finitely many linear inequalities and therefore a polytope. Hence

Theorem 6.18 *The equilibrium set of a bimatrix game is the union of a finite number of polytopes.*

Because of this finite number, we can assume that in Theorem 6.18 each of the polytopes or equivalently each of the characteristic sets is *maximal*, i.e. not properly contained in another one.

One easily checks that a characteristic set is a Nash set. Moreover

Theorem 6.19 *Let (A, B) be a bimatrix game. A maximal characteristic set is a maximal Nash set for (A, B) and vice versa.*

Proof. We have proved the theorem if we show that each Nash set is contained in a characteristic set.

Let T be a Nash set. According to Lemma 6.8, $S = \text{conv}(T)$ is also a Nash set.

Let $(\mathring{p}, \mathring{q}) \in \text{relint}(S)$. As in part (a) of the proof of Lemma 6.15, one can show that for a $(p, q) \in T$, $C(p) \subset C(\mathring{p})$, $PB_2(\mathring{p}) \subset PB_2(p)$, $C(q) \subset C(\mathring{q})$ and $PB_1(\mathring{q}) \subset PB_1(q)$. Hence (p, q) is an element of the characteristic set corresponding to the characteristic quartet of $(\mathring{p}, \mathring{q})$. By consequence, T is contained in this characteristic set. □

Thus Theorem 6.19 settles the existence of maximal Nash sets. Furthermore, this theorem implies Theorem 6.16. Note that in this approach Zorn's lemma is not used.

Obviously, a characteristic set $F(I, J, K, L)$ is maximal if and only if there is no characteristic quartet (I', J', K', L') different from (I, J, K, L) such that $I \subset I', J \supset J', K \subset K'$ and $L \supset L'$. Hence the following lemma implies that, more generally, each characteristic set is a face of a maximal Nash set and conversely.

Lemma 6.20 *Let (I, J, K, L) be a characteristic quartet for a game (A, B). Then F is a face of $F(I, J, K, L)$ if and only if $F = F(I', J', K', L')$ for some characteristic quartet with $I' \subset I, J' \supset J, K' \subset K$ and $L' \supset L$.*

Proof. (a) First let (I', J', K', L') be a characteristic quartet such that $I' \subset I, J' \supset J, K' \subset K$ and $L' \supset L$. Then $F(I', J', K', L') \subset F(I, J, K, L)$. Let G be the smallest face of $F(I, J, K, L)$ containing $F(I', J', K', L')$. We will prove that $F(I', J', K', L')$ is a face of $F(I, J, K, L)$ by showing that $G \subset F(I', J', K', L')$.

Since $F(I', J', K', L') \cap \text{relint}(G) \neq \phi$, we can take a $(\mathring{p}, \mathring{q}) \in F(I', J', K', L') \cap \text{relint}(G)$. Let $(p, q) \in G$. Arguments similar to those in the proof of Theorem 6.19 yield that $C(p) \subset C(\mathring{p})$,

$PB_2(\mathring{p}) \subset PB_2(p)$, $C(q) \subset C(\mathring{q})$ and $PB_1(\mathring{q}) \subset PB_1(q)$. Since moreover $(\mathring{p}, \mathring{q}) \in F(I', J', K', L')$, it follows that $(p, q) \in F(I', J', K', L')$. So $G \subset F(I', J', K', L')$.

(b) Secondly let F be a face of $F(I, J, K, L)$. Choose $(\mathring{p}, \mathring{q}) \in \text{relint}(F)$. As in the foregoing part one can show that $F \subset F(I', J', K', L')$, where $I' = C(\mathring{p}) \subset I$, $J' = PB_2(\mathring{p}) \supset J$, $K' = C(\mathring{q}) \subset K$ and $L' = PB_1(\mathring{q}) \supset L$.

The proof is complete if we can show that $F = F(I', J', K', L')$. Therefore we suppose that $F \neq F(I', J', K', L')$.

By part (a), $F(I', J', K', L')$ is a face of $F(I, J, K, L)$. Hence, F is a face of $F(I', J', K', L')$.

Choose $(p', q') \in \text{relint}(F(I', J', K', L'))$. It is easily shown that (I', J', K', L') is the characteristic quartet of (p', q').

Let, for $\epsilon > 0$,

$$(p^\epsilon, q^\epsilon) = \frac{1}{1-\epsilon}(\mathring{p}, \mathring{q}) - \frac{\epsilon}{1-\epsilon}(p,' q').$$

Then $(p^\epsilon, q^\epsilon) \in F(I', J', K', L')$ for small ϵ.

Since F is a face of $F(I', J', K', L') \neq F$, there are a pair $(u, v) \in \mathbb{R}^m \times \mathbb{R}^n$ and a real number c such that

$$(p, q) \cdot (u, v) = c, \text{ for } (p, q) \in F$$

and

$$(p, q) \cdot (u, v) < c, \text{ for } (p, q) \in F(I', J', K', L') \setminus F.$$

This implies that

$$c \geq (p^\epsilon, q^\epsilon) \cdot (u, v) = \frac{1}{1-\epsilon}(\mathring{p}, \mathring{q}) \cdot (u, v) - \frac{\epsilon}{1-\epsilon}(p,' q') \cdot (u, v) > c,$$

which is a contradiction. Hence $F = F(I', J', K', L')$. □

In fact, since $F(I, J, K, L) = F(I', J', K', L')$ implies that (I, J, K, L) equals (I', J', K', L'), we infer from Lemma 6.20:

Theorem 6.21 *For a bimatrix game (A, B) there is a one-to-one correspondence between the elements of $Char(A, B)$ and the set of faces of maximal Nash sets for (A, B).*

6.10 The Approach of Vermeulen and Jansen

In this section the method of Vermeulen and Jansen (1994) is described. The advantage of this method is that it can easily be adjusted to get the same structure result for perfect (cf. Vermeulen and Jansen, 1994) and proper equilibrium (cf. Jansen, 1993).

The key of this approach is the introduction of an equivalence relation for each player by identifying the strategies to which the other player has the same pure best replies. With the help of these relations the strategy spaces of both players are partitioned in a finite number of equivalence classes. The closure of each of these classes appears to be a polytope. By considering the intersection of the set of equilibria with the closure of the product of two equivalence classes (one for each player), Vermeulen and Jansen show that the set of equilibria is in fact the finite union of polytopes.

For a bimatrix game two strategies p and $\tilde{p}$ are called *best-reply equivalent*, denoted as $p \sim_{\text{BR}} \tilde{p}$, if $PB_2(p) = PB_2(\tilde{p})$. In a similar way an equivalence relation can be defined for the strategies of player 2.

Since for an $m \times n$-bimatrix game, $PB_2(p)$ is a subset of N for all $p \in \Delta_m$, the number of equivalence classes in Δ_m corresponding to the equivalence relation $\sim_{BR}$ must be finite. The equivalence classes are denoted as $\mathcal{V}_1, \ldots, \mathcal{V}_S$. Similarly, Δ_n is the finite union of equivalence classes, say $\mathcal{W}_1, \ldots, \mathcal{W}_T$. For later purposes, we choose representants p^s in $\mathcal{V}_s$ and q^t in $\mathcal{W}_t$ for all s and t.

Obviously, each equivalence class is a convex set. Furthermore,

Lemma 6.22 *For all pairs* (s, t)

$$\text{cl}(\mathcal{V}_s) = \{p \in \Delta_m |\ PB_2(p) \supset PB_2(p^s)\}$$

and

$$\text{cl}(\mathcal{W}_t) = \{q \in \Delta_n |\ PB_1(q) \supset PB_1(q^t)\}.$$

Proof. We will only give a proof of the first equality.

Obviously, for a $p \in \text{cl}(\mathcal{V}_s)$, $PB_2(p) \supset PB_2(p^s)$.

For a $p \in \Delta_m$ with $PB_2(p) \supset PB_2(p^s)$, we consider the strategy

$$p(\epsilon) = \epsilon p + (1 - \epsilon)p^s,$$

where $0 \le \epsilon < 1$. In order to show that $p(\epsilon) \in \mathcal{V}_s$ for all ϵ, first we take a $k \in PB_2(p^s) \subset PB_2(p)$. Then for all j

$$p(\epsilon)Be_k = \epsilon pBe_k + (1-\epsilon)p^sBe_k \ge \epsilon pBe_j + (1-\epsilon)p^sBe_j = p(\epsilon)Be_j.$$

For a $k \notin PB_2(p^s)$ and a $j \in PB_2(p^s) \subset PB_2(p)$

$$p(\epsilon)Be_k = \epsilon pBe_k + (1-\epsilon)p^sBe_k < \epsilon pBe_j + (1-\epsilon)p^sBe_j = p(\epsilon)Be_j.$$

Hence, for all ϵ, $PB_2(p(\epsilon)) = PB_2(p^s)$ which means that $p(\epsilon) \in \mathcal{V}_s$. Then however

$$p = \lim_{\epsilon \to 1} p(\epsilon) \in \mathrm{cl}(\mathcal{V}_s)$$

which concludes the proof. □

With the help of the representation of the closure of an equivalence class as given in the previous lemma, it is easy to prove that the closure of an equivalence class corresponding to the relation $\sim_{BR}$ is a polytope.

Next we consider the set of equilibria contained in the closure of the product of two equivalence classes (one for each player). For a pair (s,t), we consider the Nash set

$$\mathcal{N}_{s,t} = \{(p,q) \in \mathrm{cl}(\mathcal{V}_s) \times \mathrm{cl}(\mathcal{W}_t) |\ C(p) \subset PB_1(q^t) \text{ and } C(q) \subset PB_2(p^s)\}.$$

Obviously a Nash set $\mathcal{N}_{s,t}$ is a polytope and each equilibrium is contained in some Nash set $\mathcal{N}_{s,t}$. Further, if (p,q) is an element of some Nash set $\mathcal{N}_{st}$, then Lemma 6.22 implies that $PB_1(q) \supset PB_1(q^t) \supset C(p)$ and $PB_2(p) \supset PB_2(p^s) \supset C(q)$. Hence, by Lemma 6.1, (p,q) is an equilibrium. So we have the following result.

Theorem 6.23 *The set of equilibria of a bimatrix game is the finite union of polytopes.*

Since the number of Nash sets $\mathcal{N}_{s,t}$ is finite, each Nash set is contained in a maximal one and the set of equilibria of a bimatrix game is the finite union of maximal Nash sets.

References

Heuer, G.A., and C.B. Millham (1976): "On Nash subsets and mobility chains in bimatrix games," *Naval Res. Logist. Quart.*, 23, 311–319.

Jansen, M.J.M. (1981): "Maximal Nash subsets for bimatrix games," *Naval Res. Logist. Quart.*, 28, 147–152.

Jansen, M.J.M. (1993): "On the set of proper equilibria of a bimatrix game," *Internat. J. of Game Theory*, 22, 97–106.

Jurg, A.P. (1993): "Some topics in the theory of bimatrix games," Dissertation, University of Nijmegen.

Kuhn, H.W. (1961): "An algorithm for equilibrium points in bimatrix games, *Proc. Nat. Acad. Sci. U.S.A.*, 47, 1656–1662.

Mangasarian, O.L. (1964): "Equilibrium points of bimatrix games," *J. Soc. Industr. Appl. Math.*, 12, 778–780.

Mills, H. (1960): "Equilibrium points in finite games," *J. Soc. Indust. Appl. Math.*, 8, 397–402.

Nash, J.F. (1950): "Equilibrium points in n-person games," *Proc. Nat. Acad. Sci. U.S.A.*, 36, 48–49.

Nash, J.F. (1951): "Noncooperative games," *Ann. of Math.*, 54, 286–295.

Quintas, L.G. (1989): "A note on polymatrix games," *Internat. J. Game Theory*, 18, 261–272.

Vermeulen, A.J., and M.J.M. Jansen (1994): "On the set of perfect equilibria of a bimatrix game," *Naval Res. Logist. Quart.*, 41, 295–302.

von Neumann, J., and O. Morgenstern, O. (1944): *Theory of Games and Economic Behavior.* Princeton: Princeton University Press.

Vorobev, N.N. (1958): "Equilibrium points in bimatrix games," *Theor. Probability Appl.*, 3, 297–309.

Winkels, H.M. (1979): "An algorithm to determine all equilibrium points of a bimatrix game," in: O. Moeschlin and D. Pallaschke (eds.), *Game Theory and Related Topics.* Amsterdam: North-Holland, 137–148.

Chapter 7

Concave and Convex Serial Cost Sharing

BY MAURICE KOSTER

7.1 Introduction

A finite set of agents jointly own a production technology for one or more but a finite set of output goods, and to which they have equal access rights. The production technology is fully described by a *cost function* that assigns to each level of output *the* minimal necessary units of (monetary) input. Each of the agents has a certain level of demand for the good; then given the profile of individual demands the aggregate demand is produced and the corresponding costs have to be allocated. This situation is known as the *cooperative production problem.* For instance, sharing the overhead cost in a multi-divisional firm is modeled through a cooperative production problem by Shubik (1962). Furthermore, the same model is used by Sharkey (1982) and Baumol *et al.* (1982) in addressing the problem of natural monopoly. Israelsen (1980) discusses a dual problem, i.e., where each of the agents contributes a certain amount of inputs, and correspondingly the maximal output that can thus be generated is shared by the collective of agents. In this chapter I consider cost sharing rules as possible solutions to cooperative production problems, i.e. devices that assign to each instance of a cooperative production problem a unique distribution of costs. In particular the focus will be on variations of the serial rule of Moulin and Shenker (1992), the cost sharing rule that caught the most attention during the last decade

P. Borm and H. Peters (eds.), Chapters in Game Theory, 143–155.

by its excellent performance in different strategic environments. Moulin and Shenker (1992) discuss the attractive features of the serial rule in case of technologies exhibiting negative externalities, and Moulin (1996) focusses on the serial rule in presence of positive externalities. Here two new cost sharing rules are introduced: the *concave serial rule* and the *convex serial rule.* Both cost sharing rules calculate the individual costs shares by a compound of two operations: the first operation is that of a particular transformation of the cost sharing problem using methodology as in Tijs and Koster (1998), and, secondly, the serial rule is applied to this adaptation. To be more precise, the concave serial rule applies the serial rule to the cost sharing problem with the (concave) *pessimistic cost function*, whereas the convex serial rule applies the serial rule to the cost sharing problem with (convex) *optimistic cost function.* It is shown that these cost sharing rules have diametrically opposed equity features. The concave serial rule is shown to be the unique cost sharing rule that consistently minimizes the range of cost shares (being the difference between the maximal and minimal cost share) under all those cost sharing rules that satisfy the classical equity property *ranking* (see, e.g., Moulin, 2000) and the property *excess lower bound*, which considers minimal justifiable differences between agents based on differences in their demands. On the other hand, the convex serial rule maximizes the range of cost shares given *ranking* and the property *excess upper bound*, that can be seen as the dual of *excess lower bound.* In particular, it follows that the serial rule combines diametrically opposed equity properties, since it coincides with the concave serial rule on the class of cost sharing problems with concave cost functions, and it coincides with the convex serial rule on the class of cost sharing problems with convex cost function.

7.2 The Cost Sharing Model

Consider a fixed and finite set of agents $N = \{1, 2, \ldots, n\}$ sharing a production technology for the production of some divisible good. Correspondingly, a *cost sharing problem* consists of an ordered pair (d, c), where

(*i*) $d = (d_i)_{i \in N} \in \mathbb{R}_+^N$ stands for the profile of individual demands for production; d_i is the demand of agent i,

(*ii*) $c : \mathbb{R}_+ \to \mathbb{R}_+$ is the *cost function,* i.e. a nondecreasing absolutely continuous function[1] with $c(0) = 0$, that summarizes the production technology. For any output level $y \in \mathbb{R}_+$, $c(y)$ denotes the corresponding necessary amount of (monetary) input. The condition $c(0) = 0$ indicates the absence of fixed costs.

Denote the class of all cost sharing problems by $\mathcal{G}$ and the class of all cost functions is denoted $\mathcal{C}$.

For $c \in \mathcal{C}$ denote by c' the function that relates each nonnegative real t to the derivative of c at t, if it exists, and to 0 otherwise. We may unambiguously speak of the *marginal cost function* c', and $c'(t)$ is called the *marginal cost at* production level t for any $t \geq 0$. The marginal cost function c' is integrable and the total costs of production of t units can be expressed in terms of the marginal cost function, since by Lebesgue (1904) it holds for all $t \in \mathbb{R}_+$

$$c(t) = \int_0^t c'(s)\, ds. \tag{7.1}$$

Given a cost sharing problem (d, c), we seek to allocate the total costs for producing the aggregate demand, i.e. $c(d(N))$. A systematic device for the allocation of costs for the class of cost sharing problems $\mathcal{G}$ is modelled through the notion of a *cost sharing rule.* More formally, a *cost sharing rule* is a mapping $x : \mathcal{G} \to \mathbb{R}_+^N$ such that for all $G = (d, c) \in \mathcal{G}$ it holds

$$\sum_{i \in N} x_i(d, c) = c(d(N)). \tag{7.2}$$

Here $x_i(d, c)$ stands for the *cost share* of agent $i \in N$, $d(N)$ is the aggregate demand of the coalition of agents N, i.e. $\sum_{i \in N} d_i$. More generally, for any $S \subseteq N$, the aggregate demand of the coalition of agents S, $\sum_{i \in S} d_i$, is denoted $d(S)$.

In the literature many cost sharing rules are discussed, as there are, for instance, the *proportional cost sharing rule*, and the *serial cost sharing rule* of Moulin and Shenker (1992). The cost shares according to the proportional cost sharing rule for $(d, c) \in \mathcal{G}$ are given by

$$x_i^{\mathrm{P}}(d, c) = \begin{cases} d_i \frac{c(d(N))}{d(N)} & \text{if } d(N) > 0, \\ 0 & \text{else.} \end{cases}, \text{ for each } i \in N.$$

[1] A function $c : \mathbb{R}_+ \to \mathbb{R}_+$ is *absolutely continuous* if for all intervals $[a, b] \subset \mathbb{R}_+$ and all $\epsilon > 0$ there is a $\delta > 0$ such that for any finite collection of pairwise disjoint intervals $(a_k, b_k) \subset [a, b], k = 1, 2, \ldots, n$ with $\sum_{k=1}^n (b_k - a_k) < \delta$, it holds $\sum_{k=1}^n |c(b_k) - c(a_k)| < \epsilon$.

The *serial cost sharing rule*, denoted x^{S}, is defined as follows. Take $(d, c) \in \mathcal{G}$ and let π be a permutation of N that orders the demands increasingly, such that $d_{\pi(i)} \leq d_{\pi(j)}$ if $i \leq j$. Define *intermediate production levels*

$$t_1 = nd_{\pi(1)}, t_k = \sum_{i=1}^{k-1} d_{\pi(i)} + (n-k+1) \cdot d_{\pi(k)} \text{ for } k \geq 2.$$

Then the *serial cost shares* for $G = (d, c)$ are specified by

$$\begin{aligned} x^{\mathrm{S}}_{\pi(1)}(d,c) &:= \tfrac{1}{n}c(t_1), \\ x^{\mathrm{S}}_{\pi(i)}(d,c) &:= x^{\mathrm{S}}_{\pi(i-1)}(d,c) + \tfrac{c(t_i)-c(t_{i-1})}{n-i+1} \text{ for all } i \geq 2. \end{aligned} \tag{7.3}$$

7.3 The Convex and the Concave Serial Cost Sharing Rule

In the literature serial ideas in cost sharing are discussed for different types of production situations, and especially results are discussed for the extreme cases in presence of solely positive or negative externalities. Moulin and Shenker (1992) show that the serial cost sharing rule is, from a strategic point of view, an attractive allocation device in demand games related to production situations with convex cost function. Moulin (1996) discusses the serial cost sharing rule in case of economies of scale. De Frutos (1998) defines the *decreasing* serial cost sharing rule, with outstanding strategic properties in demand games in case of economies of scale. Hougaard and Thorlund-Petersen (2001) axiomatically characterize a cost sharing rule that coincides with the serial cost sharing rule if the cost function is convex, and with the decreasing serial cost sharing rule if the cost function is concave. I will show that the serial cost sharing rule has diametrically opposed equity properties in the above extreme settings of either a concave or a convex cost function. Two new cost sharing rules are introduced in this section, i.e. the *convex serial cost sharing rule* and the *concave serial cost sharing rule*, that determine cost shares according to the serial cost sharing rule for some adapted cost sharing problem. These adapted cost sharing problems rely on techniques from Tijs and Koster (1998) and Koster (2000). The convex (concave) serial cost sharing rule coincides with the serial cost sharing rule on the class of cost sharing problems with convex (concave) cost function. As will turn out, the concave serial cost sharing rule minimizes the range of cost shares subject to some constraint, while the convex serial cost sharing rule can be seen as its *dual* in the sense that it maximizes the range of cost shares subject to a corresponding *dual* constraint.

Before defining these cost sharing rules some preparations are needed. For each cost sharing problem $G = (d, c) \in \mathcal{G}$, Tijs and Koster (1998) study two cooperative games for G as an alternative for the traditional stand alone game (see e.g. Sharkey, 1982; Young, 1985; Hougaard and Thorlund-Petersen, 2000), using the notion of the *pessimistic-* and *optimistic cost function.* Given a particular cost sharing problem $G = (d, c)$, the pessimistic cost function relates each partial demanded production level y in $[0, d(N)]$ to the aggregate of highest marginal costs at which this level possibly could have been processed, whereas the optimistic cost function focusses on the lowest marginal costs in this respect.

Definition 7.1 Given $G = (d, c) \in \mathcal{G}$, the *pessimistic cost function* c_G^- is defined by

$$c_G^-(y) = \begin{cases} \sup\{\int_T c'(s)ds \mid T \in \mathcal{B}([0,d(N)]), \lambda(T) = y\} & \text{if } y \in [0, d(N)], \\ c(y) & \text{else.} \end{cases}$$

Here $\mathcal{B}([0, d(N)])$ stands for the Borel-σ-algebra on $[0, d(N)]$ and λ denotes the Lebesgue measure. The *optimistic cost function*, c_G^+, is defined by [2]

$$c_G^+(y) = \begin{cases} c(d(N)) - c_G^-(d(N) - y) & \text{if } y \in [0, d(N)], \\ c(y) & \text{else.} \end{cases}$$

◁

Calculating the pessimistic- and optimistic cost function can be quite demanding, even for simple cost functions. A useful technique to calculate the pessimistic cost function is discussed in Koster (2000). Note that c_G^- and c_G^+ indeed define cost functions and that

$$c_G^+(y) \leq c(y) \leq c_G^-(y) \text{ for all } y \geq 0. \tag{7.4}$$

Moreover, Koster (2000) shows that c_G^- is concave and c_G^+ is convex on $[0, d(N)]$, respectively. Due to these observations $c_G^-(y)$ and $c_G^+(y)$

[2] The original definition in Tijs and Koster (1998) resembles that of the pessimistic cost function where the supremum is interchanged with infimum. It is shown that the pessimistic and optimistic cost functions are duals in the sense of the first line of the present definition.

are referred to as the *pessimistic-* and *optimistic cost* for producing the amount $y \in [0, d(N)]$. The transformations of the cost sharing problem $G = (d, c)$ to (d, c_G^-) and (d, c_G^+) are used to define the concave- and convex serial cost sharing rule.

Definition 7.2 The *concave serial cost sharing rule*, denoted x^{CCS}, is defined by $x^{\mathrm{CCS}}(d,c) := x^{\mathrm{S}}(d, c_G^-)$ for all $G = (d,c) \in \mathcal{G}$. Similarly, the *convex serial cost sharing rule*, denoted x^{CVS}, is defined through $x^{\mathrm{CVS}}(d,c) = x^{\mathrm{S}}(d, c_G^+)$ for all $G = (d,c) \in \mathcal{G}$. $\triangleleft$

Note that both cost sharing rules can be seen as extensions of the serial cost sharing rule: in case of a concave cost function c it holds $(d, c) = (d, c_G^-)$ and thus $x^{\mathrm{CCS}}(d,c) = x^{\mathrm{S}}(d,c)$, and if c is convex then $(d,c) = (d, c_G^+)$ and hence $x^{\mathrm{CVS}}(d,c) = x^{\mathrm{S}}(d,c)$.

Both cost sharing rules share desirable properties with other eligible cost sharing rules. For instance, one can show that both cost sharing rules are *demand monotonic*, i.e., an agent who increases his demand will pay more in the new situation. Another feature of the above cost sharing rule is *ranking*; the natural ordering of the vector of cost shares preserves the natural ordering of the demand profile. Formally,

Axiom A cost sharing rule x satisfies *ranking* if for all $(d,c) \in \mathcal{G}$ it holds that

$$d_i \leq d_j \Rightarrow x_i(d,c) \leq x_j(d,c) \text{ for all } i, j \in N. \qquad \triangleleft$$

Thus ranking is the equity principle that requires from the larger demanders a higher contribution to the total costs of producing the aggregate demand. The property is certainly transparent within the actual setting of nondecreasing costs. In particular, ranking implies the classical *equal treatment of equals.*

Also x^{CCS} and x^{CVS} satisfy the bounds on cost shares specified by the core of the cooperative *pessimistic cost game* of Tijs and Koster (1998). Each such bound comprises the pessimistic- or optimistic costs for producing the aggregate demand of a coalition of agents as part of the total production. Instead of considering bounds on individual cost shares, the focus is on (minimal) maximal differences between the cost shares of the agents, thereby using information of the (optimistic) pessimistic costs for producing the *excess demands.*

Axiom Consider a cost sharing rule x on $\mathcal{G}$ and let $(d, c) \in \mathcal{G}$. Then x satisfies *excess lower bound for agent* $i \in N$, if

$$\sum_{k \in D_i} (x_k(d,c) - x_i(d,c)) \geq c_G^+ \left(\sum_{k \in D_i} (d_k - d_i) \right)$$
$$\text{where } D_i = \{k \in N \mid d_k > d_i\}. \tag{7.5}$$

If (7.5) holds for all $i \in N$ then x satisfies *excess lower bounds*. Similarly, x satisfies *excess upper bound for agent* $i \in N$ if

$$\sum_{k \in D_i} (x_k(d,c) - x_i(d,c)) \leq c_G^- \left(\sum_{k \in D_i} (d_k - d_i) \right). \tag{7.6}$$

If (7.6) holds for all $i \in N$, then x satisfies *excess upper bounds*.[3] ◁

The excess lower bound property ascertains that the collective of larger agents is not subsidized in the sense that they do not pay a lower price per unit of excess production that can be sustained by the production technology. A similar interpretation can be given to the excess upper bound: the larger agents do not subsidize the smaller ones by paying a price that exceeds a level that is supported by the production technology. The properties are not very restrictive. In fact, it is shown that even the combination of the two is to be considered weak. More specifically, the most popular cost sharing rules like x^{S} and x^{P} satisfy both properties, as well as the newly proposed x^{CCS} and x^{CVS}. Importantly, the inequalities (7.5) and (7.6) turn out to be tight for x^{CCS} and x^{CVS}, respectively.

Proposition 7.3 *x^{P} satisfies excess lower bounds and excess upper bounds.*

Proof. Consider $G = (d, c) \in \mathcal{G}$ with $d(N) > 0$. Then

$$\sum_{j \in D_i} \left(x_j^{\mathrm{P}}(d,c) - x_i^{\mathrm{P}}(d,c) \right) = \left(\sum_{j>i} d_j - (n-i)\, d_i \right) \frac{c(d(N))}{d(N)} \tag{7.7}$$
$$= (1-\mu)\, c(0) + \mu c(d(N)), \tag{7.8}$$

where $\mu = \left(\sum_{j>i} d_j - (n-i)\, d_i \right) d(N)^{-1}$. Recall that $c_G^-(0) = c_G^+(0) = c(0)$, $c_G^-(d(N)) = c_G^+(d(N)) = c(d(N))$. Then by equality (7.7) and

[3]The bounds are similar to those discussed in Aadland and Kolpin (1998) in the case of airport situations.

the fact that c_G^- and c_G^+ are concave and convex on $[0, d(N)]$, respectively, we get the desired inequalities, since

$$\begin{aligned}\textstyle\sum_{j\in D_i}\left(x_j^{\mathrm{P}}(d,c)-x_i^{\mathrm{P}}(d,c)\right) &= (1-\mu)\,c_G^+(0)+\mu c_G^+(d(N))\\ &\geq c_G^+((1-\mu)\cdot 0+\mu d(N))\\ &= c_G^+\left(\textstyle\sum_{j\in D_i}(d_j-d_i)\right),\end{aligned}$$

and

$$\begin{aligned}\textstyle\sum_{j\in D_i}\left(x_j^{\mathrm{P}}(d,c)-x_i^{\mathrm{P}}(d,c)\right) &= (1-\mu)\,c_G^-(0)+\mu c_G^-(d(N))\\ &\leq c_G^-((1-\mu)\cdot 0+\mu d(N))\\ &= c_G^-\left(\textstyle\sum_{j\in D_i}(d_j-d_i)\right).\end{aligned}$$

□

Proposition 7.4 $x^{\mathrm{CCS}}, x^{\mathrm{CVS}}, x^{\mathrm{S}}$ *satisfy excess lower bounds and excess upper bounds. In particular, in case of* x^{CCS} *the inequalities* (7.5), *and in case of* x^{CVS} *the inequalities* (7.6) *are tight, respectively.*

Proof. Take $G=(d,c)\in\mathcal{G}$ and assume that the demands are ordered such that $d_1\leq d_2\leq\ldots\leq d_n$. Let x be a cost sharing rule such that $x(d,c)=x^{\mathrm{S}}(d,\tilde{c})$ for some cost function $\tilde{c}$ with $\tilde{c}(d(N))=c(d(N))$. Then it holds for any $i\in N$

$$\begin{aligned}\sum_{j<i}x_j(d,c)+(n-i+1)\,x_i(d,c) &= \sum_{j<i}x_j^{\mathrm{S}}(d,\tilde{c})+(n-i+1)\,x_i^{\mathrm{S}}(d,\tilde{c})\\ &= \tilde{c}\left(\sum_{j<i}d_j+(n-i+1)\,d_i\right). \qquad (7.9)\end{aligned}$$

Then budget balance implies

$$\begin{aligned}\sum_{k\in D_i}(x_k(d,c)-x_i(d,c)) &= c(d(N))\\ &\quad-\left(\sum_{j<i}x_j(d,c)+(n-i+1)\,x_i(d,c)\right)\\ &= c(d(N))\\ &\quad-\tilde{c}\left(\sum_{j<i}d_j+(n-i+1)\,d_i\right). \qquad (7.10)\end{aligned}$$

Now distinguish between the three cases:
(a) $x = x^{\mathrm{S}}, \tilde{c} = c$. Then by (7.10), the inequalities (7.4), and the duality relation between c_G^+ and c_G^-,

$$\begin{aligned}\sum_{k \in D_i} \left(x_k^{\mathrm{S}}(d,c) - x_i^{\mathrm{S}}(d,c)\right) &= c\left(d\left(N\right)\right) - c\left(\sum_{j<i} d_j + (n-i+1)\,d_i\right) \\ &\leq c\left(d\left(N\right)\right) - c_G^+\left(\sum_{j<i} d_j + (n-i+1)\,d_i\right) \\ &= c_G^-\left(\sum_{j>i} d_j - (n-i)\,d_i\right) \\ &= c_G^-\left(\sum_{j>i} (d_j - d_i)\right).\end{aligned}$$

This proves that x^{S} satisfies excess upper bounds. Excess lower bounds follows directly from (7.10), the inequalities (7.4), and the duality relation between c_G^+ and c_G^-, by flipping the above inequality sign together with interchanging c_G^+ and c_G^-.
(b) $x = x^{\mathrm{CCS}}, \tilde{c} = c_G^-$. Then x^{CCS} satisfies excess lower bounds with equalities since the combination of equality (7.10) and the duality relation between c_G^- and c_G^+ gives

$$\begin{aligned}\sum_{k \in D_i} (x_k^{\mathrm{CCS}}(d,c) - x_i^{\mathrm{CCS}}(d,c)) &= c(d(N)) - c_G^-(\sum_{j<i} d_j + (n-i+1)d_i) \\ &= c_G^+(d(N) - \{\sum_{j<i} d_j + (n-i+1)d_i\}) \\ &= c_G^+(\sum_{j>i}(d_j - d_i)).\end{aligned}$$

Excess upper bounds follows by almost the same reasoning as for case (a).
(c) $x = x^{\mathrm{CVS}}, \tilde{c} = c_G^+$. This case resembles case (b). One only needs to interchange c_G^+ and c_G^- in the proof of case (b) in order to obtain the desired (in)equalities for x^{CVS}. □

Remark It is left to the reader to show that if a cost sharing rule x satisfies excess lower bounds and excess upper bounds, than x satisfies a property that is called *constant returns.* Constant returns is a most compelling answer to solving cost sharing problems in total absence of externalities: x satisfies constant returns when $x(d,c) = \lambda d$ in case (d,c) is such that there is $\lambda \geq 0$ with $c(y) = \lambda y$ for all $y \geq 0$. In other words, each agent pays a fixed price λ per unit of the good. $\triangleleft$

Proposition 7.4 is indicative of the special character of the cost sharing rules x^{CVS} and x^{CCS}. As I am about to show, in the universe of all cost sharing rules with the property ranking, these rules can be seen as *the* extremes of the set of cost sharing rules satisfying the excess lower- and upper bounds. Among all cost sharing rules with the properties ranking and excess upper bounds, x^{CVS} creates the highest difference between the smallest and the largest cost share in a consistent way. Similarly, among the cost sharing rules with the properties ranking and excess lower bounds, x^{CCS} is the unique rule that consistently minimizes the gap between the largest and smallest cost share. So where x^{CCS} may be perceived as a constrained egalitarian cost sharing rule, x^{CVS} is on the other side of the spectrum.

Define the *range* of a vector $y \in \mathbb{R}^N$ by $R(y) = \max_{i \in N} y_i - \min_{i \in N} y_i$. For a cost sharing rule x and cost sharing problem $G = (d,c)$ the corresponding *range of cost shares* is the number $R(x(d,c))$.

Theorem 7.5 *The concave serial cost sharing rule is the unique cost sharing rule which minimizes the range of cost shares for all cost functions among the cost sharing rules satisfying ranking and excess lower bounds.*

Proof. By Proposition 7.4 only the proof of the uniqueness part remains. Take $(d,c) \in \mathcal{G}$ and let x be a cost sharing rule with the premises as enlisted above (inclusive of range minimization). For notational convenience, put $y = x(d,c)$, $\tilde{y} = x^{\text{CCS}}(d,c)$. Concerning the uniqueness proof, suppose on the contrary $y \neq \tilde{y}$. Without loss of generality assume that $d_1 \leq d_2 \leq \ldots \leq d_n$. By ranking it holds that $y_i \leq y_j$ whenever $i \leq j$ and thus the range $R(y) = y_n - y_1$. Distinguish two cases. First consider the case that $y_n \leq \tilde{y}_n$. Since $y \neq \tilde{y}$ there is a maximal i such that $y_i \neq \tilde{y}_i$. Then excess lower bound for agent i gives

$$\sum_{k>i} y_k - (n-i)y_i \geq c_G^+\left(\sum_{k>i} d_k - (n-i)d_i\right) = \sum_{k>i} \tilde{y}_k - (n-i)\tilde{y}_i.$$

Hence $(n-i)(y_i - \tilde{y}_i) \leq \sum_{k>i}(y_k - \tilde{y}_k)$. As by the choice of i, $y_k = \tilde{y}_k$ for all $k > i$ and thus the righthandside of the latter inequality is zero.

So, $n = i$ or $y_i \leq \tilde{y}_i$ and as $y_i \neq \tilde{y}_i$ by assumption we have $y_i < \tilde{y}_i$. Suppose that $y_\ell < \tilde{y}_\ell, y_{\ell+1} < \tilde{y}_{\ell+1}, \ldots, y_{i-1} < \tilde{y}_{i-1}, y_i < \tilde{y}_i$ for some $\ell \in \{2, 3, \ldots, i\}$. Then by excess lower bound for agent $\ell - 1$

$$\begin{aligned}\sum_{k>\ell-1} y_k - (n-\ell+1)y_{\ell-1} &\geq c_G^+\left(\sum_{k>\ell-1} d_k - (n-\ell+1)d_{\ell-1}\right)\\ &= \sum_{k>\ell-1} \tilde{y}_k - (n-\ell+1)\tilde{y}_{\ell-1},\end{aligned}$$

and $y_{\ell-1} < \tilde{y}_{\ell-1}$. This shows that $y_k < \tilde{y}_k$ for all $k \leq i$. But then $\sum_{j\in N} y_j < \sum_{j\in N} \tilde{y}_j = c(d(N))$, which contradicts budget balance. So it must hold that $y_n > \tilde{y}_n$. The *excess lower bound* for 1 gives

$$\sum_{k>1} y_k - (n-1)y_1 \geq c_G^+\left(\sum_{k>1} d_k - (n-1)d_1\right).$$

Budget balance implies $c(d(N)) - ny_1 \geq c_G^+(d(N) - nd_1)$, hence $y_1 \leq \frac{1}{n}(c(d(N)) - c_G^+(d(N) - nd_1)) = \frac{1}{n}c_G^-(nd_1) = \tilde{y}_1$. Consequently $R(y) = y_n - y_1 > \tilde{y}_n - \tilde{y}_1 = R(\tilde{y})$, contradicting range minimization. □

More or less in the same way one can prove the following:

Theorem 7.6 *The convex serial rule is the unique cost sharing rule which maximizes the range of cost shares for each cost function among those rules satisfying ranking and excess upper bounds.*

Remark Always splitting costs equally among the agents yields a cost sharing rule that is usually referred to as the *equal split cost sharing rule.* This cost sharing rule minimizes the range of cost shares subject to ranking, but it does not satisfy the excess bounds previously discussed.◁

A result similar to Theorems 7.5 and 7.6 is the characterization of the constrained egalitarian solution for fixed tree cost sharing problems by Koster *et al.* (1998). This cost sharing rule uniquely minimizes the range of cost shares among those cost sharing rules satisfying some monotonicity condition. In addition it is also shown that minimization of the range of cost shares under the given monotonicity restrictions is equivalent with minimization of the highest cost share. This idea carries over to the present context.

Theorem 7.7 *The concave serial rule is the unique cost sharing rule which minimizes the largest cost share for each cost function among those rules satisfying ranking and excess lower bound.*

Proof. The same argument as in Theorem 7.5 works here. If x is a cost sharing rule that minimizes the maximal cost share then it should hold that $x_n(d,c) \leq x_n^{\mathrm{CCS}}(d,c)$ for all problems (d,c) with d_n being the largest demand. Then we are exactly in the first case in the proof of Theorem 7.5. □

Now the following result should be no surprise:

Theorem 7.8 *The convex serial rule is the unique cost sharing rule which maximizes the largest cost share for each cost function among those rules satisfying ranking and excess upper bound.*

References

Aadland, D., and V. Kolpin (1998): "Shared irrigaton cost: an empirical and axiomatical analysis," *Mathematical Social Sciences*, 35, 203–218.

Baumol, W., J. Panzar, R. Willig and E. Bailey (1998): *Contestable Markets and the Theory of Industry Structure.* San Diego, California: Harcourt Brace Jovanovich.

De Frutos, A. (1998): "Decreasing serial cost sharing under economies of scale," *Journal of Economic Theory*, 79, 245–275.

Hougaard, J., and L. Thorlund-Petersen (2000): "The stand-alone test and decreasing serial cost sharing," *Economic Theory*, 16, 355–362.

Hougaard, J., and L. Thorlund-Petersen (2001): "Mixed serial cost sharing," *Mathematical Social Sciences*, 41, 51–68.

Israelsen, D. (1980): "Collectives, communes, and incentives," *Journal of Comparative Economics*, 4, 99–124.

Koster, M. (2000): *Cost Sharing in Production Situations and Network Exploitation.* PhD Thesis, Tilburg University.

Koster, M., S. Tijs, Y. Sprumont and E. Molina (1998): "Sharing the cost of a network: core and core allocations," CentER Discussion Paper 9821, Tilburg University.

Lebesgue, H. (1904). *Leçons sur l'intégration et la recherche des fonctions primitives.* Paris: Gauthier-Villars.

Moulin, H. (1996): "Cost sharing under increasing returns: a comparison of simple mechanisms," *Games and Economic Behavior*, 13, 225–251.

Moulin, H. (2000): "Axiomatic cost and surplus-sharing," in: Arrow, S. and K. Suzumura (eds.), *Handbook of Social Choice and Welfare* (forthcoming).

Moulin, H., and S. Shenker (1992): "Serial cost sharing," *Econometrica*, 60, 1009–1037.

Moulin, H., and S. Shenker (1994): "Average cost pricing versus serial cost sharing; an axiomatic comparison," *Journal of Economic Theory*, 64, 178–201.

Sharkey, W. (1982): *The Theory of Natural Monopoly.* Cambridge, UK: Cambridge University Press.

Shubik, M (1962): "Incentives, decentralized control, the assignment of joint cost, and internal pricing," *Management Science*, 8, 325–343.

Tijs, S.H., and M. Koster (1998): "General aggregation of demand and cost sharing methods," *Annals of Operations Research*, 84, 137–164.

Young, H.P. (1985). *Cost Allocation: Methods, Principles, Applications.* Amsterdam: North-Holland.

Chapter 8

Centrality Orderings in Social Networks

BY HERMAN MONSUUR AND TON STORCKEN

8.1 Introduction

Social networks describe relationships between agents or actors in a society or community. Examples of such relations are: 'is able to communicate with', 'is in the same club as', 'has strategic alliances with', 'trades with', 'has diplomatic contacts with', 'is friend of', etc. These relations can be formalized by dyadic attributes of pairs of agents. This yields a graph where vertices or nodes play the roles of agents, and edges or arcs those of these attributes. Such a network or graph enables the study of structural characteristics describing the agents' position in the network. In literature, a variety of power or status measures have been discussed, see for example, Braun (1997) or Bonacich (1987). For measures of proximity, see Chebotarev and Shamis (1998). Also measures for centrality have been discussed, see for instance Faust (1997), Friedkin (1991), and many others. Centrality captures the potential of influencing decision making or group processes in general, of being a focal point of communication, of being strategically located and the like, see for example Gulatti and Gargiulo (1999) and Freeman (1979). Centrality, therefore, plays an important role in networks on social, inter-organizational, or communicational issues.

Let a centrality ordering f be a mapping assigning to a graph G a partial ordering, $f(G)$, on the set of vertices of that graph. This ordering

P. Borm and H. Peters (eds.), Chapters in Game Theory, 157–181.

is a reflexive and transitive relation. A pair of vertices, (x, y), is in that relation whenever x is at a position in the graph which is considered to be at least as central as the position of y in that graph. So, if the ordering is complete, then $f(G)$ constitutes a complete list of the vertices arranged from best to worst with respect to their centralities. In Monsuur and Storcken (2001), centrality positions have been studied, yielding a subset of vertices considered to be the central ones. So, in the latter case this set would consist of all best ordered vertices. In the present chapter, similar to the centrality position approach, the focus is on the conceptual issue of what makes a vertex more central than another one. This leads to an axiomatic study of centrality orderings.

Three centrality orderings defined for simple undirected connected graphs, are characterized. These are the cover, the degree and the median centrality orderings. The cover relation originates from Miller (1980): in a social network, vertex a is said to cover vertex b if all neighbours of b are also neighbours of a. If a covers b, then every social link of b can be covered by one of a. So, a weakly dominates b. Degree centrality orders the vertices according to their number of neighbours. Here, the assumption is that the more neighbours a vertex has, the more it is a focal point of communication. In the extreme case that all vertices are neighbours of a vertex a, this vertex a is at a star position, and vertex a occupies a most central position, see also Freeman (1979). Median centrality orders the vertices of a graph according to their sum of distances to all other vertices. The smaller this sum, the more central a vertex is considered. This refers to network communication, where each agent is to be reached separately, and costs are determined by distances. Another way of looking at this is as follows. A vertex is viewed central to the extent that it can avoid the control of communication of others. See Freeman (1979), who introduced closeness centrality by this. Median and closeness centrality yield the same outcome.

For cover as well as median centrality we introduce a set of characterizing conditions. For degree centrality, we provide four such sets. We strived to employ as few as possible conditions in these six characterizations. Of course, the independence of the conditions within each set is proved as well. All the conditions might be appealing from an intuitive point of view. For instance, the star condition: a star position is ordered better than a non-star position. Or neutrality: the names of the vertices are not important. A number of conditions that we use have the following general format. If in going from graph G to graph

G', the change in network environment for vertex x is similar in spirit to the change for that of vertex y, then the ordering between x and y in G is the same as that in G'. So, a centrality ordering is invariant with respect to these environment changes. In Monsuur and Storcken (2001), in two of the three characterizations, a convergence condition is used. Here such a condition is absent. Although the setting here is different from that of centrality positions, some of the conditions in Monsuur and Storcken (2001) could be adapted to the present situation.

The chapter is organized as follows. In Section 8.2, the model is spelled out and several centrality orderings are defined. Section 8.3 is on the cover relation. That is, a characterization is discussed and furthermore it is shown that many centrality orderings are just refinements of the cover relation. In Section 8.4, four characterizations of the degree centrality ordering are provided. In Section 8.5, median centrality is characterized. Finally, Section 8.6 deals with the independence of the conditions in all these characterizations.

8.2 Examples of Centrality Orderings

Let $\mathbf{V} = \{v_1, v_2, \ldots\}$ denote an infinite but countable set of potential vertices. A *graph* or network G is an ordered pair (V, E), where V is a finite, non-empty set of *vertices*, $V \subseteq \mathbf{V}$, and E is a subset of $\{\{x, y\} : x, y \in V, x \neq y\}$, the set of non-ordered pairs of V. Elements of E are called *edges* or *arcs*. Vertices represent agents while arcs represent the relation between these agents. If $\{x, y\} \in E$, then x and y are *neighbours*.

Furthermore, $N(x) = \{y \in V : \{x, y\} \in E\}$ denotes the *neighbourhood* of x and $N[x] = N(x) \cup \{x\}$ the *closed neighbourhood* of x. The number of neighbours determines the degree of a vertex x, $degree(x) = \#N(x)$.

The star of a graph G consists of all vertices which are adjacent to all other vertices, i.e. $star(G) = \{x \in V : N[x] = V\}$.

Let V' be a non-empty subset of V and let E' be a subset of $E \cap \{\{x, y\} : x, y \in V'\}$. Then a graph $G' = (V', E')$ is a *subgraph* of G, which we denote by $G' \subseteq G$. In case $E' = \{\{x, y\} \in E : x, y \in V'\}$, it is said to be the *sub*graph $G|_{V'}$ of G, induced by V'.

Let $P = (W, F)$ be a graph, where $W = \{w_0, w_1, \ldots, w_k\}$. Then P is called a *path* between w_0 and w_k if $F = \{\{w_0, w_1\}, \{w_1, w_2\}, \ldots, \{w_{k-1}, w_k\}\}$. P is a path in G, if P is a path and a subgraph of G.

The length of the path P is k, i.e. $\#W - 1$. P is also denoted by $< w_0, w_1, \ldots, w_k >$.

Let $G = (V, E)$ and $G' = (V', E')$ be graphs. Then the union $G \cup G'$ is the graph determined by the ordered pair $(V \cup V', E \cup E')$.

In the sequel we assume that a graph is *connected*, i.e. between any two vertices there is a path.

To have connected union, we take V and V' not disjoint. In view of connectedness, the geodesic distance between two vertices x and y in a graph G, i.e., the minimal length of the paths between x and y, induces a well-defined function, denoted as $d(G, x, y)$. Obviously $d(G, x, x)$ is defined to be zero. Further, the sum of distances between a given vertex x and all other vertices is denoted by $sum(G, x) = \sum_{y \in V} d(G, x, y)$.

Let V be a non-empty and finite subset of vertices. A *partial ordering* $\succeq$ on V is a reflexive and transitive relation on V. For $x, y \in V$, $(x, y) \in \succeq$ means that x is *at least as good as* y, which we write as $x \succeq y$. If $(x, y) \in \succeq$ and $(y, x) \in \succeq$, x and y are *indifferent*: $x \sim y$. If $(x, y) \in \succeq$ and $(y, x) \notin \succeq$ then x is *better than* y, $x \succ y$. If, in addition, an ordering is *complete*, then it is a weak ordering. For $X \subseteq V$, $\succeq |_X$ denotes the restriction of $\succeq$ to X, i.e. $\succeq |_X = \succeq \cap (X \times X)$.

Given a network $G = (V, E)$, where $V = \{v_1, v_2, ..., v_n\}$, we may rank the vertices from most central to least central. As we consider orderings with respect to centrality, we call all these possible orderings centrality orderings.

Centrality ordering. A *centrality ordering* is a function f, assigning to each undirected, connected graph $G = (V, E)$ a partial ordering $\succeq_{f(G)}$ on V.

In social network analysis, this ordering is often based on scores assigned to vertices, indicating the centrality of that vertex or point, resulting in a complete ordering. That is, the higher the score, the more central a vertex is considered. First, we introduce measures that are discussed in the existing literature on this subject, see for example Faust (1997), Freeman (1979) or Friedkin (1991).

Let G be a graph. Then $C_{degree}(v_i)$ [or $C_{degree}(G, v_i)$], the **degree centrality**, is equal to

$$C_{degree}(G, v_i) = degree(v_i) \tag{8.1}$$

It is clear that, intuitively speaking, the point at the center of a star is the most central one. The measure C_{degree} assigns to this center the highest centrality. With respect to communication, a point with highest degree centrality is visible in this network, or is a focal point of communication, see for example Freeman (1979).

Let G be a graph. Then $C_{between}(v_i)$ [or $C_{between}(G, v_i)$], the **betweenness centrality**, is

$$C_{between}(v_i) = \frac{1}{2} \sum_{s \neq t, s, t \neq i} \frac{g_{st}(v_i)}{g_{st}} \tag{8.2}$$

where $g_{st}(v_i)$ is the number of geodesics from v_s to v_t containing v_i, i.e. paths in G from v_s to v_t of minimal length, that contain v_i, while g_{st} is the total number of geodesics from v_s to v_t.
This view of centrality is based upon the frequency with which a vertex falls on a shortest path in G, connecting two other vertices. A vertex linking pairs of other vertices can influence the transmission of information.

Let G be a graph. Then $C_{close}(v_i)$ [or $C_{close}(G, v_i)$], the **closeness centrality**, is defined by :

$$C_{close}(v_i) = \left(\frac{sum(G, v_i)}{n - 1} \right)^{-1} \tag{8.3}$$

So, $C_{close}(v_i)$ is the inverse of the average distance of v_i to other vertices. Interpreting each vertex as a point that controls the flow of information which passes through, a vertex is viewed as central to the extent that it can avoid this control of communication of others, (Freeman, 1979).

Next, let the $n \times n$ adjacency matrix M be defined for an element m_{ij} at position (i, j) by $m_{ij} = 1$ if $(v_i, v_j) \in E$, otherwise $m_{ij} = 0$. This means that the square matrix M is a nonnegative and symmetric matrix. From the theory of nonnegative matrices (see for example Berman and Plemmons, 1979) we deduce the following results. An $n \times n$ nonnegative matrix A is *reducible* if these exists a permutation of its rows and its columns, such that we obtain $A = \begin{pmatrix} B & 0 \\ C & D \end{pmatrix}$ with B and D square matrices, or $n = 1$ and $A = 0$. Otherwise A is irreducible. It can be shown that a nonnegative matrix A is irreducible if and only if for every i, j there exists a natural number q such that $a_{ij}^{(q)} > 0$, where $a_{ij}^{(q)}$ is the element of A^q at position (i, j). If a nonnegative matrix is irreducible then $r(A)$, the spectral radius of A, is a simple eigenvalue, any eigenvalue of A of the same modulus is also simple, A has a positive eigenvector $\mathbf{x}$ corresponding to $r(A)$ and any nonnegative eigenvector of A is a multiple of $\mathbf{x}$. If there exists a natural number q such that A^q is positive, in which case A is a primitive matrix, then A is irreducible and $r(A)$ is greater in magnitude than any other eigenvalue. If A is nonnegative and primitive with $r(A) = r$, then $\lim_{p\to\infty}(\frac{A}{r})^p$ is a positive matrix whose columns are positive eigenvectors corresponding to r.

If we assume that the graph G is connected then M^n is positive, so M is primitive. We deduce that $r(M)$ is a simple eigenvalue, M has a positive eigenvector $\mathbf{w}$ corresponding to $r(M)$, and $\lim_{p\to\infty}(\frac{M}{r})^p$ exists and gives positive copies of $\mathbf{w}$.

Let G be a graph. Then $C_{eigen}(v_i)$ [or $C_{eigen}(G, v_i)$], the **eigenvector centrality**, is defined as

$$C_{eigen}(v_i) = \mathbf{w}_i \tag{8.4}$$

We give a motivation for this measure. In determining the centrality of a vertex, one may take into account the centrality of its neighbours: being connected to a highly central vertex adds to the centrality of a vertex. Since in turn, the centrality of the neighbours also depends upon the centralities of other vertices, this process looks circular. The eigenvector approach proves to be useful in solving this problem. Indeed, if we let $\mathbf{w}$ represent the centralities of the vertices, then $M\mathbf{w}$ contains, for each vertex, the sum of the centralities of its neighbours. Since, for arbitrary vertices x and y $M\mathbf{w}_x/M\mathbf{w}_y = r(M)\mathbf{w}_x/r(M)\mathbf{w}_y = \mathbf{w}_x/\mathbf{w}_y$, this process of assigning to a vertex the sum of centralities of its neighbours, does not change the (relative) centralities. The (iterative) procedure of computing the eigenvector, as described above, is also implemented in a prototype search engine for the Web, see the Clever Project (1999).

We next introduce a new measure, which uses the Shapley value of a game, see Shapley (1953). For a graph G, let the cooperative (transferable utility) game (V, ν) be defined by letting $\nu(W)$ be the number of unordered pairs v_i, v_j such that all shortest paths from v_i to v_j pass through W, where W is any subset of V. The term 'passes through' refers to the situation that at least one vertex of the path P is also an element of W. Seeing paths as communication links, $\nu(W)$ measures the mediator role of W in these communication links.

If we consider the dual game (V, ν^*) that is defined by $\nu^*(W) = \nu(V) - \nu(V \setminus W)$, then it is easy to verify that $\nu^*(W)$ equals the number of unordered pairs (v_i, v_j), v_i and v_j in W, such that there is a path from v_i to v_j that is contained entirely in $G|_W$.

The Shapley value of a game ν is defined as

$$Sh(\nu)(v_i) = \sum_{F \subset X, v_i \in F} \frac{(f-1)!(n-f)!}{n!}[\nu(F) - \nu(F \setminus \{v_i\}] \tag{8.5}$$

where $f = \#F$, $n = \#V$. It is known that the Shapley values of ν and ν^* coincide.

Let G be a graph. Then $C_{Shapley}(v_i)$ [or $C_{shapley}(G, v_i)$], the **Shapley centrality**, is defined by

$$C_{Shapley}(v_i) = Sh(\nu)(v_i). \tag{8.6}$$

Further, $C_{center}(v_i)$ [or $C_{center}(G, v_i)$], the **center centrality** is defined by

$$C_{center}(v_i) = -max[d(G, v_i, v_j) : v_j \in V]. \tag{8.7}$$

This measure of centrality is based on the geometrical notion of centrality: minimizing the eccentric distances. Here the score is based on the eccentricity.

The **median centrality** $C_{median}(v_i)$ [or $C_{median}(G, v_i)$] is defined by

$$C_{median}(v_i) = -sum(G, v_i). \tag{8.8}$$

This measure is based on the sum. This refers to a communication network, where each agent is to be reached individually and costs are determined by the distances. So, the smaller the sum, the more central an agent's position.

Note that C_{close} and C_{median} yield coinciding orderings. Of course, the underlying ideas are the same.

All these centrality measures induce corresponding centrality orderings f_{degree}, $f_{between}$, f_{close}, f_{eigen}, $f_{Shapley}$, f_{center} and f_{median}.

The centrality orderings introduced so far, are based on scores, which can depend on the local structure around a vertex, like degree centrality, or can depend on global structures, like median, center and eigenvector centrality orderings.

Next, we define the cover centrality ordering, which is not based on scores but on the very local structure around vertices. Actually, we discuss the cover relation, which is not necessarily complete. See also Miller (1980), Moulin (1986), Fishburn (1977), Peris and Subiza (1999) and Dutta and Laslier (1999). In a graph G, a vertex a covers vertex b if either $a = b$ or $N(b) \subseteq N[a]$. It is straightforward to prove that this cover relation is transitive (and reflexive). Therefore, it is a (partial) ordering.

The the **cover centrality** ordering f_{cover}, assigns to each graph G its cover relation:

$$a \succeq_{f_{cover}(G)} b \text{ if and only if } a \text{ covers } b. \tag{8.9}$$

8.3 Cover Centrality Ordering

In this section, we characterize the cover centrality ordering as the inclusion minimal centrality ordering satisfying four independent conditions. Furthermore, we show that degree, closeness, hence median, eigenvector and Shapley centrality orderings are refinements of this cover centrality ordering, while betweenness and center centrality orderings are not.

A centrality ordering f satisfies the **star condition** if for each graph $G = (V, E)$,

$$\text{for each } x, y \in star(G) \text{ and } z \in V \setminus star(G): \quad x \sim_{f(G)} y \text{ and } x \succ_{f(G)} z. \tag{8.10}$$

As it is natural that star vertices are the most central positions, this condition is intuitively clear.

A centrality ordering f satisfies **partial independence** if for every graph $G = (V, E)$ and *sub*graph $G' = (V, E')$ such that $E = E' \cup \{\{v, w\}\}$ for some $v, w \in V$:

$$\text{for every } x, y \in V \setminus \{v, w\} : x \succeq_{f(G)} y \text{ if and only if } x \succeq_{f(G')} y. \tag{8.11}$$

So, partial independence means that the centrality ordering between x and y only depends on the local network environment of x and y in a graph. It is therefore clear that degree and cover centrality orderings satisfy this condition, while, for example, the median centrality ordering does not.

A centrality ordering f is said to be **equable of equal distance connected arc additions**, if for each graph $G = (V, E)$ and *sub*graph $G' = (V, E')$ and vertices $x, y \in V$:

$$x \succeq_{f(G)} y \text{ if and only if } x \succeq_{f(G')} y, \tag{8.12}$$

whenever there are v,v',w,$w' \in V$ such that $\{v, v'\} \cap \{w, w'\} \neq \emptyset$, $E \setminus E' = \{\{v, v'\}, \{w, w'\}\}$, $d(G, x, v) = d(G, y, w)$, $d(G, x, v) = d(G', x, v) - 1$,

$d(G, y, w) = d(G', y, w) - 1$, $d(G, x, t) = d(G', x, t)$ for all $t \in V \setminus \{v\}$ and $d(G, y, t) = d(G', y, t)$ for all $t \in V \setminus \{w\}$.

To illustrate this condition, take $v = w$ and let $G' = (V, E)$, where $V = \{x, v', v, w, w', b, y\}$ and $E = \{\{x, v'\}, \{v', b\}, \{b, w'\}, \{w', y\}, \{b, v\}, \{v', v\}, \{w, w'\}\}$. So, in going from G' to G, we add connected arcs $\{v, v'\}$ and $\{w, w'\}$ such that the distances between x and v and between y and w decrease with 1 and all other distances between x and y and the other vertices remain unchanged. Furthermore, the added arc $\{v, v'\}$ has the same distance to x as arc $\{w, w'\}$ has to y. In this case, equability of equal distance connected arc addition requires that this addition has no effect on the ordering between x and y. Loosely speaking, it means that the preference between x and y remains unchanged whenever we only decrease the distance between x and v by 1 and the distance between y and w by 1 and v and w have the same distance to respectively x and y and either $v = w$ or $\{x, y\}$ is the added arc. It is straightforward to prove that f_{degree}, f_{median} and f_{cover} satisfy this equability condition.

A centrality ordering f is said to be **appendix dominating** if for graph $G = (V, E)$, with $\#V \geq 3$, and all vertices $x, y \in V$, $x \neq y$

$$x \succ_{f(G)} y \text{ if for all } z \in V \setminus \{y\}, x \text{ is on every path from } y \text{ to } z. \qquad (8.13)$$

For a graph $G = (V, E)$, with $\#V \geq 3$, let $appendix(G) = \{y \in V :$ there is a vertex $x \in V \setminus \{y\}$ such that for all $z \in V \setminus \{y\}$, x is on all paths from y to $z\}$.

These four conditions characterize the cover centrality ordering:

Theorem 8.1 *Let f be a centrality ordering that satisfies the star condition, partial independence, equability of equal distance connected arc addition and the appendix domination. Then $f_{cover}(G) \subseteq f(G)$ for all connected graphs G.*

Proof. It is straightforward to show that the cover centrality ordering satisfies these four conditions. Conversely, let f be a centrality ordering satisfying these four conditions. Further, let $G = (V, E)$ be a graph and $x, y \in V$, such that $x \neq y$. To prove the theorem it is sufficient to prove the following two implications:

$$x \sim_{f_{cover}(G)} y \text{ implies } x \sim_{f(G)} y \qquad (8.14)$$

and

$$x \succ_{f_{cover}(G)} y \text{ implies } x \succ_{f(G)} y \qquad (8.15)$$

We consider two cases.

Case 1: $y \in appendix(G)$ and $\{x,y\} \in E$. Then by the appendix domination, we have $x \succ_{f(G)} y$, proving (8.14) and (8.15).

Case 2: $y \notin appendix(G)$ or $\{x,y\} \notin E$. By partial independence, we obtain a sequence of graphs $G^0 = G, G^1 = (V, E^1), \ldots, G^k = (V, E^k)$, such that

$$\{\{s,t\} : s,t \in V \setminus \{x,y\}\} \subseteq E^k, \tag{8.16}$$

$$f(G^i)|_{\{x,y\}} = f(G^{i+1})|_{\{x,y\}} \text{ for } i \in \{0,1,\ldots,k-1\}. \tag{8.17}$$

Note that for any $v,w \in V$, $d(G^k,v,w) \leq 2$. Now, let $V \setminus (N[x] \cup N[y]) = \{v_1, v_2, \ldots, v_p\}$. Applying equability of equal distance connected arc additions p times, yields a sequence of graphs G^k, $G^{k+1} = (V, E^{k+1}), \ldots, G^{k+p} = (V, E^{k+p})$, such that

$$E^{k+p} = E^k \cup \{\{z,v\} : v \in V \setminus (N[x] \cup N[y]) \text{ and } z \in \{x,y\}\} \tag{8.18}$$

and

$$f(G^i)|_{\{x,y\}} = f(G^{i+1})|_{\{x,y\}} \text{ for } i \in \{k,\ldots,k+p\}. \tag{8.19}$$

Now, let $G^{k+p+1} = (V, E^{k+p+1})$, with $E^{k+p+1} = E^{k+p} \cup \{\{x,y\}\}$. Now, we either have $G^{k+p} = G^{k+p+1}$, or G^{k+p+1} is obtained from G^{k+p} by an equal distance connected arc addition. So, we may conclude that

$$f(G)|_{\{x,y\}} = f(G^{k+p+1})|_{\{x,y\}} \tag{8.20}$$

Note that $x \in star(G^{k+p+1})$, so that the two implications (8.14) and (8.15) follow evidently from (8.20). □

The independence of the four characterizing conditions will be discussed in Section 8.6.

The centrality ordering f satisfies the **cover principle** if for all graphs $G = (V, E)$, and all vertices a and b in V: if a covers b then $a \succeq_{f(G)} b$.

If, in addition, b does not cover a and $a \succ_{f(G)} b$, we say that f satisfies the **strict cover principle**. If f satisfies the strict cover principle, f is a refinement of f_{cover}: $\succeq_{f_{cover}} \subseteq \succeq_f$.

Proposition 8.2 *The centrality orderings induced by $C_{between}$ and C_{center} satisfy the cover principle, but do not satisfy the strict cover principle.*

Proof. We first consider the measure $C_{between}$. Suppose a covers b. We consider two cases.

Case 1. $a, b \neq v, y$. Let $< v, c_1, c_2, ..., c_k, b, c_{k+1}, \ldots, c_n, y >$ be a geodesic between v and y. Since a covers b, also $< v, c_1, c_2, \ldots, c_k, a, c_{k+1}, \ldots, c_n, y >$ is a geodesic between v and y.

Case 2: $a = v$, $b \neq v, y$. Consider a path from a to y of the following type: $< a, c_1, c_2, \ldots, c_k, b, c_{k+1}, \ldots, c_n, y >$. Because a covers b, we have a shorter path $< a, c_{k+1}, \ldots, c_n, y >$. This means that b cannot be part of a geodesic from a to y.

The conclusion is that for each geodesic path from v to y containing b, we also have a geodesic path from v to y including a, or $C_{between}(a) \geq C_{between}(b)$. The betweenness centrality measure does not satisfy the strict cover-principle. To show this, consider G with $E = \{\{a, c\}, \{a, d\}, \{d, c\}, \{d, b\}\}$. Then a strictly covers b. But $C_{between}(a) = C_{between}(b) = 0$. The proof that C_{center} satisfies the cover principle is left to the reader. To show that it does not satisfy the strict cover principle, let $E = \{\{a, b\}, \{a, c\}, \{a, e\}, \{b, c\}, \{c, d\}\}$. Then a strictly covers b, while $C_{center}(a) = C_{center}(b)$. □

Proposition 8.3 *The centrality orderings induced by C_{degree}, C_{close}, C_{eigen} and $C_{Shapley}$ satisfy the strict cover principle.*

Proof. Let $G = (V, E)$ be a graph with M its adjacency matrix and let a and b be distinct vertices. Since the assertion is obvious for C_{degree}, we only consider the other measures.

Suppose that a covers b. Since for every c, the distance from b to c is larger than or equal to the distance from a to c, $C_{close}(a) \geq C_{close}(b)$. If, in addition, b does not cover a, there is an element c such that $\{a, c\} \in E$ while $\{b, c\} \notin E$. So the distance between a and c is strictly smaller than the distance between b and c, resulting in $C_{close}(a) > C_{close}(b)$.

Next, let $\mathbf{w}$ be the vector containing the eigenvector centralities. Then $\lambda \mathbf{w}_a = (M\mathbf{w})_a = \sum_{k:\{a,k\} \in E} \mathbf{w}_k \geq \sum_{k:\{b,k\} \in E} \mathbf{w}_k = (M\mathbf{w})_b = \lambda \mathbf{w}_b$, where the inequality is strict whenever the covering is strict. Since $\lambda > 0$, the result easily follows.

For the proof for $C_{Shapley}$, we consider the dual game ν^*. Let a cover b in the graph G. We first prove that $C_{Shapley}$ satisfies the cover principle: $C_{Shapley}(a) \geq C_{Shapley}(b)$. We consider two cases.

Case 1: let F be such that $a, b \in F$. We show that $\nu^*(F \setminus \{a\}) \leq \nu^*(F \setminus \{b\})$. Let P be a path along G from vertex i to j that is contained in $F \setminus \{a\}$. If $b \notin P$ then P is entirely contained in $F \setminus \{b\}$. If $b \in P$, then $P =< k_0, k_1, \ldots, k_t, b, k_{t+1}, \ldots, k_m >$, where $i = k_0$ and $j = k_m$. Since a covers b in graph G, we also have another path P' from i to j that is contained in $F \setminus \{b\}$: $P' =< k_0, k_1, \ldots, k_t, a, k_{t+1}, \ldots, k_m >$. Hence $\nu^*(F \setminus \{a\}) \leq \nu^*(F \setminus \{b\})$. So we have $\nu^*(F) - \nu^*(F \setminus \{a\}) \geq \nu^*(F) - \nu^*(F \setminus \{b\})$.

Case 2: let F be such that $b \in F$ while $a \notin F$. Then there is a unique F' with the same cardinality as F and $a \in F'$ while $b \notin F'$: $F' = (F \setminus \{b\}) \cup \{a\}$. It is evident that $\nu^*(F \setminus \{b\}) = \nu^*(F' \setminus \{a\})$ since $F' \setminus \{a\} = F \setminus \{b\}$. Next we proceed by showing that $\nu^*(F') \geq \nu^*(F)$. Let P be a path from vertex i to j contained in F. If $b \notin P$ then P is contained entirely in F'. Now suppose that $b \in P =< k_0, k_1, \ldots, k_t, b, k_{t+1}, \ldots, k_m >$, where $i = k_0$ and $j = k_m$. If $b \neq i, j$ then, as before, we may exchange b by a, resulting in a path from i to j that is contained in F'. If $i = b$, so that $P =< b, k_{t+1}, \ldots, k_m >$ then $< a, k_{t+1}, \ldots, k_m >$ is a path in F'. So, we have $\nu^*(F') - \nu^*(F' \setminus \{a\}) \geq \nu^*(F) - \nu^*(F \setminus \{b\})$. Altogether, we have showed that $C_{Shapley}(a) \geq C_{Shapley}(b)$.

To verify that $C_{Shapley}$ does satisfy the strict cover principle, we consider the case that a covers b while b does not cover a. This means that there exists an element c such that $\{a, c\} \in E$ while $\{b, c\} \notin E$. We show that there exists a subset F of the set of vertices such that $\nu^*(F) - \nu^*(F \setminus \{b\}) < \nu^*(F) - \nu^*(F \setminus \{a\})$. To this end, take $F = \{a, b, c\}$. Then $\nu^*(F) = 1$ if $\{a, b\} \notin E$, otherwise it equals 3. Furthermore $\nu^*(F \setminus \{a\}) = 0$, while $\nu^*(F \setminus \{b\}) = 1$. This proves that $\nu^*(F) - \nu^*(F \setminus \{b\}) < \nu^*(F) - \nu^*(F \setminus \{a\})$, completing the proof of the proposition. □

8.4 Degree Centrality Ordering

In this section, four characterizations of the degree centrality ordering are presented. The notion of degree centrality for undirected graphs is similar to that of the Copeland score in tournaments. See also Freeman (1979), Moulin (1980), Rubinstein (1980) and Delver et al. (1991). Some of the characterizing conditions stem from this literature on Copeland scores. First we discuss the various conditions used in the characterizations.

The following condition is slightly stronger than equability of equal distance connected arc additions, see equation (8.12), in that we do not require that the arcs are connected.

A centrality ordering f is said to be **equable of equal distance arc additions**, if for graph $G = (V, E)$ and *sub*graph $G' = (V, E')$ and vertices $x, y \in V$

$$x \succeq_{f(G)} y \text{ if and only if } x \succeq_{f(G')} y, \tag{8.21}$$

whenever there are $v, v', w, w' \in V$ such that $E \setminus E' = \{\{v, v'\}, \{w, w'\}\}$, $d(G, x, v) = d(G, y, w)$, $d(G, x, v) = d(G', x, v) - 1$, $d(G, y, w) = d(G', y, w) - 1$, $d(G, x, t) = d(G', x, t)$ for all $t \in V \setminus \{v\}$ and $d(G, y, t) = d(G', y, t)$ for all $t \in V \setminus \{w\}$.

Note that adding connected arcs at equal distances in the neighbourhood of x , hence $v' = x$, implies that $w' = y$ and either $\{x, y\} = \{v, v'\} = \{w, w'\}$ or $v = w$. So, this addition does not affect the cover relation between x and y. If connectedness is dropped, arc additions, even at equal distances may influence the cover relation. Therefore, f_{cover} does not satisfy this new condition. On the other hand, it is straightforward to see that f_{degree} and f_{median} do satisfy this condition. In fact, if we substitute this equability condition for the connected version in Theorem 8.1 and drop the appendix domination, we obtain a set of characterizing conditions for f_{degree} as is shown in Theorem 8.4(i).

The following condition requires the notion of a lenticular graph. Let $P^1 = (V^1, E^1), \ldots, P^k = (V^k, E^k)$ be paths from vertex x to y. Then the union $L = P^1 \cup P^2 \cup \ldots \cup P^k$ is called a lenticular graph between x and y, if $V^i \cap V^j = \{x, y\}$ for all $i, j \in \{1, 2, \ldots, k\}$, with $i \neq j$. Hence the paths only meet at x and y.

A centrality ordering f is called **invariant at lenticular additions** if for graphs $G = (V, E)$, all vertices $x, y \in V$ and lenticular graphs $L = (V', E')$ between x and y:

$$x \succeq_{f(G)} y \text{ if and only if } x \succeq_{f(G \cup L)} y, \tag{8.22}$$

whenever $V' \cap V = \{x, y\}$ and $d(G, v, w) = d(G \cup L, v, w)$ for all $v, w \in V$ with $\{x, y\} \neq \{v, w\}$.

So, if we add a number of disjoint paths from x to y such that the distances between all pairs of vertices $\{v, w\}$ different from $\{x, y\}$ is not changed, then this addition does not affect the centrality ordering between x and y. Theorem 8.4(ii) shows that by substituting this condition for equability of equal distance arc additions yields a new set of characterizing conditions for the degree centrality ordering. It is easy to see that adding lenticular graphs may ruin the cover relation.

A centrality ordering f is said to be **complete** if for all graphs G, $f(G)$ is a complete ordering.

A centrality ordering f is said to be **neutral** if for all graphs $G = (V, E)$ and all permutations σ on V,

$$f(\sigma G) = \sigma f(G), \tag{8.23}$$

where $\sigma G = (V, E')$, such that $E' = \{\{\sigma(x), \sigma(y)\}\} : \{x, y\} \in E\}$ and $\sigma f(G) = \{(\sigma(x), \sigma(y)) : (x, y) \in f(G)\}$.

Neutrality means that the centrality ordering does not depend on the actual numbering of the vertices. It guarantees a similar ordering of the vertices if these are at similar positions in a graph.

A centrality ordering f is **monotonic** if for all graphs $G = (V, E)$ and *sub*graphs $G' = (V, E')$ with $E = E' \cup \{\{x, z\}\}$ for some vertices $x, z \in V$ with $x \neq z$, we have for every $y \notin \{x, z\}$

$$x \succeq_{f(G')} y \text{ implies } x \succ_{f(G)} y. \tag{8.24}$$

Going from G' to G, the arc $\{x, z\}$ is added where z is not equal to y. Then monotonicity implies that if x is at least as good as y at $f(G')$, then the relative position of x with respect to y improves, meaning that x is better than y at $f(G)$. Clearly f_{degree} satisfies this condition. Replacing the star and equability conditions by the latter three defined conditions yields yet another characterization of f_{degree} in Theorem 8.4(3).

A centrality ordering f is **swap invariant** if for all graphs $G = (V, E)$ and $G' = (V, E')$ and all vertices $x, y \in V$

$$x \succeq_{f(G)} y \text{ if and only if } x \succeq_{f(G')} y, \tag{8.25}$$

whenever there are $a, b \in V$ such that $\{a, x\}, \{b, y\} \in E, \{a, y\}, \{b, x\} \notin E$ and $E' = (E \setminus \{a, x\}, \{b, y\}) \cup \{\{a, y\}, \{b, x\}\}$.

Going from G to G', we swap neighbour $a \in N(x) \setminus N[y]$ of x, with neighbour $b \in N(y) \setminus N[x]$ of y. Swap invariance means that this type of neighbour swapping has no effect on the ordering between x and y. Replacing partial independence with swap invariance, we obtain our fourth characterization of f_{degree}.

It is straightforward to check that f_{degree} satisfies all the conditions defined in this section.

Theorem 8.4 *The degree centrality ordering f_{degree} is the only centrality ordering that satisfies either of the following four sets of conditions. (i) partial independence, equability of equal distance arc additions and the star condition, (ii) partial independence, invariance at lenticular additions and the star condition, (iii) partial independence, neutrality, monotonicity and completeness, (iv) swap invariance, neutrality, monotonicity and completeness.*

Proof. It is straightforward to prove that f_{degree} satisfies all these conditions. In order to complete the proof of these four characterizations, let f be a centrality ordering satisfying one of these four sets of conditions. We proceed by showing that $f = f_{degree}$. Let $G = (V, E)$ be a graph and $x, y \in V$ two distinct vertices, $x \neq y$. It is sufficient to prove that

$$C_{degree}(G, x) = C_{degree}(G, y) \text{ implies } x \sim_{f(G)} y \tag{8.26}$$

and

$$C_{degree}(G, x) > C_{degree}(G, y) \text{ implies } x \succ_{f(G)} y. \tag{8.27}$$

Case 1. Let f satisfy partial independence. By adding arcs which neither involve vertex x nor vertex y, we obtain a graph $G' = (V, E')$ with subgraph G, such that $\{v, w\} \in E'$ for all $v, w \in V \setminus \{x, y\}$, $v \neq w$. By partial independence,

$$f(G)|_{\{x,y\}} = f(G')|_{\{x,y\}}. \tag{8.28}$$

Let f satisfy the set of conditions (i) of the theorem. First, consider the case where $N(y) \setminus \{x\} \neq \emptyset$.

Then, since $d(G', v, w) \leq 2$ for all $v, w \in V$, by equability of equal distance arc additions, it is without loss of generality to assume that $\{x, y\} \in E'$ and that $V \setminus (N[x] \cup N[y])$ is empty. Now, let $a \in N(x) \setminus N[y]$ and $b \in N(y) \setminus N[x]$. Invoking equability of equal distance arc additions, we may assume that $\{y, a\}, \{x, b\} \in E'$. This holds for all such a and b. So, if $C_{degree}(G, x) = C_{degree}(G, y)$ then $x, y \in star(G')$. By (8.28) and the star property it follows that $x \sim_{f(G)} y$. If $C_{degree}(G, x) > C_{degree}(G, y)$ then $x \in star(G')$, $y \notin star(G')$. So, by (8.28) and the star property, we find $x \succ_{f(G)} y$.

Now, consider the special case of $N(y) \setminus \{x\} = \emptyset$. Since G is connected, we have $N(y) = \{x\}$. If $V = N[x]$, we are done with the star condition. Suppose $V \neq N[x]$. Then by partial independence, it is sufficient to prove $x \succ_{f(G'')} y$, where $G'' = (V, E'')$ is the path graph $< y, x, w, v_1, v_2, \ldots, v_t >$. Now, apply the previous case to G'', w and y. This yields $w \succ_{f(G'')} y$. Application of the previous case to G'', w and x yields $w \sim_{f(G'')} x$. Then, by transitivity of the ordering $f(G'')$, we obtain $x \succ_{f(G'')} y$. As we proved the implications (8.26) and (8.27), we showed that $f = f_{degree}$ if f satisfies the set of conditions (i).

Let f satisfy the set of conditions (ii). By invariance at lenticular additions, it is without loss of generality to assume that $N(y) \setminus \{x\} \neq \emptyset$.

Hence, $d(G', v, w) \leq 2$ for all $v, w \in V$. Now, by invariance at lenticular additions, it is without loss of generality to assume that $\{x, y\} \in E'$. Let $v \in V \setminus (N[x] \cup N[y])$. Let $u, w \in \mathbf{V} \setminus V$ and consider the path $P =< x, u, w, y >$. Adding P to G', we have by invariance at lenticular additions that $f(G')|_{\{x,y\}} = f(P \cup G')|_{\{x,y\}}$. Now, by partial independence, we may delete arc $\{u, w\}$, delete arcs $\{v, s\}$, where $s \in V$ and add the two arcs $\{u, v\}$ and $\{w, v\}$, yielding graph G'', such that $f(P \cup G')|_{\{x,y\}} = f(G'')|_{\{x,y\}}$. Now, $< x, u, v, w, y >$ is a path in G'' and by invariance at lenticular additions, it follows that $f(G'')|_{\{x,y\}} = f(G''|_{V \setminus \{u,v,w\}})|_{\{x,y\}}$. As $G''|_{V \setminus \{u,v,w\}} = G'|_{V \setminus \{u,v,w\}}$, the previous reasoning shows that it is without loss of generality to assume that $V \setminus (N[x] \cup N[y])$ is empty. Next, consider $a \in N(x) \setminus N[y]$ and $b \in N(y) \setminus N[x]$. Now $< x, a, b, y >$ is a path in G'. By partial independence and invariance at lenticular additions, we have $f(G')|_{\{x,y\}} = f(G'|_{V \setminus \{a,b\}})|_{\{x,y\}}$. This holds for all such a and b. Hence it is without loss of generality to assume that $N(x) \setminus N[y]$ is empty if $C_{degree}(G, x) \leq C_{degree}(G, y)$ and $N(y) \setminus N[x]$ is empty if $C_{degree}(G, x) \geq C_{degree}(G, y)$. Now, similar as to condition set (1), the star condition can be used to prove implications (8.26) and (8.27). So, $f = f_{degree}$ if f satisfies the conditions mentioned in (ii).

Let f satisfy the conditions in (iii). In order to prove implication (8.26), let $C_{degree}(G, x) = C_{degree}(G, y)$. Then, the cardinality of $N(x) \setminus N[y]$ equals that of $N(y) \setminus N[x]$. Hence there is a bijection τ from $N(x) \setminus N[y]$ to $N(y) \setminus N[x]$. Consider a permutation σ on V such that $\sigma^{-1} = \sigma$, $\sigma(x) = y$ and $\sigma(a) = \tau(a)$ for $a \in N(x) \setminus N[y]$, $\sigma(b) = \tau^{-1}(b)$ for $b \in N(y) \setminus N[x]$, and $\sigma(z) = z$ for all $z \in (N(x) \cap N(y)) \setminus \{x, y\}$ or $z \in V \setminus (N[x] \cup N[y])$. Now, $\sigma G' = G'$ and by neutrality $\sigma f(G') = f(\sigma G') = f(G')$. As $\sigma(x) = y$ and f is complete, it follows that $x \succeq_{f(G')} y$ and $y \succeq_{f(G')} x$. Hence by (8.28) $x \succeq_{f(G)} y$ and $y \succeq_{f(G)} x$ which proves implication (8.26). In order to prove implication (8.27), let $C_{degree}(G, x) > C_{degree}(G, y)$. So, $\#(N(x) \setminus N(y)) > \#(N(y) \setminus N(x))$. Let G'' be a subgraph of G', which is obtained by deleting some of the arcs $\{x, a\}$ for $a \in N(x) \setminus N[y]$, such that $C_{degree}(G'', x) = C_{degree}(G'', y)$. Then, by implication (8.26), we have $x \succeq_{f(G'')} y$ and by monotonicity we have $x \succ_{f(G')} y$. So, by (8.28) this yields $x \succ_{f(G)} y$, which proves implication (8.27). So, $f = f_{degree}$ if f satisfies the conditions mentioned (iii).

Case 2. f satisfies swap invariance, neutrality and completeness. First, we prove implication (8.26). Let $C_{degree}(G, x) = C_{degree}(G, y)$. By a sequence of swaps, while maintaining connectedness, we may swap all neighbours of x with those of y yielding a graph $G' = (V, E')$ such that $E' = (E \setminus \{z, v\} : z \in \{x, y\}, v \in V\}) \cup \{\{x, b\} : \{y, b\} \in E\} \cup \{\{y, a\} : \{x, a\} \in E\}$. By swap invariance, we have $f(G)|_{\{x,y\}} = f(G')|_{\{x,y\}}$. Consider a permutation σ on V such that $\sigma^{-1} = \sigma$, $\sigma(x) = y$, $\sigma(y) = x$ and $\sigma(z) = z$ for all other vertices. Then $\sigma G = G'$ and by neutrality $\sigma f(G) = f(\sigma G) = f(G')$. So by (8.28) we have $\sigma f(G)|_{\{x,y\}} = f(G)|_{\{x,y\}}$. As f is complete, this yields that $x \sim_{f(G)} y$, which proves implication (8.26). Now, similar as in the case of (iii), monotonicity gives implication (8.27). So $f = f_{degree}$ if f satisfies the set of conditions (iv). □

In Section 8.6, we discuss the independence of the conditions in Theorem 8.4.

8.5 Median Centrality Ordering

In the existing literature, one may find axiomatic characterizations of locations on networks. For example, see Foster and Vohra (1998), Holzman (1990) or McMorris *et al.* (2000). There are, however, a few differences between their work and our approach. Firstly, they consider tree networks (or median graphs), while in our model a network may be arbitrarily cyclic. Secondly, in social networks, the problem is to find the set of central vertices among the set of all vertices. This contrasts with theories concerning location or consensus, where one generally works with profiles, which are k-tuples, $k \geq 1$, of locations (not necessarily vertices). The problem then is to find a compromise location on the network. Thirdly, we assume that the distance between two adjacent vertices equals one, which is not the case in location theory. Altogether, this means that our necessary and sufficient conditions used in the following axiomatic characterization of the median for arbitrary, social networks, do not compare to the axioms used in location theory.

Here the median centrality ordering will be characterized by three conditions: the star property, invariance at lenticular additions and yet another equability condition. This latter condition is a strengthening of equal distance arc additions, see equation (8.21), in that equal distance of the added arcs is not required.

A centrality ordering f is said to be **equable of arc addition** if for all graphs $G = (V, E)$ and *sub*graphs $G' = (V, E')$ and all vertices $x, y \in V$:

$$x \succeq_{f(G)} y \text{ if and only if } x \succeq_{f(G')} y, \tag{8.29}$$

whenever there are $v, v', w, w' \in V$ such that $E \setminus E' = \{\{v, v'\}, \{w, w'\}\}$, $d(G, x, v) = d(G', x, v) - 1, d(G, y, w) = d(G', y, w) - 1, d(G, x, t) = d(G', x, t)$ for all $t \in V \setminus \{v\}$ and $d(G, y, t) = d(G', y, t)$ for all $t \in V \setminus \{w\}$.

Theorem 8.5 *The median centrality ordering f_{median} is the only centrality ordering that satisfies equability of arc addition, invariance at lenticular addition and the star condition.*

Proof. It is straightforward to prove that the median centrality ordering satisfies these three conditions. In order to prove that it is the only centrality ordering that does so, let $G = (V, E)$ be a graph and $x, y \in V, x \neq y$, two vertices. Let f be a centrality ordering that satisfies the conditions. It is sufficient to prove the following two implications:

$$sum(G, x) = sum(G, y) \text{ implies } x \succeq_{f(G)} y \tag{8.30}$$

and

$$sum(G, x) > sum(G, y) \text{ implies } x \succ_{f(G)} y. \tag{8.31}$$

The proof of these implications is based on the construction of a special sequence of graphs: $G^0 = G, G^1, \ldots, G^p$ is a *confined sequence of graphs* if, for each $i \in \{0, \ldots, p-1\}$, G^{i+1} is obtained from G^i by a lenticular addition or equable arc additions with respect to x and y. Note that

$$sum(G, x) \geq sum(G, y) \text{ if and only if } sum(G^p, x) \geq sum(G^p, y) \tag{8.32}$$

and by equability of arc additions and invariance at lenticular additions,

$$f(G)|_{\{x,y\}} = f(G^p)|_{\{x,y\}}. \tag{8.33}$$

Eventually, we construct a confined sequence of graphs, such that $star(G^p) \cap \{x, y\} \neq \emptyset$. Applying the previous two remarks and the star condition, then yield implications (8.30) and (8.31). The following Claim prepares the construction of such a confined sequence of graphs.

Claim. *Let $G = (V, E)$ be a graph. Let $x, y \in V$ be two distinct vertices and let $d(G, x, y) = k$. Then there is a confined sequence of graphs $G^0 = G, G^1, \ldots, G^p$ such that $d(G^p, x, z) \leq max\{2, k\}, d(G^p, y, z) \leq max\{2, k\}$ and $d(G^p, x, z) + d(G^p, z, y) \leq k + 2$ for all $z \in V^p$.*

Proof of Claim. First, we construct a confined sequence of graphs, say $G = G^0, G^1 = (V^1, E^1), \ldots, G^q = (V^q, E^q)$ such that $d(G^q, x, z) + d(G^q, y, z) \leq k + 2$ for all $z \in V^q$, where $k = d(G, x, y)$. If this is true for $q = 0$, then this desired sequence consists of $G = G^0 = G^q$ only. So, we assume that this is not the case. Let $W = \{z \in V : d(G,x,z) + d(G,y,z) \geq d(G,x,v)+d(G,y,v)$ for all $v \in V\}$. Further, let W_x and W_y be subsets of W defined by $W_x = \{z \in W : d(G,x,z) \geq d(G,x,w)$ for all $w \in W\}$ and $W_y = \{z \in W : d(G,y,z) \geq d(G,y,w)$ for all $w \in W\}$. We proceed by arbitrarily choosing two vertices $a \in W_x$ and $b \in W_y$ ($a = b$ is possible). Let $d(G,x,a) = t, d(G,y,a) = s, d(G,x,b) = t'$ and $d(G,y,b) = s'$. Note that $t' + s' = t + s \geq k + 3$.

Next, let $G^1 = (V^1, E^1)$ be obtained from $G = G^0$ by lenticular addition of two distinct paths $< x, a_1, a_2, \ldots, a_{t+s-3}, y >$ and $< x, b_{t'+s'-3}, \ldots, b_1, y >$. Moreover, let
$G^2 = (V^1, E^1 \cup \{\{a, a_{t-2}\}, \{b, b_{s'-2}\}\})$. It is clear that for any $z \in V^1$

$$t' + s' = t + s \geq d(G^1,x,z) + d(G^1,y,z) \geq d(G^2,x,z) + d(G^2,y,z) \quad (8.34)$$

where the latter inequality is strict for $z \in \{a, b\}$. So, if it is the case that G^2 is obtained from G^1 by equable arcs addition, then repeating this procedure yields the desired sequence $G = G^0, \ldots, G^q$. So, we have to prove that G^2 is obtained from G^1 by equable arcs addition.

Obviously, $d(G^2,x,a) = d(G^1,x,a)-1$ and $d(G^2,y,b) = d(G^1,y,b)-1$. Take $z \in V^1 \setminus \{a\}$ and $u \in V^1 \setminus \{b\}$. In order to show that $\{a, a_{t-2}\}$ and $\{b, b_{s'-2}\}$ are equable arc additions, it is sufficient to prove that $d(G^2,x,z) = d(G^1,x,z)$ and $d(G^2,y,u) = d(G^1,y,u)$. As $E^1 \subseteq E^2$, we have $d(G^2,x,z) \leq d(G^1,x,z)$ and $d(G^2,y,u) \leq d(G^1,y,u)$.

We prove that the assumption $d(G^2,x,z) < d(G^1,x,z)$ for any $z \in V^1 \setminus \{a\}$ yields a contradiction.

Case 1. Suppose that any shortest path from x to z in G^2 contains $\{a, a_{t-2}\}$. In view of the construction of G^1 and G^2, it is without loss of generality to assume $\{z, a\} \in E^2$. Since $d(G^2,x,a) = d(G^1,x,a) - 1 = t - 1$, and $\{a, a_{t-2}\}$ is on a shortest path from x to z in G^2, we have $d(G^2,x,z) = t$. Therefore $d(G,x,z) = t + 1$. Next, since $d(G,x,z) + d(G,y,z) \leq t + s$, we obtain $d(G,y,z) \leq s - 1$. Because $d(G,y,a) = s$, we have $d(G,y,z) = s - 1$. But this means that $z \in W$. But then, since $d(G,x,z) > d(G,x,a)$, a is not in W_x, a contradiction.

Case 2. Suppose that there is a shortest path from x to z in G^2, not containing $\{a, a_{t-2}\}$. So it necessarily contains $\{b_{s'-2}, b\}$. But then we obtain $d(G^2,x,b) = d(G,x,b_{s'-2}) + 1 = t' + 1$, contradicting $d(G,x,b) = t'$.

So, we may conclude that $d(G^2,x,z) < d(G,x,z)$ yields a contradiction. Similarly, $d(G^2,y,u) < d(G^1,y,u)$ for any $u \in V^1 \setminus \{b\}$ yields a contradiction.

Hence, we have a confined sequence $G = G^0, G^1, \ldots, G^q$ with the property that $d(G^q, x, z) + d(G^q, y, z) \leq k+2$, proving the second part of the Claim.

If $k = 1$, then the conclusion that $d(G^q, x, z) + d(G^q, y, z) \leq k + 2$ for all $z \in V^q$ yields $d(G^q, x, z) \leq 2$ and $d(G^q, y, z) \leq 2$ for all $z \in V^q$. So, the Claim is proved for the case $k = 1$. Therefore, we now consider the case $k \geq 2$. Let $z \in V^q$ such that $d(G^q, x, z) + d(G^q, y, z) = k + 2$ and $d(G^q, y, z) = k + 1$. So $d(G^q, x, z) = 1$, meaning that we have $\{x, z\} \in E^q$ and furthermore, since $d(G^q, x, y) = k$, we have a path $< x, a_1, a_2, \ldots, a_{k-1}, y >$. Then add a (lenticular) path $< x, b_1, b_2, \ldots, b_k, y >$ resulting in G^{q+1}. Next, we have the following equable arc additions giving G^{q+2}: if $k = 2$ then $\{x, b_k\}$, reducing $d(G^q, x, b_k)$ by one, if $k > 2$ then $\{a_{k-2}, b_k\}$, reducing $d(G^q, x, b_k)$ by one, and $\{a_1, z\}$, reducing $d(G^q, y, z)$ by one. Now we have $d(G^{q+2}, y, z) = k$. By repeating these last path and arc additions, we obtain a sequence as desired in the lemma. This completes the proof of the Claim. $\lhd$

Now, let $d(G, x, y) = k$. Then, by the Claim, there exists a sequence of graphs $G^0 = G, G^1, \ldots, G^p$, where G^{i+1} is obtained from G^i by means of equable arc additions or lenticular additions, such that

$$d(G^p, x, z) \leq max\{2, k\}, d(G^p, y, z) \leq max\{2, k\}, \tag{8.35}$$

and

$$d(G^p, x, z) + d(G^p, y, z) \leq k + 2, \text{ for all } z \in V^p. \tag{8.36}$$

Consider the case $d(G, x, y) = 1$. Then $V = \{x, y, v_1, \ldots, v_t, w_1, \ldots, w_s\}$, where $N(x) \setminus N[y] = \{v_1, \ldots, v_t\}$ and $N(y) \setminus N[x] = \{w_1, \ldots, w_s\}$. Now, suppose that $sum(G^p, x) > sum(G^p, y)$. Then $t < s$. Add $2t$ (equable) arcs $\{y, v_i\}, \{x, w_i\}$ resulting in G^{p+t}, where $y \in star(G^{p+t})$. By the star condition, $y \succ_{f(G^{p+t})} x$. Using Remark (8.33) then also $y \succ_{f(G)} x$. If $sum(G^p, x) = sum(G^p, y)$ then, by similar reasoning, $y \succeq_{f(G)} x$. This proves implication (8.30) and (8.31) for this case.

Next, consider the case $d(G, x, y) = k > 1$. By virtue of (8.35) and (8.36), the new path $< x, z_1, z_2, \ldots, z_{k-1} >$, with $z_{k-1} = y$, of length $k - 1$ is an allowed lenticular addition, giving G'. Now, by a simple induction argument, we obtain that (8.30) and (8.31) hold for G'. But then Remarks (8.33) yields (8.30) and (8.31) for graph G. $\square$

The independence of the characterizing conditions will be investigated in the following section.

8.6 Independence of the Characterizing Conditions

In the foregoing sections, three centrality orderings are characterized involving in total six sets of conditions. In this section, we discuss the independence of the conditions within each set. To prove the independence and thereby completing the characterizations, in the sense that these characterizing condition sets are inclusion minimal, we introduce six centrality orderings. These are just centrality orderings to fit the independence proofs and most likely will have no further practical use.

Stap centrality ordering, f_{stap}, is a centrality ordering defined for all graphs $G = (V, E)$ by $f_{stap}(G) = (V \times V) \setminus \{\{y, x\} \in V \times V : y \notin star(G)$ and $x \in star(G)$ or $y \in appendix(G)$ and $x \notin appendix(G)\}$

Degmed centrality ordering, f_{degmed}, is a centrality ordering, where indifferences according to f_{degree} may be resolved according to f_{median}. For a graph $G = (V, E)$ and vertices $x, y \in V$, it is defined by $x \succ_{f_{degmed}(G)} y$, whenever $x \succ_{f_{degree}(G)} y$ or whenever $x \sim_{f_{degree}(G)} y$ and $x \succ_{f_{median}(G)} y$, $x \sim_{f_{degmed}(G)} y$, whenever $x \sim_{f_{degree}(G)} y$ and $x \sim_{f_{median}(G)} y$.

Smaller than centrality ordering, $f_<$, is based on a numbering of all potential vertices in $\mathbf{V}$. Let $v_i < v_j$ whenever $i < j$ for all $i, j \in \{1, 2, 3, \ldots\}$. Now $f_<$ is defined for all graphs $G = (V, E)$ by $f_<(G) = < |_V$.

Centrality ordering f_1 is defined for all graphs $G = (V, E)$ by $f_1(G) = f_{median}(G)$ if $G \neq G^1$ and $f_1(G) = v_3 \succ v_2 \sim v_4 \succ v_1 \succ v_5$ if $G = G^1$, where G^1 is the path graph $< v_1, v_2, v_3, v_4, v_5 >$.

Centrality ordering $f_{hierdeg}$ is defined for all graphs $G = (V, E)$ and $v_i, v_j \in V$ by $v_i \succ_{f_{hierdeg}(G)} v_j$ if $v_i \succ_{f_{degree}(G)} v_j$ or $v_i \sim_{f_{degree}(G)} v_j$ and $i < j$. This centrality ordering refines the degree centrality by tie-breaking indifferences according to the numbering of the vertices.

Centrality ordering f_{app}, is defined for all graphs G by $f_{app}(G) = f_{cover}(G)$ if $G \neq G^2$ and $f_{app}(G^2)$ is the partial ordering $v_2 \succ v_4, v_3 \succ v_1$ where G^2 is the path graph $< v_1, v_2, v_3, v_4 >$.

Consider the characterizing conditions for f_{cover} in Theorem 8.1. The independence of equability of equal distance connected arc additions from the other three conditions is shown by f_{stap}, that of partial independence by f_{median}, that of the star condition by $f_{hierdeg}$ and that of appendix domination by f_{app}.

Consider the first characterization of f_{degree} in Theorem 8.4(i). The independence of equability of equal distance arc additions from the other two conditions is shown by f_{stap}, that of partial independence by f_{median} and that of the star condition by $f_{hierdeg}$.

Consider the second characterization of f_{degree} in Theorem 8.4(ii). The independence of partial independence from the other two conditions, is shown by f_{median}, that of invariance of lenticular addition by f_{stap} and that of the star condition by $f_{<}$.

Consider the third characterization of f_{degree} in Theorem 8.4(iii). The independence of partial independence from the other three conditions, is shown by f_{degmed}, that of neutrality by $f_{<}$, that of monotonicity by f_{stap} and that of completeness by f_{cover}.

Consider the fourth characterization of f_{degree} in Theorem 8.4(iv). The independence of neutrality from the other three conditions is shown by $f_{<}$ that of swap invariance by f_{degmed}, that of monotonicity by f_{stap} and that of completeness by f_{cover}.

Consider the characterization of f_{median} in Theorem 8.5. The independence of equability of arc additions from the other two conditions is shown by f_{degree}, that of invariance of lenticular addition by f_1 and that of the star condition by $f_{<}$.

This shows the independence of all conditions in each characterization. The following tables indicate which centrality ordering satisfies which condition. They are straightforward, although cumbersome, to check.

Table 1	f_{cover}	f_{degree}	f_{median}	f_{degmed}	f_{stap}
equability (Sect. 8.3)	$+_1$	$+$	$+$	$+$	$-$
equability (Sect. 8.4)	$-$	$+_2$	$+$	$+$	$-$
equability (Sect. 8.5)	$-$	$-$	$+_6$	$-$	$-$
part. independence	$+_1$	$+_{2,3,4}$	$-$	$-$	$+$
invar. lent. add.	$-$	$+_3$	$+_6$	$+$	$-$
star condition	$+_1$	$+_{2,3}$	$+_6$	$+$	$+$
cover principle	$+$	$+$	$+$	$+$	$-$
app. domination	$+_1$	$+$	$+$	$+$	$+$
neutrality	$+$	$+_{4,5}$	$+$	$+$	$+$
swap invariance	$+$	$+_5$	$-$	$-$	$+$
monotonicity	$+$	$+_{4,5}$	$-$	$+$	$-$
completeness	$-$	$+_{4,5}$	$+$	$+$	$+$

$+_1$ characterize cover centrality ordering

$+_i$ where $i \in \{2, 3, 4, 5\}$, are characterizations of degree centrality ordering

$+_6$ characterize median centrality ordering

Table 2	$f_<$	f_1	$f_{hierdeg}$	f_{app}
equability (Sect. 8.3)	+	+	+	+
equability (Sect. 8.4)	+	+	+	−
equability (Sect. 8.5)	+	+	−	−
part. independence	+	−	+	+
invar. lent. add.	+	−	+	−
star condition	−	+	−	+
cover principle	−	+	−	−
app. domination	−	+	+	−
neutrality	−	−	−	+
swap invariance	+	−	+	+
monotonicity	+	−	+	+
completeness	+	+	+	−

References

Berman, A., and R.J. Plemmons (1979): *Nonnegative matrices in the mathematical sciences.* New York: Academic Press.

Bonacich, P. (1987): "Power and centrality: a family of measures," *American Journal of Sociology*, 92, 1170–1182.

Braun, N. (1997): "A rational choice model of network status," *Social Networks*, 19, 129–142.

Chebotarev, P.Yu., and E. Shamis (1998): "On proximity measures for graph vertices," *Automation and Remote Control*, 59, 1443–1459.

Clever Project (1999): "Hypersearching the Web," *Scientific American*, June, 44–52.

Danilov, V.I. (1994): "The structure of non-manipulable social choice rules on a tree," *Mathematical Social Sciences*, 27,123–131.

Delver, R., H. Monsuur, and A.J.A. Storcken (1991): "Ordering pairwise comparison structures," *Theory and Decision*, 31, 75–94.

Delver, R., and H. Monsuur (2001): "Stable sets and standards of behaviour," *Social Choice and Welfare*, 18, 555–570.

Dutta, B., and J.-F. Laslier (1999): "Comparison functions and choice correspondences," *Social Choice and Welfare*, 16, 513–532.

Faust, K. (1997): "Centrality in affiliation networks," *Social Networks*, 19, 157–191.

Fishburn, P.C. (1977): "Condorcet social choice functions," *SIAM Journal of Applied Mathematics*, 33, 469–489.

Foster, D., and R. Vohra (1998): "An axiomatic characterization of a class of locations on trees," *Operations Research*, 46,347–354.

Freeman, L.C. (1979): "Centrality in social networks, conceptual clarifications," *Social Networks*, 1, 15–239.

Friedkin, N.E. (1991): "Theoretical foundations for centrality measures," *American Journal of Sociology*, 96, 1478–1504.

Gulati, R., and M. Gargiulo (1999): "Where do inter-organizational networks come from?" *American Journal of Sociology*, 104, 1439–1493.

Haynes, T.W., S.T. Hedetniemi, and P.J. Slater (1998). *Fundamentals of domination in graphs.* Marcel Dekker, Inc.

Holzmann, R. (1990): "An axiomatic approach to location on networks," *Mathematics of Operations Research*, 15, 553–563.

Miller, N.R. (1980): "A new solution set for tournament and majority voting: Further graph-theoretical approaches to the theory of voting," *American Journal of Political Science*, 24, 68–96.

McMorris, F.R., H.M. Mulder, R.C. and Powers (2000): "The median function on median graphs and semilattices," *Discrete Applied Mathematics*, 101, 221–230.

Moulin, H. (1986): "Choosing from a tournament," *Social Choice and Welfare*, 3, 271–291.

Papendieck, B., and P. Recht (2000): "On maximal entries in the principle eigenvector of graphs," *Linear Algebra and its Applications*, 310, 129–138.

Peris, J.E., and B. Subiza (1999): "Condorcet choice correspondences for weak tournaments," *Social Choice and Welfare*, 16, 217–231.

Rubinstein, A. (1980): "Ranking the participants in a tournament," *SIAM J. Appl. Math.*, 38, 108–11.

Ruhnau, B. (2000): "Eigenvector centrality, a node centrality?" *Social Networks*, 22, 357–365.

Seidman, S.B. (1985): "Structural consequences of individual position in nondyadic social networks," *Journal of Mathematical Psychology*, 29, 367–386.

Shapley, L.S. (1953): "A value for n-person games," in: H.W. Kuhn and A.W. Tucker (eds.), *Annals of Mathematics Studies*, 28, 307–317.

Storcken, T., and H. Monsuur (2001): "An axiomatic theory of centrality in social networks," Meteor Research Memorandum, RM/01/009, University of Maastricht.

Vohra, R. (1996): "An axiomatic characterization of some locations on trees," *European Journal of Operations Research*, 90, 78–84.

Chapter 9

The Shapley Transfer Procedure for NTU-Games

BY GERT-JAN OTTEN AND HANS PETERS

9.1 Introduction

A cooperative game is described by sets of feasible utility vectors, one set for each coalition. Such a game may arise from each situation where involved parties can achieve gains from cooperation. Examples range from exchange economies to cost allocation between divisions of multinationals or power distribution within political systems. The two central questions are: which coalitions will form; and on which payoffs will each formed coalition agree. Since an answer to the latter question seems a prerequisite to study the former question of coalition formation, most of the literature has concentrated on the question of payoff distribution. Specifically, the usual assumption is that the grand coalition of all players will form and then the question is which payoff vector(s) this coalition will agree upon.

This question has been studied extensively for two special cases: games with transferable utility, and pure bargaining games.

In a game with transferable utility, what each coalition can do is described by just one number: the total utility or payoff, which that coalition can distribute among its members in any way it wants. The underlying assumption is the presence of a common medium of exchange in which the players' utilities are linear. For instance, the payoff is in monetary units and the players have linear utility for money.

P. Borm and H. Peters (eds.), Chapters in Game Theory, 183–203.

In a pure bargaining game intermediate coalitions—coalitions other than the grand coalition or individual players—play no role. Because of these simplifying features, both types of games are easier to analyse than general cooperative games, also called games with nontransferable utility or NTU-games.

A solution is a map that assigns to every game within a certain subclass of NTU-games a feasible payoff vector or set of feasible payoff vectors for the grand coalition. In an important article, Shapley (1969) proposed a procedure to extend single-valued solutions defined on the class of games with transferable utility and satisfying a few minimal conditions, to NTU-games. This procedure works as follows. For a given NTU-game consider any vector of nonnegative weights for the players. For every coalition, maximize the correspondingly weighted sum of the utilities of its members over the set of feasible payoffs of that coalition. Regard these coalitional maxima as a game with transferable utility, and apply the given solution for transferable utility games to this game: those payoff vectors of the original NTU-game that, when similarly weighted, belong to the solution of the TU-game, are defined to be in the solution of the NTU-game. The complete solution of the NTU-game is then obtained by repeating this procedure for every possible weight vector.

This Shapley transfer procedure has been applied in particular to the Shapley value for games with transferable utility (Shapley, 1953), resulting in the 'nontransferable utility value' (Aumann, 1985) for NTU-games. For pure bargaining games, this solution coincides with the Nash bargaining solution, proposed by Nash (1950) for the case of two players. As indicated, however, by Shapley (1969) the procedure can be applied to a variety of solutions; also the existence result established by Shapley holds under quite mild conditions. An explicit example of this is the NTU τ-value studied in Borm *et al.* (1992). Another example is the so called 'inner core' studied by Qin (1994) but proposed earlier by Shapley (1984).

The main objective of the present contribution is twofold. First we review and extend the Shapley transfer procedure; the extension is to any compact and convex valued, continuous solution. Shapley's existence result will be re-established for this extension. Second, we characterize solutions that are obtained by this procedure. This characterization can be seen as an alternative description of the Shapley transfer procedure. The price for obtaining existence and characterization within the same framework is the occurrence of zero weights.

Section 2 introduces the Shapley transfer procedure. Section 3 contains the existence result and a digression on a well known and earlier procedure proposed by Harsanyi (1959,1963)—to which the existence result applies equally. In Section 4 the announced characterization is presented. Section 5 discusses applications to several TU-solutions: the Shapley value, the core, the nucleolus, and the τ-value. Section 6 concludes.

Notations. For a finite subset S of the natural numbers let $\mathbb{R}^S_+$ denote the nonnegative orthant of $\mathbb{R}^S$. For $x, y \in \mathbb{R}^S$ denote $x > y$ if $x_i > y_i$ for every $i \in S$, and denote $x \geq y$ if $x_i \geq y_i$ for every $i \in S$. The vector inequalities $<, \leq$ are defined analogously. The $\cdot$ denotes the usual inner product: $x \cdot y = \sum_{i \in S} x_i y_i$. The product xy denotes the vector in $\mathbb{R}^S$ with ith coordinate equal to $x_i y_i$ $(i \in S)$. For $Y \subseteq \mathbb{R}^S$, $xY := \{xy : y \in Y\}$; for a real number a, $aY := (a, \ldots, a)Y$. For another finite set of natural numbers M with $M \cap S \neq \emptyset$, $x \in \mathbb{R}^S$ and $Y \subseteq \mathbb{R}^S$ let $x_M \in \mathbb{R}^{M \cap S}$ be defined by $(x_M)_i := x_i$ for every $i \in M \cap S$; let $Y_M := \{y_M \in \mathbb{R}^{M \cap S} : y \in Y\}$. Thus, the set Y_M is the projection of Y on the $M \cap S$-plane.

9.2 Main Concepts

Let $N = \{1, 2, \ldots, n\}$ denote the set of *players*. A *coalition* is a nonempty subset of N. A subset $D \subseteq \mathbb{R}^S_+$ is *comprehensive* if $x \in D$ and $y \leq x$ imply $y \in D$ for all $x, y \in \mathbb{R}^S_+$. The *Pareto optimal* subset of D is the set

$$PO(D) := \{x \in D : \text{for all } y \in D, \text{ if } x \leq y \text{ then } x = y\}$$

and the *weakly Pareto optimal* subset of D is the set

$$WPO(D) := \{x \in D : \text{for all } y \in \mathbb{R}^S_+, \text{ if } x < y \text{ then } y \notin D\}.$$

A *nontransferable utility game* or NTU-game is a pair (N, V) where V assigns to every coalition S a *feasible set* $V(S)$ such that

(N1) $V(\{i\}) = \{0\} \subseteq \mathbb{R}$ for every $i \in N$;

(N2) $V(S)$ is a nonempty compact convex and comprehensive subset of $\mathbb{R}^S_+$ for every coalition S;

(N3) $PO(V(S)) = WPO(V(S))$ for every coalition S.

These assumptions, though restrictive, are still quite standard. The normalization in (N1) is mainly for convenience; it is not innocent because together with (N2) it implies, for instance, that every coalition can at least as good as singleton coalitions. The convexity assumption in (N2) may arise from the players having von Neumann-Morgenstern utility functions over uncertain outcomes, or concave ordinal utility functions over bundles of goods. It is essential to what follows. Condition (N3) means that, for every coalition, every weakly Pareto optimal point is also Pareto optimal: there are no flat segments in the weakly Pareto optimal boundary of $V(S)$. One consequence is that if $x \in PO(V(S))$ with $x_i > 0$ for some $i \in S$, then there is a $y \in PO(V(S)$ with $0 < y_i < x_i$ and $y_j > x_j$ for all $j \neq i$; note that in that case $y > 0$. It follows, in particular, that either $V(S) = \{0\}$ or there is a $y \in V(S)$ with $y > 0$.

If $V(S) = \{0\} \subseteq \mathbb{R}^S_+$ for every $S \neq N$, then (N, V) is called a *pure bargaining game.* If for every coalition S there is a nonnegative real number $v(S)$ such that $V(S) = \{x \in \mathbb{R}^S_+ : \sum_{i \in S} x_i \leq v(S)\}$, then (N, V) is called a *game with transferable utility* or TU-game. Such a TU-game is sometimes also denoted by (N, v). Our definition deviates from the usual one in that all payoff vectors are restricted to the nonnegative orthant.

Instead of (N, V) or (N, v) we will usually write V or v, with the understanding that the player set is N.

The class of NTU-games [TU-games, pure bargaining games] with player set N is denoted by $\mathcal{G}^N$ [$\mathcal{T}^N$, $\mathcal{B}^N$]. Often the superscript 'N' is omitted. Subclasses are denoted by $\mathcal{G}'$, $\mathcal{T}'$, etc.

Let $\mathcal{G}' \subseteq \mathcal{G}$ be a subclass of NTU-games. An *NTU-solution* is a correspondence $\Phi : \mathcal{G}' \twoheadrightarrow \mathbb{R}^N_+$ that assigns to each NTU-game $V \in G'$ a set $\Phi(V) \subseteq V(N)$. (We use $\twoheadrightarrow$ to denote a correspondence, i.e., a set-valued function.) If $\mathcal{G}' \subseteq \mathcal{T}$ then Φ is also called a TU-solution. Usually TU-solutions are denoted by small characters, e.g., φ.

A TU-solution φ defined on a class $\mathcal{T}' \subseteq \mathcal{T}$ is *regular* if it satisfies the following three conditions. In condition (T3), for an NTU-game V and a real number $a > 0$ we denote by aV the NTU-game with $(aV)(S) := aV(S)$ for every coalition S.

(T1) $\varphi(V)$ is a nonempty, compact, and convex subset of $PO(V(N))$ for every $V \in \mathcal{T}'$;

(T2) φ is continuous on $\mathcal{T}'$;

(T3) φ is *homogeneous*, that is, for every $V \in \mathcal{T}'$ and real number $a > 0$,

$aV \in \mathcal{T}'$ implies $\varphi(aV) = a\varphi(V)$.

Here, continuity is meant with respect to the restriction to $\mathcal{T}'$ of the Euclidean metric on $\mathbb{R}^{2^N \setminus \{\emptyset\}} \cong \mathcal{T}^N$ and the Hausdorff metric for compact sets in $\mathbb{R}^N$. Conditions (T1), (T2) and (T3) are not very restrictive. Most known single-valued solutions (e.g., Shapley value, nucleolus, τ-value) are continuous and homogeneous on the classes of TU-games on which they are defined. The best known multi-valued concept, the core, satisfies (T1), (T2), and (T3) on the class of balanced games. See Section 5 for some of the details.

For an arbitrary NTU-game V and an arbitrary vector $\lambda \in \mathbb{R}^N_+ \setminus \{0\}$ the associated *λ-transfer game* V^λ is the transferable utility game defined by

$$V^\lambda(S) := \{x \in \mathbb{R}^S_+ : \sum_{i \in S} x_i \leq v^\lambda(S)\}$$

where

$$v^\lambda(S) := \max\{\lambda_S \cdot x : x \in V(S)\}$$

for every coalition S. By (N1) and (N2) these numbers $v^\lambda(S)$ are well defined. For a class $\mathcal{G}'$ of NTU-games denote by

$$T(\mathcal{G}') := \{V^\lambda \in \mathcal{T}^N : V \in \mathcal{G}',\ \lambda \in \mathbb{R}^N_+ \setminus \{0\}\}$$

the class of all TU-games that arise as transfer games associated with NTU-games in $\mathcal{G}'$. Let $\varphi : \mathcal{T}' \twoheadrightarrow \mathbb{R}^N$ be a regular TU-solution defined on $T(\mathcal{G}')$. We extend φ to an NTU-solution $\bar{\varphi} : \mathcal{G}' \twoheadrightarrow \mathbb{R}^N$, as follows. For each $V \in \mathcal{G}'$ and $x \in PO(V(N))$,

$$x \in \bar{\varphi}(V) :\Leftrightarrow \exists \lambda \in \mathbb{R}^N_+ \setminus \{0\}, y \in \varphi(V^\lambda) : y = \lambda x.$$

One way to understand this procedure (cf. Qin, 1994) is to think of the players as countries and of the coordinates of λ as exchange rates between these countries. Then the transfer game V^λ expresses what coalitions of countries can do in real monetary terms, and the vector y represents a payoff distribution in real monetary terms. If this payoff distribution is feasible in terms of the original individual currencies, then it is a solution of the game.

We call $\bar{\varphi}$ the *transfer solution* associated with φ. Observe that not all of the properties in (T1) and (T2) are trivially inherited by $\bar{\varphi}$. In this contribution we will be mainly concerned with existence, i.e., nonemptiness.

The transfer procedure can also be applied to TU-games, resulting in an extension of a TU-solution to the associated transfer solution on TU-games. We observe:

Lemma 9.1 *Let φ be a regular TU-solution on a class $\mathcal{T}'$. Let $V \in \mathcal{T}'$ with $T(\{V\}) \subseteq \mathcal{T}'$. Then $\varphi(V) \subseteq \bar{\varphi}(V)$.*

Proof. Let $x \in \varphi(V)$ and $\lambda = (1, 1, \ldots, 1)$. Then $V^\lambda(S) = V(S)$ for every coalition S, so $\lambda x = x \in \varphi(V) = \varphi(V^\lambda)$. Hence, $x \in \bar{\varphi}(V)$. □

Observe that actually we do not need regularity for Lemma 9.1 to hold. The inclusion in the lemma, however, can be strict, even for regular solutions, as the following example shows.

Example 9.2 Let $N = \{1, 2, 3\}$ and for every $v \in \mathcal{T}^N$ let $\epsilon(v) := \frac{1}{3}\min\{v(N)/2v(\{2,3\}), 1\}$ if $v(\{2,3\}) > 0$ and $\epsilon(v) := \frac{1}{3}$ otherwise. Define $\varphi : \mathcal{T}^N \twoheadrightarrow I\!R^N$ by $\varphi(v) := \{(\frac{1}{3} + 2\epsilon(v), \frac{1}{3} - \epsilon(v), \frac{1}{3} - \epsilon(v))v(N)\}$ for every $v \in \mathcal{T}^N$. Observe that φ is a regular TU-solution. Consider the game w with $w(S) = 1$ for all coalitions S with more than one player. Then $\epsilon(w) = \frac{1}{6}$ and $\varphi(w) = \{(\frac{2}{3}, \frac{1}{6}, \frac{1}{6})\}$. Let $\lambda = (4, 1, 1)$. Then $w^\lambda(N) = 4$, $w^\lambda(\{2,3\}) = 1$, $\epsilon(w^\lambda) = \frac{1}{3}$, and $\varphi(w^\lambda) = \{(4, 0, 0)\}$. Since $\lambda(1, 0, 0) = (4, 0, 0)$, it follows that $(1, 0, 0) \in \bar{\varphi}(w)$. Hence, $\bar{\varphi}(w)$ is strictly larger than $\varphi(w)$. ◁

Note that the possibility of the strict inclusion $\varphi(w) \subset \bar{\varphi}(w)$ is not due to the possibility of zero coordinates of λ, but to the occurrence of boundary solution points with zero coordinates.

In Section 5 we present examples showing that also for TU-solutions such as the core and the nucleolus the inclusion in Lemma 9.1 can be strict: hence, the transfer procedure applied to TU-games may add additional solution outcomes. This will not be the case for the Shapley value or the τ-value.

We conclude this section by reviewing a well known application of this procedure. Let, as above, $\mathcal{B} = \mathcal{B}^N$ be the class of pure bargaining games. The class $T(\mathcal{B})$ of associated transfer games consists of all TU-games v with $v(S) = 0$ for all $S \neq N$. Consider the equal-split solution ε on $T(\mathcal{B})$, that is,

$$\varepsilon(v) := \{\frac{1}{n}(v(N), \ldots, v(N))\} \text{ for every } v \in T(\mathcal{B}).$$

Let V be a pure bargaining game such that $x \in V(N)$ for some $x > 0$. (This is without loss of generality since the only alternative case is $V(N) = \{0\}$, see above.) Then $v^\lambda(N) > 0$ for every $\lambda \in I\!R^N_+\backslash\{0\}$. Let $z \in PO(V(N))$, then $z \in \bar{\varepsilon}(V)$ if, and only if, there is a $\lambda \in I\!R^N_+\backslash\{0\}$ such that $\lambda z = (v^\lambda(N)/n)(1,\ldots,1)$. Since $v^\lambda(N) > 0$ for every λ it follows that both z and λ are positive and hence $\lambda = k(1/z_1,\ldots,1/z_n)$, where $k = v^\lambda(N)/n$, and by definition of $v^\lambda(N)$, there is a hyperplane supporting $V(N)$ at z with normal λ. Consider the product $\prod_{i\in N} x_i$ on $V(N)$. At z, the supporting hyperplane to the level curve of this product has normal $(1/z_1,\ldots,1/z_n)$ as follows straightforwardly by partial differentiation; hence this hyperplane also supports $V(N)$. It follows that the product $\prod_{i\in N} x_i$ is maximized on $V(N)$ at z. So z is the (n-person) Nash bargaining solution outcome for V (Nash, 1950). Thus, $\bar{\varepsilon}(V)$ contains exactly one point, which is the Nash bargaining solution outcome.

9.3 Nonemptiness of Transfer Solutions

In this section we show that under the imposed conditions a transfer solution assigns a nonempty set of payoff vectors to any NTU-game for which it is defined. This result extends Shapley's (1969) existence result to the case where the TU-solution under consideration may be a correspondence. The proof closely follows Shapley's proof.

In order to obtain a compact set we normalize the λ-transfer vectors. Specifically, let $\Delta^N := \{\lambda \in I\!R^N_+ : \sum_{i\in N}\lambda_i = 1\}$ denote the $(n-1)$-dimensional simplex in $I\!R^N$. Elements of Δ^N are also called *weight vectors*.

Theorem 9.3 *Let $V \in \mathcal{G}^N$ and let $\mathcal{T}' := \{V^\lambda : \lambda \in \Delta^N\}$. Let $\varphi : \mathcal{T}' \twoheadrightarrow I\!R^N$ be a regular TU-solution. Then $\bar{\varphi}(V) \neq \emptyset$.*

Proof. Define the correspondence $P : \Delta^N \twoheadrightarrow I\!R^N$ by

$$P(\lambda) := \{\pi \in I\!R^N : \sum_{i\in N} \pi_i = 0,\ \exists x \in \lambda PO(V(N)), y \in \varphi(V^\lambda)[\pi = y - x]\}.$$

So P assigns to a weight vector the set of all 'sidepayments' by which solution payoff vectors of the transfer game are carried over to feasible elements of the correspondingly weighted NTU-game. Note that P is nonempty, convex and compact valued. Moreover, it is upper semicontinuous since $\lambda \mapsto \varphi(V^\lambda)$ is continuous in the Hausdorff metric, and

$\lambda \mapsto PO(V^\lambda(N)) \cap \lambda V(N)$ is upper semicontinuous. These properties are inherited by the correspondence $L : \Delta^N \twoheadrightarrow I\!R^N$ defined by

$$L : \lambda \mapsto \lambda + P(\lambda) := \{\lambda + \pi : \pi \in P(\lambda)\}.$$

In particular this implies compactness of the set $\bigcup_{\lambda\in\Delta^N} L(\lambda)$. Therefore we can find a compact and convex subset D of $\{x \in I\!R^N : \sum_{i\in N} x_i = 1\}$ such that $D \supseteq \Delta^N \cup (\cup_{\lambda\in\Delta^N} L(\lambda))$. Extend the correspondence L to D by defining, for every $d \in D$

$$\begin{aligned} L(d) &:= L(f(d)) \\ &\quad \text{where } f_i(d) := \max\{0, d_i\}/\textstyle\sum_{j\in N}\max\{0, d_j\} \text{ for every } i \in N. \end{aligned}$$

Since the projection f is also continuous, Kakutani's fixed point theorem implies that there is a $d^* \in D$ with $d^* \in L(d^*)$. Let $\lambda^* := f(d^*) \in \Delta^N$.

If $d^* = \lambda^*$, then $0 \in P(\lambda^*)$, hence by definition of P there exists $y \in \varphi(V^{\lambda^*}) \cap \lambda^* PO(V(N))$. So $y \in \bar{\varphi}(V)$ and the proof is complete.

Suppose $d^* \neq \lambda^*$. We will show that this is not possible. In this case there is an $i \in N$ with $\lambda_i^* = 0 > d_i^*$. Since $d^* \in L(d^*) = L(f(d^*)) = L(\lambda^*) = \lambda^* + P(\lambda^*)$, there exists a $\pi^* \in P(\lambda^*)$ with $\pi_i^* < 0$. Moreover, by definition we have $x \geq 0$ for all $x \in \varphi(V^{\lambda^*})$; take $x^* \in \varphi(V^{\lambda^*})$ with $x^* - \pi^* \in \lambda^* PO(V(N))$ then in particular $x_i^* - \pi_i^* > 0$, in contradiction with $\lambda_i^* = 0$. □

By checking the proof of this theorem one observes that it would be valid for any other transfer procedure as long as the resulting TU-games for a specific NTU-game depend continuously on the weight vector. One example of such a procedure is the one underlying the definition of the Harsanyi NTU-value (Harsanyi, 1959, 1963).

In order to define this procedure let V be an NTU-game and let λ be a weight vector. We first assume that λ has only positive coordinates. We recursively define the *dividends* $d_S \in I\!R$ $(\emptyset \neq S \subseteq N)$ as follows: $d_{\{i\}} := 0$ for every $i \in N$, and for S with more than one player:

$$d_S := \max\{t \in I\!R : x^S(t) \in V(S)\},$$

where $x_i^S(t)$ for every $i \in S$ is given by

$$x_i^S(t) = \lambda_i^{-1}(t + \sum_{T: i\in T\subseteq S,\ T\neq S} d_T).$$

Observe that $x_i^{\{i,j\}}(t) = t/\lambda_i$ for all $i, j \in N$, so that $d_{\{i,j\}}(\lambda_i^{-1}, \lambda_j^{-1})$ is the maximal element in the direction reciprocal to $\lambda_{\{i,j\}}$ on the boundary of $V(\{i,j\})$. For coalitions of three and more players the idea is similar, but the starting point is, generally speaking, no longer the origin. Now define, for each coalition S, the vector $x^S := x^S(d_S)$, and let v_d^λ be defined by

$$v_d^\lambda(S) := \begin{cases} \lambda_S \cdot x^S & \text{if } S \neq N \\ \max\{\lambda \cdot x : x \in V(N)\} & \text{if } S = N. \end{cases}$$

Note that there is an asymmetry in this definition between the grand coalition and the smaller coalitions. We comment on this below.

That v_d^λ is actually a TU-game follows immediately since $x^S \in V(S) \subseteq \mathbb{R}_+^S$ by definition. For positive λ the game v_d^λ depends continuously on λ. We still have to define the games v_d^λ for the case where λ has one or more coordinates equal to zero. The following lemma shows that this can be done by taking limits.

Lemma 9.4 *Let $\lambda \in \Delta^N$ and let V be an NTU-game. Then there is a TU-game v such that $v_d^{\lambda^k} \to v$ for any sequence $\lambda^1 > 0, \lambda^2 > 0, \ldots$ in Δ^N with $\lambda^k \to \lambda$.*

Proof. Let $\lambda^1 > 0, \lambda^2 > 0, \ldots$ in Δ^N with $\lambda^k \to \lambda$. Then obviously $v_d^{\lambda^k}(N)$ converges: call the limit $v(N)$. For $S \neq N$ and $k \geq 1$, let $(x^S)^k$ be the vector as defined above associated with λ^k. Since $(x^S)^k$ is in the compact set $V(S)$ for every k, we have $\lim_{k\to\infty} \lambda_S^k \cdot (x^S)^k = \lim_{k\to\infty} \lambda_{S'}^k \cdot (x^S)_{S'}^k$, where $S' := \{i \in S : \lambda_i > 0\}$. Since $(x^S)_{S'}^1, (x^S)_{S'}^2, \ldots$ converges to $x_{S'}^S$ for some $x^S \in V(S)$, the proof is complete by defining $v(S) := \lambda_S \cdot x^S$. □

In view of this lemma we can define a collection of *Harsanyi transfer games* V_d^λ for every $\lambda \in \Delta^N$. The extension of a TU-solution can be defined completely analogous as in the case of the Shapley transfer procedure. Since the Harsanyi transfer games again depend continuously on λ the existence result, Theorem 9.3, also holds for this case:

Theorem 9.5 *Under the assumptions in Theorem 9.3, $\hat{\varphi}(V) \neq \emptyset$, where $\hat{\varphi}$ is the extension of φ according to the Harsanyi procedure.*

One application is to take the Shapley value as a TU-solution: this results in the so-called Harsanyi NTU-value under the Harsanyi transfer procedure. Specifically, if in the definition of the TU-game v_d^λ above we would take $v_d^\lambda(N) = \lambda \cdot x^N$, then x^N would be the Shapley value of the resulting TU-game. If, additionally, x^N maximizes $\lambda \cdot x$ on $V(N)$ then x^N is said to be in the Harsanyi solution. Hence, such points x^N result by applying the Harsanyi transfer procedure to the Shapley TU-solution. See Hart (1985) for a characterization.

9.4 A Characterization

In this section we present a general characterization of NTU-solutions that are obtained by extending regular TU-solutions through the Shapley transfer procedure.

Let Φ be an NTU-solution defined on a class $\mathcal{G}'$ of NTU-games. We list the following possible properties of Φ.

Property 9.6 Φ is *Pareto optimal* if $\Phi(V) \subseteq PO(V(N))$ for every $V \in \mathcal{G}'$.

This property needs no further explanation.

In the following property, for a game V and a positive vector $\alpha \in \mathbb{R}^N$, αV denotes the game defined by $(\alpha V)(S) := \alpha_S V(S)$ for every nonempty coalition $S \subseteq N$.

Property 9.7 Φ is *scale covariant* if for every $V \in \mathcal{G}'$, every $x \in \Phi(V)$, and every $\alpha \in \mathbb{R}^N$ with $\alpha > 0$ we have: if $\alpha V \in \mathcal{G}'$, then $\alpha x \in \Phi(\alpha V)$.

One possible interpretation of this property is that the players have cardinal utility functions that are unique only up to a positive affine transformation.

For the next two properties and the ensuing characterization result we need to introduce some additional notation. For a game V, a nonempty coalition S and a nonnegative vector $\alpha \in \mathbb{R}^N$ denote by $H(V, S, \alpha)$ the halfspace of $\mathbb{R}^S$ of all points on or below the hyperplane with normal α_S supporting $V(S)$ from above, and by $\partial H(V, S, \alpha)$ its boundary. Thus, for any point of tangency $y \in PO(V(S)) \cap H(V, S, \alpha)$,

$$H(V, S, \alpha) = \{x \in \mathbb{R}^S : \alpha_S \cdot x \leq \alpha_S \cdot y\},$$

and

$$\partial H(V, S, \alpha) = \{x \in \mathbb{R}^S : \alpha_S \cdot x = \alpha_S \cdot y\}.$$

For $\alpha \in \mathbb{R}^N$ with $\alpha > 0$, an NTU-game V is called an (α-)*hyperplane game* if for every nonempty coalition S the set $PO(V(S))$ coincides with the nonnegative part of a hyperplane with normal α_S. Note that every TU-game is a hyperplane game.

Property 9.8 Φ is *expansion independent* if for every $V \in \mathcal{G}'$ and $x \in \Phi(V) \cap PO(V(N))$ there is a $\lambda \in \mathbb{R}^N_+ \setminus \{0\}$ with $x \in \partial(H(V, N, \lambda))$ such that, for $L := \{i \in N : \lambda_i > 0\}$ and for all $V' \in \mathcal{G}'$ satisfying $V(S)_L \subseteq V'(S)_L \subseteq H(V, S, \lambda)_L$ for all S with $S \cap L \neq \emptyset$, we have: $x \in \Phi(V')$.

This property captures the essence of the Shapley transfer procedure. If a game is extended by allowing sidepayments that preserve the utility comparison ratios between the players at a certain solution outcome, then that outcome should still belong to the solution of the extended game. The property, naturally, requires this to hold also for games in between the original game and the game extended by sidepayments, although this is not needed for the characterization theorem below. If the utility comparison ratios are not uniquely determined—which is the case if there is no unique supporting hyperplane at the solution outcome—then the requirement applies to at least one set of ratios. A proviso is made for the zero components of the vector λ in the formulation of Property 9.8; by property (N3) of an NTU-game it follows that players outside the set L must have zero at x, and then Property 9.8 implies that these players will stay at zero in the solution.

The expansion independence property was first introduced by Thomson (1981) in the context of pure bargaining problems.

The fourth property is in a sense the 'dual' of Property 9.8.

Property 9.9 Φ is *contraction independent* if for every hyperplane game $V \in \mathcal{G}'$, every $x \in \Phi(V)$, and every $V' \in \mathcal{G}'$ with $x \in V'(N)$ such that, for $L := \{i \in N : x_i > 0\}$, the set $PO(V(S)_L)$ supports $V'(S)_L$ from above for every coalition S with $S \cap L \neq \emptyset$, we have: $x \in \Phi(V')$.

This property is a variant of the well known 'independence of irrelevant alternatives' condition proposed by Nash (1950).

The characterization result is as follows.

Theorem 9.10 *Let Φ be an NTU-solution defined on a class $\mathcal{G}'$ of NTU-games containing all hyperplane games, and let φ be a regular TU-solution on $T(\mathcal{G}')$. The following two statements are equivalent:*

(i) Φ satisfies Properties 9.6–9.9 and $\Phi(V) = \bar{\varphi}(V)$ for every $V \in T(\mathcal{G}')$.

(ii) $\Phi(V) = \bar{\varphi}(V)$ for every $V \in \mathcal{G}'$.

Proof. In this proof we use the following fact, the proof of which is left to the reader.

Fact. Let $V \in \mathcal{G}'$ and let $\alpha, \lambda \in \mathbb{R}^N_+$ with $\alpha > 0$. Then $(\alpha V)^\lambda = V^{\alpha\lambda}$.

Proof of (ii) ⇒ (i). Assume that (ii) holds. To prove (i), we have to show that $\bar{\varphi}$ satisfies Properties 9.6–9.9.

Pareto optimality of $\bar{\varphi}$ follows by definition.

For scale covariance, let $V \in \mathcal{G}'$ and $\alpha \in \mathbb{R}^N$ with $\alpha > 0$ and $\alpha V \in \mathcal{G}'$. Let $x \in \bar{\varphi}(V)$ and $\mu \in \mathbb{R}^N_+ \backslash \{0\}$ with $\mu x \in \varphi(V^\mu)$. Define $\lambda \in \mathbb{R}^N_+ \backslash \{0\}$ by $\lambda_i := \mu_i / \alpha_i$ for every $i \in N$. Then $\lambda\alpha = \mu$, so $\lambda\alpha x \in \varphi(V^{\lambda\alpha}) = \varphi((\alpha V)^\lambda)$ by Fact (i). Hence, $\alpha x \in \bar{\varphi}(\alpha V)$. This implies scale covariance of $\bar{\varphi}$.

For expansion independence, take $V \in \mathcal{G}'$ and $x \in \bar{\varphi}(V)$. Let $\lambda \in \mathbb{R}^N_+ \backslash \{0\}$ with $\lambda x \in \varphi(V^\lambda)$. Then $x \in \partial(H(V, N, \lambda))$. Let L and V' as in the definition of Property 9.8. Then by (N3) $x_{N \backslash L} = 0$, $(V')^\lambda = V^\lambda$ and $x \in PO(V'(N))$. Hence, $x \in \bar{\varphi}(V')$, which proves expansion independence.

Finally, for contraction independence, let $V \in \mathcal{G}'$ be an α-hyperplane game ($\alpha > 0$). Let $x \in \bar{\varphi}(V)$, hence there is a $\lambda \in \mathbb{R}^N_+ \backslash \{0\}$ with $\lambda x \in \varphi(V^\lambda)$. Observe that for $i \in L := \{j \in N : x_j > 0\}$ we have $\lambda_i > 0$, otherwise there would be a y in $V(N)$ with $\lambda \cdot y > \lambda \cdot x$, contradicting $\lambda x \in \varphi(V^\lambda)$. Then for V' as in Property 9.9 it follows that $(V')^\lambda = V^\lambda$. Together with $x \in V'(N)$ this implies $x \in \bar{\varphi}(V')$.

Proof of (i) ⇒ (ii). Assume that (i) holds. Let $V \in \mathcal{G}'$ and $x \in \Phi(V)$. We prove that $x \in \bar{\varphi}(V)$ and, thus, that $\Phi(V) \subseteq \bar{\varphi}(V)$. By Pareto optimality and expansion independence of Φ we can take λ and L as in Property 9.8. Define $\lambda' \in \mathbb{R}^N$ by $\lambda'_i := \lambda_i$ if $i \in L$ and $\lambda'_i := 1$ if $i \in N \backslash L$. For V' as in Property 9.8 take a λ'-hyperplane game. Property 9.8, expansion independence, implies $x \in \Phi(V')$. By scale covariance, $(\lambda')^{-1}x \in \Phi((\lambda')^{-1}V')$, where $(\lambda')^{-1}_i := 1/\lambda'_i$ for every $i \in N$. Since $(\lambda')^{-1}V'$ is a TU-game, this implies $(\lambda')^{-1}x \in \bar{\varphi}((\lambda')^{-1}V')$. Hence, by scale covariance, $x \in \bar{\varphi}(V')$, and by Property 4: $x \in \bar{\varphi}(V)$.

For the converse implication, let $x \in \bar{\varphi}(V)$. We show that $x \in \Phi(V)$, which completes the proof of the theorem. Let $\lambda \in \mathbb{R}^N_+ \backslash \{0\}$ with $\lambda x \in \varphi(V^\lambda)$. By Lemma 9.1, $\varphi(V^\lambda) \subseteq \bar{\varphi}(V^\lambda)$; and since $\bar{\varphi}(V^\lambda) = \Phi(V^\lambda)$, we have $\lambda x \in \Phi(V^\lambda)$. Define $\mu \in \mathbb{R}^N$ by $\mu_i := \lambda_i^{-1}$ if $\lambda_i > 0$ and $\mu_i := 1$ if $\lambda_i = 0$. By scale covariance and noting that $L := \{i \in N : x_i > 0\} \subseteq \{i \in N : \lambda_i > 0\}$, we obtain $x = \mu\lambda x \in \Phi(\mu V^\lambda)$. Now the game μV^λ is a μ^{-1}-hyperplane game and V satisfies the requirements for V' with respect to this hyperplane game as in Property 9.9, contraction independence. Hence, this property implies $x \in \Phi(V)$. □

9.5 Applications

The Shapley transfer procedure and the corresponding results on existence and characterization can be applied to most known solutions for TU-games. Here, we consider applications to the Shapley value, the core, the nucleolus, and the τ-value.

First we state a lemma characterizing the transfer games associated with TU-games. The proof is straightforward and left to the reader. For a vector $\lambda \in I\!R^N_+ \backslash \{0\}$ and a coalition $S \subseteq N$ denote $\bar{\lambda}_S := \max\{\lambda_i : i \in S\}$.

Lemma 9.11 *Let $v \in \mathcal{T}$ and $\lambda \in I\!R^N_+ \backslash \{0\}$. Then $v^\lambda(S) = \bar{\lambda}_S v(S)$ for every coalition $S \subseteq N$. If x is efficient in v, i.e., $\sum_{i \in N} x_i = v(N)$, and $\lambda \cdot x = v^\lambda(N)$, then $x_i = 0$ for every $i \in N$ for which $\lambda_i < \bar{\lambda}_N$.*

9.5.1 The Shapley Value

For a TU-game v the Shapley value σ (Shapley, 1953) is defined by

$$\sigma_i(v) := \sum_{S \subseteq N : i \notin S} \frac{|S|!(n - |S| - 1)!}{n!} [v(S \cup \{i\}) - v(S)]$$

for every $i \in N$, where $v(\emptyset) := 0$, and $|\cdot|$ denotes the cardinality of a finite set. The Shapley TU-solution assigns the set $\{\sigma(v)\}$ to a game v. With some abuse of notation we use σ for the Shapley solution and omit the set-brackets. An alternative definition using dividends (cf. Section 3) is also possible. Within our framework, the Shapley value is well defined as long as $\sigma(v) \geq 0$. Call an NTU-game V *monotonic* if $V(S) \subseteq V(T)_S$ whenever $S \subseteq T$. Hence, a TU-game v is monotonic if $v(S) \leq v(T)$ whenever $S \subseteq T$. Denote the corresponding classes of games by $\mathcal{G}_{\text{mon}}$ and $\mathcal{T}_{\text{mon}}$. Then the Shapley value is well defined on $\mathcal{T}_{\text{mon}}$, and $\mathcal{T}_{\text{mon}} = T(\mathcal{G}_{\text{mon}})$. It is also straightforward to check that σ is a regular TU-solution on $\mathcal{T}_{\text{mon}}$. Moreover, the inclusion in Lemma 9.1 turns out to be an equality, as the following lemma shows.

Lemma 9.12 *Let $v \in \mathcal{T}_{\text{mon}}$. Then $\bar{\sigma}(v) = \sigma(v)$.*

Proof. Let $\lambda \in I\!R^N_+ \backslash \{0\}$ and $x \in I\!R^N_+$ with $\sum_{i \in N} x_i = v(N)$ such that $\lambda x = \sigma(v^\lambda)$. In view of Lemma 9.1 it is sufficient to prove that $x = \sigma(v)$.

Let $M := \{i \in N : \lambda_i = \bar{\lambda}_N\}$. Then, by Lemma 9.11, $v^\lambda(S) = \bar{\lambda}_N v(S)$ for every coalition S with $S \cap M \neq \emptyset$, and $x_i = 0$ for all $i \in N \backslash M$. By monotonicity of v^λ we have $v^\lambda(S) \geq v^\lambda(S \cap M)$ for every coalition S with $S \cap M \neq \emptyset$. Suppose $v^\lambda(S) > v^\lambda(S \cap M)$ for some S, then by definition of the Shapley value there must be an $i \in N \backslash M$ with $\sigma_i(v^\lambda) > 0$ and hence $\lambda_i x_i > 0$, a contradiction. Hence $v^\lambda(S) = v^\lambda(S \cap M)$ for every coalition S with $S \cap M \neq \emptyset$.

We claim that $v(S) = 0$ for every coalition S with $S \subseteq N \backslash M$. Suppose not, then there is a $T \subseteq N \backslash M$ with $v(T) > 0$. Take $i \in M$ arbitrary. By monotonicity, $v(T \cup \{i\}) > 0$, hence $v^\lambda(T \cup \{i\}) = \bar{\lambda}_N v(T \cup \{i\}) > 0$. Therefore, $v^\lambda(\{i\}) = v^\lambda((T \cup \{i\}) \cap M) = v^\lambda(T \cup \{i\}) > 0$, a contradiction since $v(\{i\}) = 0$ by definition. This proves our claim.

Since $v(S) = 0$ whenever $S \cap M = \emptyset$ and $v^\lambda(S) = \bar{\lambda}_N v(S)$ whenever $S \cap M \neq \emptyset$, we have $v^\lambda = \bar{\lambda}_N v$. Also, $\lambda x = \bar{\lambda}_N x$ since $x_i = 0$ for all $i \in N \backslash M$. Hence $\bar{\lambda}_N x = \sigma(\bar{\lambda}_N v)$, so by homogeneity of σ we have $x = \sigma(v)$. □

Application of the results in the preceding sections now yields:

Corollary 9.13 *$\bar{\sigma}(V) \neq \emptyset$ for every $V \in \mathcal{G}_{\text{mon}}$. Moreover, $\bar{\sigma}$ is the unique NTU-solution on $\mathcal{G}_{\text{mon}}$ that satisfies Pareto optimality, scale covariance, expansion and contraction independence and coincides with the Shapley value on $\mathcal{T}_{\text{mon}}$.*

Proof. Theorem 9.3 implies nonemptiness of $\bar{\sigma}$ on $\mathcal{G}_{\text{mon}}$. The second part follows from Theorem 9.10 and Lemma 9.12. □

On the subclass of pure bargaining games $\mathcal{B} \subseteq \mathcal{G}_{\text{mon}}$, $\bar{\sigma}$ coincides with the Nash bargaining solution: see the last part of Section 2.

An earlier characterization of $\bar{\sigma}$ (also called the Shapley NTU-value) was given by Aumann (1985). This characterization presumes existence and makes use of specific properties of the Shapley value; it also uses the standard concept of unbounded TU-games.

9.5.2 The Core

The *core* of a TU-game v is defined by

$$\begin{aligned} c(v) \ := \ & \{x \in \mathbb{R}^N : \textstyle\sum_{i \in N} x_i = v(N) \\ & \text{and } \textstyle\sum_{i \in S} x_i \geq v(S) \text{ for every coalition } S\}. \end{aligned}$$

More generally, the *core* of an NTU-game V is defined by

$$C(V) := \{x \in PO(V(N)) : \forall \emptyset \neq S \neq N[x_S \notin \text{int}V(S)]\}$$

where 'int' denotes the topological interior. Nonemptiness of the core is closely connected to the idea of balancedness. A collection of nonnegative numbers $\{\gamma_S : S \text{ a coalition}\}$ is called *balanced* if $\sum_{S:i\in S} \gamma_S = 1$ for every player $i \in N$. An NTU-game V is called *balanced* if for every balanced collection $\{\gamma_S\}$ we have $\sum_S \gamma_S \tilde{V}(S) \subseteq V(N)$, where $\tilde{V}(S) \subset I\!R^N$ is constructed from $V(S)$ by adding zeros for players outside S. It is well known (Bondareva, 1963; Shapley, 1967) that a TU-game has a nonempty core if and only if it is balanced. For an NTU-game, balancedness—and even a weaker balancedness condition, cf. Scarf (1967)—implies nonemptiness of the core, but not the other way around.

Let $\mathcal{T}_{\text{bal}}$ denote the class of balanced TU-games, i.e., TU-games with nonempty cores. The TU-solution c is regular, as is easy to verify. Let $\mathcal{G}_{\text{bal}}$ denote the class of balanced NTU-games. By slightly adapting an argument of Qin (1994)[1] it can be shown that $V \in \mathcal{G}_{\text{bal}}$ if and only if $T(\{V\}) \subseteq \mathcal{T}_{\text{bal}}$. In words, an NTU-game is balanced if and only if all associated transfer games are balanced.

Corollary 9.14 *$\bar{c}(V) \neq \emptyset$ for every $V \in \mathcal{G}_{\text{bal}}$. Moreover, $\bar{c}$ is the unique NTU-solution on $\mathcal{G}_{\text{bal}}$ that satisfies Pareto optimality, scale covariance, expansion and contraction independence and coincides with $\bar{c}$ on $\mathcal{T}_{\text{bal}}$.*

Proof. Theorem 9.3 implies nonemptiness of $\bar{c}$ on $\mathcal{G}_{\text{bal}}$. The second part follows from Theorem 9.10. □

In this case, applying the transfer procedure on TU-games may add solution outcomes, as the following example shows.

Example 9.15 Consider the four-person TU-game v with player set $N = \{1, 2, 3, 4\}$ and with $v(N) = 2$, $v(S) = 1$ if $|S| = 3$ or $S \in \{\{1, 2\}, \{3, 4\}\}$, and $v(S) = 0$ otherwise. This is a monotonic game with core equal to

$$c(v) = \{x \in I\!R^N_+ : x_1 + x_2 = 1, x_3 + x_4 = 1\}.$$

Let $\lambda = (1, 1, 0, 0)$, then v^λ is equal to v except that $v^\lambda(\{3, 4\}) = 0$. Now, for $x = (1, 1, 0, 0)$, $\lambda x = (1, 1, 0, 0) \in c(v^\lambda)$, hence $x = (1, 1, 0, 0) \in \bar{c}(v)$ but $x \notin c(v)$. ◁

[1] Attributed to Shapley.

Example 9.15 also implies that the core $C(V)$ of an NTU-game V does not have to contain $\bar{c}(V)$. Also the converse is not true:

Example 9.16 Consider the three-person NTU-game V with player set $N = \{1,2,3\}$ and with $V(N) = \{x \in I\!R^N_+ : x_1+x_2+x_3 \leq 3\}$, $V(\{1,2\}) = \{x \in I\!R^{\{1,2\}}_+ : x_1 + 2x_2 \leq 3\}$, and $V(S) = \{0\}$ otherwise. Note that $(1,1,1) \in C(V)$. The only possible transfer game through which we could obtain $(1,1,1) \in \bar{c}(V)$ would be one corresponding to $\lambda = (1,1,1)$ (or a positive multiple of that vector). For this transfer game we have $v^\lambda(N) = 3$ and $v^\lambda(\{1,2\}) = 3$, so that $\lambda(1,1,1) = (1,1,1) \notin c(v^\lambda)$, hence $(1,1,1) \notin \bar{c}(V)$. ◁

Example 9.16 still works if we replace the game V by V' with as the only difference that now $V'(\{1,2\}) = \{x \in I\!R^{\{1,2\}}_+ : x_1 + 3x_2 \leq 4\}$. In that case, however, the resulting $(1,1,1)$-transfer game has an empty core and therefore $V' \notin \mathcal{G}_{\text{bal}}$. The latter fact follows also directly by considering the collection $\gamma_{\{1,2\}} = 1$, $\gamma_{\{3\}} = 1$, and $\gamma_S = 0$ otherwise. This shows that if an NTU-game has a nonempty core, then this property is not necessarily inherited by the associated transfer games.[2]

9.5.3 The Nucleolus

The *nucleolus* (Schmeidler, 1969) for a TU-game v is defined as follows. For every Pareto optimal payoff vector x, arrange the so-called excesses $v(S) - \sum_{i \in S} x_i$ $(\emptyset \neq S \neq N)$ in a nonincreasing order. Then compute x such that the thus associated vector of excesses is lexicographically minimal: the resulting payoff vector is the nucleolus of the game. If the game has a nonempty core, then the nucleolus is in the core.

The nucleolus on $\mathcal{T}$ is a regular TU-solution, so Theorems 9.3 and 9.10 apply again. Consequently, denoting the nucleolus by ν, we have:

Corollary 9.17 *$\bar{\nu}(V) \neq \emptyset$ for every $V \in \mathcal{G}$. Moreover, $\bar{\nu}$ is the unique NTU-solution on $\mathcal{G}$ that satisfies Pareto optimality, scale covariance, expansion and contraction independence and coincides with $\bar{\nu}$ on $\mathcal{T}$.*

Just as was the case with the Shapley value, also the transfer solution associated with the nucleolus coincides with the Nash bargaining solution

[2] Qin (1994) also studies an extension of the core to NTU-games by applying the concept of transfer games. These transfer games are (λ-) hyperplane games rather than TU-games, and his approach also differs from ours since in the transfer game the feasible set of a coalition of players with zero weights is unbounded. In particular, such a game will always have an empty core.

on the subclass of pure bargaining games (see the last part of Section 2).

Like in the case of the core the transfer procedure may add outcomes to TU-games, as is illustrated by the next example.

Example 9.18 Consider the four-person TU-game with $N = \{1,2,3,4\}$, $v(N) = v(\{1,2\}) = (\{3,4\}) = 1$, and $v(S) = 0$ otherwise. Then $\nu(v) = \frac{1}{4}(1,1,1,1)$ as is easily derived by symmetry. Take $\lambda = (1,1,0,0)$, then v^λ is equal to v except that now $v^\lambda(\{3,4\}) = 0$. By symmetry and the fact that the nucleolus is in the core, $\nu(v^\lambda) = \frac{1}{2}(1,1,0,0)$. Hence $\nu(v^\lambda) = \lambda\frac{1}{2}(1,1,0,0)$ so that $\frac{1}{2}(1,1,0,0) \in \bar{\nu}(v)$. ◁

Observe that the game v in this example is not balanced. It is an open question to find an example with a balanced TU-game.

9.5.4 The τ-Value

The τ-value for TU-games (Tijs, 1981, Borm *et al.*, 1992) is defined as follows. For a TU-game v, define the 'utopia vector' $M(v) \in \mathbb{R}^N$ by $M_i(v) = v(N) - v(N\backslash\{i\})$ and the 'minimal right vector' $m(v) \in \mathbb{R}^N$ by $m_i(v) = \max_{S:i\in S}(v(S) - \sum_{j\in S\backslash\{i\}} M_j(v))$ for every $i \in N$. Then the τ-value $\tau(v)$ is the unique Pareto optimal point on the line segment with $m(v)$ and $M(v)$ as endpoints, if such a point exists and if $m(v) \leq M(v)$. Games for which these two conditions are satisfied are called quasi-balanced. It can be shown that that every balanced game is quasi-balanced. By $\mathcal{T}_{\text{qbal}}$ we denote the class of quasi-balanced TU-games.

We will show that transfer games associated with quasi-balanced TU-games are again quasi-balanced. First, we derive some inequalities concerning the utopia and minimal right vectors of transfer games.

Lemma 9.19 *Let v be a TU-game and $\lambda \in \mathbb{R}^N_+\backslash\{0\}$. Then*

$M_i(v^\lambda) \geq \bar{\lambda}_N M_i(v)$ *for all* $i \in N$

and

$m_i(v^\lambda) \leq \bar{\lambda}_N m_i(v)$ *for all* $i \in N$.

Proof. Let $i \in N$. Then, by Lemma 9.11,

$$\begin{aligned} M_i(v^\lambda) &= v^\lambda(N) - v^\lambda(N\backslash\{i\}) \\ &= \bar{\lambda}_N v(N) - \bar{\lambda}_{N\backslash\{i\}} v(N\backslash\{i\}) \\ &\geq \bar{\lambda}_N[v(N) - v(N\backslash\{i\})] \\ &= \bar{\lambda}_N M_i(v) \end{aligned}$$

and

$$\begin{aligned} m_i(v^\lambda) &= \max_{S:S\ni i}[v^\lambda(S) - \sum_{j\in S\setminus\{i\}} M_j(v^\lambda)] \\ &\leq \max_{S:S\ni i}[\bar{\lambda}_S v(S) - \sum_{j\in S\setminus\{i\}} \bar{\lambda}_N M_j(v)] \\ &\leq \max_{S:S\ni i} \bar{\lambda}_N[v(S) - \sum_{j\in S\setminus\{i\}} M_j(v)] \\ &= \bar{\lambda}_N m_i(v). \end{aligned}$$

Here, the before-last inequality follows from the first part of the proof. □

Lemma 9.20 *Let $v \in \mathcal{T}_{\text{qbal}}$ and $\lambda \in \mathbb{R}^N_+\setminus\{0\}$. Then $v^\lambda \in \mathcal{T}_{\text{qbal}}$.*

Proof. Let $i \in N$. Then by Lemma 9.19 and the fact that $v \in \mathcal{T}_{\text{qbal}}$,

$$m_i(v^\lambda) \leq \bar{\lambda}_N m_i(v) \leq \bar{\lambda}_N M_i(v) \leq M_i(v^\lambda),$$

and

$$\sum_{j\in N} m_j(v^\lambda) \leq \bar{\lambda}_N \sum_{j\in N} m_i(v) \leq \bar{\lambda}_N v(N) \leq \bar{\lambda}_N \sum_{j\in N} M_j(v) \leq \sum_{j\in N} M_i(v^\lambda),$$

hence $\sum_{j\in N} m_j(v^\lambda) \leq v^\lambda(N) \leq \sum_{j\in N} M_i(v^\lambda)$. So $v^\lambda \in \mathcal{T}_{\text{qbal}}$. □

We next show that the Shapley transfer procedure does not add solution outcomes to TU-games. Cf. Lemma 9.12, where we prove this for the Shapley value.

Lemma 9.21 *Let $v \in \mathcal{T}_{\text{qbal}}$. Then $\bar{\tau}(v) = \tau(v)$.*

Proof. Let $\lambda \in \mathbb{R}^N_+\setminus\{0\}$ and $x \in \mathbb{R}^N_+$ with $\sum_{i\in N} x_i = v(N)$ such that $\lambda x = \tau(v^\lambda)$. In view of Lemma 9.1 it is sufficient to prove that $x = \tau(v)$. Let $M := \{i \in N : \lambda_i = \bar{\lambda}_N\}$ then, by Lemma 9.11, $x_i = 0$ for all $i \in N\setminus M$. Hence,

$$\tau_i(v^\lambda) = 0 \text{ for all } i \in N\setminus M. \tag{9.1}$$

Let $\alpha \in [0, 1]$ such that $\tau(v^\lambda) = \alpha M(v^\lambda) + (1 - \alpha)m(v^\lambda)$.
Case (a): $\alpha = 0$.
Then

$$\tau(v^\lambda) = m(v^\lambda), \tag{9.2}$$

and

$$\bar{\lambda}_N v(N) \geq \bar{\lambda}_N \sum_{j \in N} m_j(v) \geq \sum_{j \in N} m_j(v^\lambda) = v^\lambda(N) = \bar{\lambda}_N v(N), \tag{9.3}$$

where the second inequality follows from Lemma 9.19, the second equality from Lemma 9.11, and the first equality from (9.2). Hence, all inequalities in (9.3) are equalities. In particular, $m(v)$ is efficient in v, so $\tau(v) = m(v) = (1/\bar{\lambda}_N)\tau(v^\lambda)$. So by (9.1), $x_i = 0$ and $\tau_i(v) = 0$ for all $i \in N \backslash M$, so $x_i = \tau_i(v)$ for these i. For $i \in M$, $x_i = (1/\lambda_i)\tau_i(v^\lambda) = (1/\bar{\lambda}_N)\tau_i(v^\lambda)$, hence $x_i = \tau_i(v)$. Altogether, $x = \tau(v)$.

Case (b): $\alpha > 0$.
Then for $i \in N\backslash M$, by (9.1), $m_i(v^\lambda) = M_i(v^\lambda) = 0$, hence by Lemma 9.11, $\bar{\lambda}_N v(N) = v^\lambda(N) = v^\lambda(N\backslash\{i\}) = \bar{\lambda}_N v(N\backslash\{i\})$ so that $M_i(v) = 0$. Thus,

$$M_i(v) = \tau_i(v) = 0 = x_i \text{ for all } i \in N\backslash M. \tag{9.4}$$

Now let $i \in M$. First suppose $|M| > 1$. Then $M_i(v^\lambda) = \bar{\lambda}_N v(N) - \bar{\lambda}_N v(N\backslash\{i\}) = \bar{\lambda}_N M_i(v)$ and
$m_i(v^\lambda) = \max_{S:S\ni i}[v^\lambda(S) - \sum_{j\in S\backslash\{i\}} M_j(v^\lambda)] = \max_{S:S\ni i}[\bar{\lambda}_N v(S) - \sum_{j\in S\backslash\{i\}} \bar{\lambda}_N M_j(v)] = \bar{\lambda}_N m_i(v)$. Hence $\bar{\lambda}_N x_i = \tau_i(v^\lambda) = \bar{\lambda}_N \tau_i(v)$, so that $x_i = \tau_i(v)$. This concludes the proof for $|M| > 1$. If $M = \{i\}$, then by (9.4) and efficiency of the τ value, $\tau_i(v) = v(N)$. Because of $\lambda x = \tau(v^\lambda)$, (9.4) and efficiency, we have $\tau_i(v^\lambda) = v^\lambda(N) = \bar{\lambda}_N x_i$, hence $x_i = v(N) = \tau_i(v)$. This concludes the proof of the lemma. □

Denote by $\mathcal{G}_{\text{qbal}}$ the class of NTU-games such that $T(\mathcal{G}_{\text{qbal}}) = \mathcal{T}_{\text{qbal}}$. By Lemma 9.20 this class contains $\mathcal{T}_{\text{qbal}}$. Moreover, $\mathcal{G}_{\text{qbal}}$ contains the class $\mathcal{G}_{\text{bal}}$ of all balanced NTU-games, since every transfer game associated with a balanced NTU-game is balanced and therefore also quasi-balanced.

We have:

Corollary 9.22 *$\bar{\tau}(V) \neq \emptyset$ for every $V \in \mathcal{G}_{\text{qbal}}$. Moreover, $\bar{\tau}$ is the unique NTU-solution on $\mathcal{G}_q$ that satisfies Pareto optimality, scale covariance, expansion and contraction independence and coincides with the τ-value on $\mathcal{T}_{\text{qbal}}$.*

Proof. Theorem 9.3 implies nonemptiness of $\bar{\tau}$ on $\mathcal{G}_{\text{qbal}}$. The second part follows from Theorem 9.10 and Lemma 9.21. □

Since on $T(\mathcal{B})$, where as before $\mathcal{B}$ is the class of pure bargaining games, the τ-value coincides with the equal-split solution, it follows again that the transfer solution $\bar{\tau}$ coincides with the Nash bargaining solution on the class of pure bargaining games.

9.6 Concluding Remarks

The main objective of this contribution was to provide existence and characterization of NTU-solutions obtained from TU-solutions by the Shapley transfer procedure within one and the same framework. The price paid for this is the allowance of zero weights and the associated technical problems. The benefit is that the results can be applied to many TU-solutions: see Corollaries 9.13–9.22.

The approach followed above can be modified in many ways. If existence is less of an issue then we may restrict attention to only positive weights and consider other classes of games: this is the approach usually adopted in the literature. Also the transfer procedure may be varied: cf. the Harsanyi procedure as discussed in Section 3, or the procedure used to extend the so-called consistent value—which coincides with the Shapley value on TU-games—to NTU-games (see Maschler and Owen, 1992).

References

Aumann, R.J. (1985): "An axiomatization of the non-transferable utility value,", *Econometrica*, 53, 599–612.

Bondareva, O.N. (1963): "Some applications of linear programming methods to the theory of cooperative games," *Problemy Kibernetiki*, 10, 119–139.

Borm, P., H. Keiding, R.P. McLean, S. Oortwijn, and S.H. Tijs (1992): "The Compromise Value for NTU-Games," *International Journal of Game Theory*, 21, 175–189.

Harsanyi, J.C. (1959): "A bargaining model for the cooperative n-person game," *Annals of Mathematics Studies*, Princeton University Press, Princeton, 40, 325–355.

Harsanyi, J.C. (1963): "A simplified bargaining model for the n-person cooperative game," *International Economic Review*, 4, 194–220.

Hart, S. (1985): "An axiomatization of Harsanyi's nontransferable utility solution," *Econometrica*, 53, 1295–1313.

Maschler, M., and G. Owen (1992): "The consistent value for games without side payments," in: R. Selten (ed.), *Rational Interaction*, 5–12. New York: Springer Verlag.

Nash, J.F. (1950): "The bargaining problem," *Econometrica*, 18, 155–162.

Qin, C.-Z. (1994): "The inner core of an n-person game," *Games and Economic Behavior*, 6, 431–444.

Scarf, H. (1967): "The core of an n-person game," *Econometrica*, 35, 50–67.

Schmeidler, D. (1969): "The nucleolus of a characteristic function game," *SIAM Journal of Applied Mathematics*, 17, 1163–1170.

Shapley, L.S. (1953): "A value for n-person games," in: H. Kuhn, A.W. Tucker (eds.), *Contributions to the Theory of Games*, Princeton University Press, Princeton, 307–317.

Shapley, L.S. (1967): "On balanced sets and cores," *Naval Research Logistics Quarterly*, 14, 453–460.

Shapley, L.S. (1969): "Utility comparison and the theory of games," in: G.Th. Guilbaud (ed.), *La Décision*. Editions du CNRS, Paris.

Shapley, L.S. (1984): "Lecture notes on the inner core," Department of Mathematics, University of California, Los Angeles.

Thomson, W. (1981): "Independence of irrelevant expansions," *International Journal of Game Theory*, 10, 107–114.

Tijs, S.H. (1981): "Bounds for the core and the τ-value," in: O. Moeschlin and D. Pallasche (eds.), *Game Theory and Mathematical Economics*, 123–132. Amsterdam: North-Holland.

Chapter 10

The Nucleolus as Equilibrium Price

BY JOS POTTERS, HANS REIJNIERSE, AND ANITA VAN GELLEKOM

10.1 Introduction

The exchange economies studied in this chapter find their origins in Debreu (1959). They have a finite set of *agents* N and a finite set of *indivisible goods* Q. Besides there is an infinitely divisible good referred to as 'money'. It can be used to 'transfer utility' from one agent to another agent: the marginal utility of money does not depend on the agent nor his wealth.

We introduce the notions of a *stable* equilibrium (with respect to a price vector) and a *regular* price. Stable equilibria are robust in the sense that they are not affected by any increase of the money supply. A price vector is regular if it can be considered to be a shadow-price of the linear program corresponding to the economy. We show that price vectors that support the stability of an equilibrium are regular. Furthermore, conditions on the economy are provided such that reallocations maximizing so-called *social welfare* can be extended to a stable equilibrium by any regular price.

Economies of the considered type do not necessarily have equilibria, but regular prices can always be found. A particular one will be defined by means of the *nucleolus*. The existence of algorithms to calculate the nucleolus facilitates the task to find a price vector.

P. Borm and H. Peters (eds.), Chapters in Game Theory, 205–222.

The nucleolus is introduced by Schmeidler (1969), primarily as a tool to prove the nonemptiness of the *bargaining set* (Aumann and Maschler, 1964). Gradually, the nucleolus became a one-point solution rule in its own right for TU-games with nonempty *imputation set*. It shares with the Shapley value the properties of *efficiency, dummy player*, *symmetry* and *anonymity* but does not satisfy some other properties of the Shapley value, like *additivity* (Shapley, 1953) and *strong monotonicity* (Young, 1985). On the other hand, the nucleolus is a *core allocation* whenever the core is nonempty, and it satisfies the *reduced game property* in the sense of Snijders (1992).

As a solution rule the nucleolus is an expression of *one-sided egalitarianism on coalition level.* It is an attempt to treat all coalitions equally in the sense that they exceed or fall short to their coalition values with the same amount. When this is not possible, it exhibits the tendency to help the 'poorer' coalitions (coalitions with a high *excess*).

When computing the nucleolus, it turns out that only a few coalition values have an influence on the position of the nucleolus: apart from the grand coalition there is a collection of at most $2(n-1)$ coalitions that determine the nucleolus (Reijnierse and Potters, 1998). Here, n represents the number of players.

In the early nineties (cf., among other papers, Potters and Tijs, 1992; Maschler *et al.*, 1992) several other types of (pre-)nucleolus concepts were introduced. One of these is the $\mathcal{B}$-nucleolus. The difference with the standard pre-nucleolus is that only the excesses of coalitions $S \in \mathcal{B}$ are taken into account. A consequence of the restriction to coalitions in $\mathcal{B}$ is that the $\mathcal{B}$-nucleolus can be empty or can consist of more than one point. By applying a result of Derks and Reijnierse (1998), we provide necessary and sufficient conditions for the $\mathcal{B}$-nucleolus to be a singleton.

In Potters and Reijnierse (1998) the idea of the $\mathcal{B}$-nucleolus is used to simplify the computation of the nucleolus. For certain classes of TU-games one can find relatively small collections $\mathcal{B}$ such that the $\mathcal{B}$-nucleolus coincides with the nucleolus. When the size of $\mathcal{B}$ is much smaller than 2^n, (e.g., polynomial in n) the computation of the $\mathcal{B}$-nucleolus has a lower complexity than the computation of the nucleolus. E.g., for *assignment games* the one-coalitions and the mixed two-coalitions 'determine the nucleolus'.

The present contribution shows another application of the $\mathcal{B}$-nucleolus. It will be shown that in economies with indivisible goods, money and—what is called—*quasi-linear* utility functions a (potential) equilibrium price can be computed by computing a $\mathcal{B}$-nucleolus in an 'associated' TU-game. The collection $\mathcal{B}$ will be polynomial in the number of agents, but exponential in the number of goods.

With such an economy a partial TU-game $(\bar{N}, \mathcal{B}, \bar{v})$ is associated with the following features:

- the player set $\bar{N}$ equals $N \cup Q$,
- the collection $\mathcal{B}$ of participating coalitions is $\{T \subseteq \bar{N} : |T \cap N| = 1\} \cup \{\bar{N}\}$,
- $\bar{v}: \mathcal{B} \to \mathbb{R}$ is defined by: $\bar{v}(C \cup \{i\})$ is the utility of agent i for bundle C,
- $\bar{v}(\bar{N})$ is taken sufficiently large to guarantee the nonemptiness of the $\mathcal{B}$-core of $(\bar{N}, \bar{v})$.

If $(x, y) \in \mathbb{R}^Q \times \mathbb{R}^N$ is the $\mathcal{B}$-nucleolus of $(\bar{N}, \mathcal{B}, \bar{v})$, the part $x \in \mathbb{R}^Q$ can be understood as a price vector. If the exchange economy has a (stable) price equilibrium, the vector x is one of the possible equilibrium price vectors.

The chapter consists of the following sections. The preliminaries introduce the type of exchange economies to be studied and repeat the basic definitions in this area; it also contains a short recapitulation of the main concepts from the theory of TU-games.

Section 10.3 provides two properties economies can have. One of them, the *SW-condition,* is necessary for the existence of stable price equilibria, together they are sufficient. This part is a generalization of the results of Bikhchandani and Mamer (1997). It will be shown how *regular price vectors*, potential equilibrium price vectors, look like. Section 10.4 gives the proofs of the theorems of Section 10.3.

Section 10.5 considers the partial TU-game $(\bar{N}, \mathcal{B}, \bar{v})$ and proves that x is a regular price vector if (x, y) is the $\mathcal{B}$-nucleolus of $(\bar{N}, \mathcal{B}, \bar{v})$. If $\mathcal{B}_{\max}$ is defined to be the collection of coalitions in $\mathcal{B}$ with maximal excess, the SW-condition holds if and only if $\mathcal{B}_{\max}$ *contains a partition of* $\bar{N}$.

10.2 Preliminaries

This section consists of two parts. The first part introduces the type of exchange economies that will be considered. The second part recalls the definitions of some concepts of the theory of TU-games.

10.2.1 Economies with Indivisible Goods and Money

The economies we consider in this chapter have the following features:

- There is a finite set of *agents* N, $n = |N|$ and $i, j, k, \ldots \in N$.
- There is a finite set of *indivisible goods* Q, $q = |Q|$ and $\alpha, \beta, \gamma, \ldots \in Q$.
- Each agent $i \in N$ has an *initial endowment* (A_i, m_i); $A_i \subseteq Q$ denotes the set of goods initially held by agent i and $m_i \geq 0$ is the amount of money agent i has in the beginning. We assume that $\{A_i\}_{i \in N}$ is a *distribution* of Q, i.e., $A_i \cap A_j = \emptyset$ whenever $i \neq j$, and $\bigcup_{i \in N} A_i = Q$. We allow, however, that $A_i = \emptyset$ for some agents i.
- Each agent $i \in N$ has a *preference relation* $\preceq_i$ on the set of *commodity bundles* (B, x) with $\emptyset \subseteq B \subseteq Q$ and $x \in \mathbb{R}_+$. We assume that $\preceq_i$ can be represented by a *utility function* U_i of the form ($i \in N$, $x \in \mathbb{R}_+$):
  $U_i(B, x) := V_i(B) + x$, ($\emptyset \subseteq B \subseteq Q$) (separability of money),
  $V_i(B) \leq V_i(C)$, whenever $B \subseteq C$ and $i \in N$ (monotonicity),
  $V_i(\emptyset) := 0$.

Because of the last assumption, an economy $\mathcal{E}$ is determined by N, Q and $\{A_i, m_i, V_i\}_{i \in N}$:

$$\mathcal{E} := (N, Q, \{A_i, m_i, V_i\}_{i \in N}).$$

Comment: Separability of money is the most restrictive condition. In fact, it induces four properties, namely:

- *separability per se*, saying that $U_i(B, x) = V_i(B) + W_i(x)$ for some function W_i, defined on $\mathbb{R}_+$,
- *strict monotonicity in money*, saying that W_i is strictly monotonic,
- the possibility of *interpersonal comparison of utility,* which is expressed by $W_i = W_j$, and
- the property that money can be used as a physical means to *transfer utility* because the marginal utility for money is constant (and, by scaling, set to be 1).

By trading, a *coalition* $T \subseteq N$ can realize any redistribution of the goods in $\bigcup_{j\in T} A_j$ and any redistribution of the money supply $\sum_{j\in T} m_j$. We call such a twofold redistribution a *T-reallocation*. So,
a T-reallocation $\{(C_j, y_j)\}_{j\in T}$ must satisfy:

$$\bigcup_{j\in T} C_j = \bigcup_{j\in T} A_j, \qquad \sum_{j\in T} y_j = \sum_{j\in T} m_j \qquad \text{and} \qquad y \geq 0.$$

An N-reallocation $\{(B_i, x_i)\}_{i\in N}$ is a *price equilibrium*, if there exists a *price vector* $p \in I\!R^Q$ with the following properties:

$$\begin{array}{rcl}
\text{(i) } p(B_i) + x_i & \leq & p(A_i) + m_i \text{ for all } i \in N \text{ (budget constraints)} \\
\text{(ii) if } U_i(C, y) & > & U_i(B_i, x_i) \text{ for some } C \subseteq Q,\ y \in I\!R_+ \text{ and } i \in N, \text{ then} \\
p(C) + y & > & p(A_i) + m_i \text{ (maximality conditions).}
\end{array}$$

Here, $p(C)$ is an abbreviation of $\sum_{\alpha\in C} p_\alpha$. By the strict monotonicity of the utility functions U_i, the maximality conditions imply that the budget constraints are, in fact, equalities. Furthermore, by the monotonicity of the reservation prices $V_i(C)$, an equilibrium price is nonnegative.

10.2.2 Preliminaries about TU-Games

A *transferable utility game* or *TU-game* is a pair (N, v) consisting of a finite player set N and a map v: $2^N \to I\!R$ with $v(\emptyset) = 0$. In a *partial TU-game* $(N, \mathcal{B}, v)$ the map v is only defined on a collection of coalitions $\mathcal{B}$ containing N. The *$\mathcal{B}$-core* of a partial game $(N, \mathcal{B}, v)$ consists of all vectors $x \in I\!R^N$ with $x(N) = v(N)$ and $x(S) \geq v(S)$ for all $S \in \mathcal{B}$. For partial TU-games the pre-imputation set $I^*(N, v)$ consists of all vectors $x \in I\!R^N$ with $x(N) = v(N)$. The *excess* of a pre-imputation x with respect to a coalition $S \in \mathcal{B}$ and the partial game v: $\mathcal{B} \to I\!R$ is:

$$\text{exc}(S, x \mid v) := v(S) - x(S).$$

For $x \in I^*(N, v)$ we define $E(x \mid v) := \{\text{exc}(S, x \mid v)\}_{S\in\mathcal{B}} \in I\!R^{\mathcal{B}}$. Let $b := |\mathcal{B}|$ and let θ: $I\!R^{\mathcal{B}} \to I\!R^b$ be the map that orders the coordinates of each vector of $I\!R^{\mathcal{B}}$ in a weakly decreasing order. Let $\preceq_{\text{lex}}$ be the lexicographic order[1] on $I\!R^b$. The *$\mathcal{B}$-nucleolus of* $(N, \mathcal{B}, v)$ consists of all pre-imputations that are lexicographically optimal:

$\text{Nu}(N, \mathcal{B}, v) := \{x \in I^*(v) : \theta \circ E(x \mid v) \preceq_{\text{lex}} \theta \circ E(y \mid v)$ for all $y \in I^*(N, v)\}$.

[1] I.e. $x \preceq_{\text{lex}} y$ if $x = y$ or if $x_j < y_j$, j being the *first* coordinate at which x and y differ.

In Maschler *et al.* (1992) also the nucleolus $\mathrm{Nu}(N, \mathcal{B}, v, \Pi)$ is defined in a similar way, with the exception that only allocations in a closed subset Π of the pre-imputation set are possible candidates.

Unlike the regular nucleolus, the $\mathcal{B}$-nucleolus can be empty or consist of more than one point. However, by applying a result due to Derks and Reijnierse (1998) we will prove that the $\mathcal{B}$-nucleolus consists of one point for all partial games $(N, \mathcal{B}, v)$ if and only if:

$\mathcal{B}$ is complete: for every $z \in \mathbb{R}^N$, the equation $\sum_{S \in \mathcal{B}} y_S\, e_S = z$ has a solution,

$\mathcal{B}$ is balanced: the equation $\sum_{S \in \mathcal{B}} y_S\, e_S = e_N$ has a *positive* solution.

Here, e_S denotes the indicator vector of coalition S $(S \in \mathcal{B})$. Maschler *et al.* (1992) proved that $\mathrm{Nu}(N, \mathcal{B}, v, \Pi)$ is nonempty when Π is a *compact* subset of $I^*(N, v)$ and that all excess-functions $\mathrm{exc}(S, \cdot \mid v)$ are constant on $\mathrm{Nu}(N, \mathcal{B}, v, \Pi)$ $(S \in \mathcal{B})$.

10.3 Stable Equilibria

Equilibria can arise by a lack of money. This will be illustrated in Example 10.4. Such equilibria are *unstable* in the sense that a (sufficiently high) increase of the initial money supply upsets the equilibrium character of the reallocation. We are interested in equilibria for which this does not occur. Let $\mathcal{E}$ be an exchange economy with indivisible goods and money.

- An N-reallocation $\{(B_i, x_i)\}_{i \in N}$ is a *stable* price equilibrium of $\mathcal{E}$ if there exists a price vector $p \in \mathbb{R}^Q$ such that for every $\Delta \in \mathbb{R}^N_+$ the reallocation $\{(B_i, x_i + \Delta_i)\}_{i \in N}$ is a price equilibrium with equilibrium price p, if the initial money supply becomes $m + \Delta$.

So, a price equilibrium is stable if it remains a price equilibrium when the initial endowment of money is increased. If an equilibrium is stable, not necessarily every equilibrium price supports its stability, as the following example shows:

Example 10.1 Let $N = \{1, 2\}$ and $Q = \{\alpha, \beta\}$. Let $A_1 = \{\alpha\}$, $A_2 = \{\beta\}$ and $m = (1, 1)$. The reservation values are given by:

	α	β	$\alpha\beta$
1	7	0	10
2	0	7	10

If we set the price to be $p := (2,2)$, both agents would like to have both goods (yielding profits of value (10-7)-2=1), but their budgets are not sufficient. Therefore, they just keep their own goods. Hence, the initial endowment $\{(A_i, m_i)\}_{i\in N}$ is an equilibrium. Price p fails to be an equilibrium price if we increase the money endowments to $(3,3)$.

Another price leading to the same equilibrium is $\hat{p} := (4,4)$. This price remains an equilibrium price at any increase of the money supply. Therefore, $\{(A_i, m_i)\}_{i\in N}$ is a stable equilibrium. We say that $\hat{p}$ *supports* the stability of the equilibrium (and p does not). $\lhd$

This section provides necessary conditions and sufficient conditions for the existence of stable price equilibria. As will be proved, the existence of stable price equilibria requires two conditions

(1) a condition on the *reservation prices* $\{V_i(C)\}_{i\in N, C\subseteq Q}$, and
(2) a condition on the *money supply*.

Because of the separability of the utility functions these conditions can to a large extent be handled separately, as we shall see. The first condition does not depend on the initial endowments (we call it the *social welfare condition* or *SW-condition*); the second condition (this is called the *abundance condition* or *AB-condition*, for short) depends on initial endowments. To formulate these conditions we need the following concepts:

- An N-redistribution $\{B_i\}_{i\in N}$ *maximizes social welfare* or *is efficient* if $\sum_{i\in N} V_i(B_i)$ is maximal among all N-redistributions. The maximal social welfare is denoted by $\mathcal{SW}_{\max}(\mathcal{E})$.

- A *stochastic redistribution* consists of a set of numbers $y_{C,i} \geq 0$, one for each agent i in N and each subset C of Q, with the property that $\sum_{C\subseteq Q} y_{C,i} = 1$ for all $i \in N$ and $\sum_{C:\alpha\in C} \sum_{i\in N} y_{C,i} = 1$ for every commodity $\alpha \in Q$. So, a stochastic redistribution is a nonnegative solution of the vector equation:

$$\sum_{C,i} y_{C,i}\,(e_i \oplus e_C) = e_N \oplus e_Q.$$

Here e_i and e_C are the characteristic vectors of $\{i\} \subseteq N$ and $C \subseteq Q$ and $\oplus$ denotes the direct sum: $e_i \oplus e_C \in I\!R^N \oplus I\!R^Q$.

The numbers $\{y_{C,i}\}_{C\subseteq Q}$ can be understood as a lottery for agent i. The number $y_{C,i}$ is the chance that agent i obtains bundle C. The second condition says that the probability that object α will be assigned is also one. Note that the integer-valued stochastic redistributions are exactly the N-redistributions: each agent obtains with probability one a bundle C_i and $\{C_i\}_{i\in N}$ is a redistribution.
Expected social welfare realized by the stochastic redistribution $\{y_{C,i}\}_{C\subseteq Q, i\in N}$ is, by definition,
$\sum_{i\in N}\sum_{C\subseteq Q} y_{C,i}\, V_i(C)$.
Now we can formulate the *SW-condition*:

- An economy $\mathcal{E}$ satisfies the **SW-condition** if no stochastic redistribution has a higher expected social welfare than $\mathcal{SW}_{\max}(\mathcal{E})$.

As maximal expected social welfare is determined by the following linear program (LP):

$$\begin{array}{lrl} \text{maximize:} & \displaystyle\sum_{i\in N}\sum_{C\subseteq Q} y_{C,i}\, V_i(C) & \text{subject to:} \\ & y_{C,i} \geq 0 & \text{for } i\in N \text{ and } \emptyset\subseteq C\subseteq Q, \\ & \displaystyle\sum_{C\subseteq Q} y_{C,i} = 1 & \text{for all } i\in N, \\ & \displaystyle\sum_{C:\alpha\in C}\sum_{i\in N} y_{C,i} = 1 & \text{for all } \alpha\in Q, \end{array}$$

the SW-condition says that (LP) has an integer-valued optimal solution. Note that the SW-condition is not dependent on the initial endowments.

- An economy $\mathcal{E}$ satisfies the **AB-condition** if there is an N-redistribution $\{B_i\}_{i\in N}$ that maximizes social welfare and satisfies the inequalities

$$V_i(B_i) \leq V_i(A_i) + m_i \qquad (i\in N).$$

The AB-condition depends on initial endowments. If, e.g., the initial distribution of the indivisible goods $\{A_i\}_{i\in N}$ maximizes social welfare, it is even an empty condition.

The AB-condition stipulates that each agent has enough money to sell A_i for the price $V_i(A_i)$ (the lowest price for which he is willing to sell A_i) and to buy B_i for the price $V_i(B_i)$ (the highest price he is willing to pay for B_i). This makes clear that the AB-condition might be too restrictive. If the price for A_i is higher than $V_i(A_i)$ or the price of B_i is lower than $V_i(B_i)$, a smaller amount of money is sufficient. We will frequently use the phrase '$\{B_i\}_{i\in N}$ satisfies the AB-condition'.

Note that the AB-condition is much weaker than the abundance conditions that are found in the literature, namely:

$$V_i(C) \leq V_i(A_i) + m_i, \quad (i \in N, C \subseteq Q)$$

or even:

$$V_i(C) \leq m_i, \quad (i \in N, C \subseteq Q)$$

see Beviá *et al.* (1999) and Bikhchandani and Mamer (1997). These conditions are, in our opinion, unreasonably restrictive: *every* agent must be able to buy *all* indivisible goods for the highest price he is willing to pay.

- A vector $p \in I\!R^Q$ is called a *regular price vector* if there is a vector $q \in I\!R^N$ such that (p, q) is an optimal solution of the dual linear program (LP)*:

$$\begin{array}{lrl} \text{minimize:} & p(Q) + q(N) & \text{subject to:} \\ & p(C) + q_i \geq V_i(C) & \text{for all } i \in N \text{ and } C \subseteq Q. \end{array}$$

Let us formulate the two theorems concerning the existence of price equilibria. The proofs will be postponed untill the next section.

Theorem 10.2 [cf. Bikhchandani and Mamer (1997)] *An exchange economy $\mathcal{E}$ with quasi-linear utility functions, indivisible goods and money has a price equilibrium if the SW-condition and the AB-condition are satisfied.*

In fact, the proof of Theorem 10.2 shows that every redistribution $\{B_i\}_{i \in N}$ maximizing (expected) social welfare for which the AB-condition holds can be extended to a *stable* price equilibrium $\{(B_i, x_i)\}_{i \in N}$ and that the set of prices supporting its stability contains all regular price vectors.

In Example 10.4, we shall see that an economy can have equilibria that do not maximize social welfare and that equilibrium prices need not be regular.

In the following theorem we prove that the SW-condition is a necessary condition for the existence of *stable* price equilibria.

Theorem 10.3 [cf. Bikhchandani and Mamer (1997)] *If an economy $\mathcal{E}$ with quasi-linear utility functions, indivisible goods and money has a price equilibrium, then the SW-condition holds. Every stable equilibrium allocation maximizes (expected) social welfare and every equilibrium price supporting its stability is regular.*

Combining the two theorems we see that, if the AB- and SW-conditions are satisfied, the set of prices supporting some stable price equilibrium consists of all regular prices. The following simple examples show what can happen in economies with indivisibilities.

Example 10.4 Let $N = \{1, 2\}$ and $Q = \{\alpha, \beta\}$. Let $A_1 = \{\alpha\}$ and $A_2 = \{\beta\}$. The reservation values are *additive*

	α	β
1	10	70
2	20	50

It is easy to see that social welfare is optimized if the agents switch their endowments. A price vector (p_α, p_β) supporting this exchange obeys $10 \leq p_\alpha \leq 20$ and $50 \leq p_\beta \leq 70$. By solving the linear program (LP)*, one can verify that the set of regular prices vectors is given by these inequalities.

To support the redistribution $\beta \to 1$ and $\alpha \to 2$ by a regular price vector, player 1 has, after payment, $m_1 + p_\alpha - p_\beta$. This amount lies between $m_1 - 60$ and $m_1 - 30$, so *lack of money* may block the existence of regular equilibrium prices (if $m_1 < 30$) or may block some regular equilibrium prices (if $30 \leq m_1 < 60$).

Let us consider the case that $m = (10, 20)$ and the price vector is $p = (15, 40)$. Then the reallocation $(\emptyset, 25) \to 1$ and $(\{\alpha, \beta\}, 5) \to 2$ is a price equilibrium that *does not maximize social welfare* and the equilibrium price is *not regular*. The better assignment $\beta \to 1$ and $\alpha \to 2$ cannot be realized because agent 1 does not have enough money to buy β. $\triangleleft$

The next example originates from Beviá *et al.* (1999). They show that if the money supply is sufficiently large, the reservation values exclude the existence of equilibrium prices at all.

Example 10.5 Let $N = \{1, 2, 3\}$ and $Q = \{\alpha, \beta, \gamma\}$. The reservation values $\{V_i(B)\}_{i \in N,\ B \subseteq Q}$ are given in the table below. The initial endowments are $A_1 = \{\alpha\}$, $A_2 = \{\beta\}$ and $A_3 = \{\gamma\}$.

	α	β	γ	$\alpha\beta$	$\alpha\gamma$	$\beta\gamma$	$\alpha\beta\gamma$
1	10	8	2	13	11	9	13
2	8	5	10	13	14	13	15
3	1	1	8	2	9	9	10

The authors show that the unique social optimum $[\alpha \to 2,\ \beta \to 1$ and $\gamma \to 3]$ is not supported by regular equilibrium prices. The reason is that a *stochastic redistribution* has a higher value. If agent 1 obtains α and β each with chance $\frac{1}{2}$, agent 2 obtains γ or $\alpha\beta$ with equal chances and agent 3 obtains γ or $\emptyset$ each with chance $\frac{1}{2}$, the total *expected* utility is 24.5, higher than the social optimum 24. And, indeed, if we increase e.g. $V_2(\alpha)$ to 8.5, the price vector $p = (6.5, 4.5, 8)$ supports the socially optimal redistribution, if there is enough money ($m_2 \geq 2$ is sufficient). Note that, in the original economy (with $V_2(\alpha) = 8$) the social optimal redistribution $\alpha \to 2$, $\beta \to 1$ and $\gamma \to 3$ is a price equilibrium, if the prices are $(7, 5, 8)$ and $m = (0, 2, 0)$. $\triangleleft$

These examples show that if the money supply is sufficiently restrictive, non-stable equilibria or equilibrium prices not supporting stability can exist. We end this section with a scheme overviewing these phenomena. Let $(B_i, x_i)_{i\in N}$ be an equilibrium with equilibrium price p.

Does the SW-condition hold?	$\xrightarrow{\text{no}}$ no stability (Ex. 10.5)

$\downarrow$ yes

Does $(B_i)_{i\in N}$ maximize social welfare?	$\xrightarrow{\text{no}}$ no stability (Ex. 10.4)

$\downarrow$ yes

Is p regular?	$\xrightarrow{\text{no}}$

p does not support the stability of $(B_i, x_i)_{i\in N}$ (cf. Example 10.1)

$\downarrow$ yes

p supports the stability of $(B_i, x_i)_{i\in N}$ (see next section, Comments (iii)).

10.4 The Existence of Price Equilibria: Necessary and Sufficient Conditions

This section provides the proofs of the theorems in the previous one. We start by giving a proof of the fact that the SW-condition and the AB-condition guarantee the existence of price equilibria (Theorem 10.2).

Proof of Theorem 10.2. Let $\{B_i\}_{i \in N}$ be *any* redistribution of the indivisible goods that maximizes social welfare having the property $V_i(B_i) \leq V_i(A_i) + m_i$, $(i \in N)$.

Let (p, q) be *any* optimal solution of the linear program (LP)*:

$$\begin{array}{lrl} \text{minimize:} & p(Q) + q(N) & \text{subject to:} \\ & p(C) + q_i \geq V_i(C) & \text{for all } i \in N \text{ and } C \subseteq Q. \end{array}$$

Define $x_i := p(A_i) + m_i - p(B_i)$ for every agent $i \in N$. In order to show that $\{(B_i, x_i)\}_{i \in N}$ is a price equilibrium, the inequality $x \geq 0$ and the maximality conditions have to be checked.

By the SW-condition, the integer-valued stochastic reallocation

$$\begin{array}{ll} y_{C,i} := 1 & \text{if } C = B_i \text{ and} \\ y_{C,i} := 0 & \text{else} \end{array}$$

is an optimal solution of (LP). By complementary slackness, we find:

$$p(B_i) + q_i = V_i(B_i), \ (i \in N).$$

Since:

$$p(A_i) \geq V_i(A_i) - q_i = V_i(A_i) + p(B_i) - V_i(B_i),$$

we find:

$$x_i = p(A_i) + m_i - p(B_i) \geq V_i(A_i) + m_i - V_i(B_i),$$

which is nonnegative by the AB-condition $(i \in N)$. So, $x \geq 0$.

If $V_i(C) + y > V_i(B_i) + x_i$ for some agent i, commodity bundle C and $y \in I\!R_+$, then:

$$p(C)+y \geq V_i(C)-q_i+y > V_i(B_i)+x_i-q_i = p(B_i)+x_i = p(A_i)+m_i.$$

The first inequality follows from the feasibility condition $p(C) + q_i \geq V_i(C)$. As this inequality is an equality for $C = B_i$, we find the third relation. The last equality follows from the definition of x_i. Hence, the N-reallocation $\{(B_i, x_i)\}_{i \in N}$ is a price equilibrium with equilibrium price p. □

Comments. (i) If we reconsider the proof of Theorem 10.2, we see that, if the SW-condition holds, *every* N-reallocation $\{B_i\}_{i \in N}$ satisfying the AB-condition and *every* regular price vector can be matched to a price equilibrium.

(ii) Furthermore, if we increase the initial money supply by $\Delta \in I\!R^N_+$, p remains regular (since LP and LP^* are independent of m), moreover $(B_i)_i \in N$ still maximizes social welfare and still supports the AB-condition. Therefore, in the new situation the proof above again shows that $(B_i, x_i + \Delta_i)_{i \in N}$ is an equilibrium. Hence, $(B_i, x_i)_{i \in N}$ is a *stable* equilibrium of the original economy, supported by p.

(iii) Finally, the proof above can be used to verify that if $(B_i, x_i)_{i \in N}$ is an equilibrium supported by *regular* price p, then the AB-condition is not necessary to prove the stability of price p. In this case we have by the definition of a price equilibrium:

$$x_i = p(A_i) + m_i - p(B_i) \geq 0.$$

If m is raised by $\Delta \in I\!R^N_+$, p remains regular and the (second part of the) proof once more shows that $(B_i, x_i + \Delta_i)_{i \in N}$ is an equilibrium of the new situation. ◁

The SW-condition is necessary for the existence of *stable* price equilibria (Theorem 10.3).

Proof of Theorem 10.3. Let $\{(B_i, x_i)\}_{i \in N}$ be a stable price equilibrium with equilibrium price p. Define $q_i := \max_{C \subseteq Q} [V_i(C) - p(C)]$ for each agent $i \in N$. The pair (p, q) is a feasible point of (LP)*. We prove that $q_i = V_i(B_i) - p(B_i)$ for each agent $i \in N$.

Let $C \subseteq Q$ be any commodity bundle and let $i \in N$ be any agent. Let y be the real number $y := V_i(B_i) + x_i - V_i(C)$ and let δ be any positive number. Then $V_i(C) + y + \delta > V_i(B_i) + x_i$.

If y is nonnegative, the maximality condition and the budget constraint generate the inequality $p(C)+y+\delta > p(A_i)+m_i = p(B_i)+x_i$. So, $p(C)+y \geq p(B_i)+x_i$. Substitution of y gives $p(C)+V_i(B_i)+x_i-V_i(C) \geq p(B_i) + x_i$ and therefore, $V_i(C) - p(C) \leq V_i(B_i) - p(B_i)$.

If y is negative, we use the fact that $\{(B_i, x_i+\Delta_i)\}_{i \in N}$ is also an equilibrium if the initial amount of money is $m + \Delta$. This time, we redefine $y := V_i(B_i)+x_i+\Delta_i-V_i(C)$ and y is nonnegative for Δ_i sufficiently large. We can proceed as before and find again $V_i(C)-p(C) \leq V_i(B_i)-p(B_i)$.

Hence, in both cases, $p(B_i) + q_i = V_i(B_i)$ for all $i \in N$. Define the *integer-valued* stochastic redistribution $(y_{C,i})_{C \subseteq Q, i \in N}$ by:

$y_{C,i} := 1$ if $C = B_i$,
$y_{C,i} := 0$ else.

The vectors $(y_{C,i})_{C \subseteq Q, i \in N}$ and (p, q) are feasible vectors in the primal and dual programs respectively, leading to the same value, i.e., $p(Q) + q(N)$. Hence, this is the value of the programs and the vectors are optimal solutions. Because $(y_{C,i})_{C \subseteq Q, i \in N}$ is integer valued, the SW-condition is satisfied. Finally, the redistribution $\{B_i\}_{i \in N}$ maximizes social welfare. □

Summarizing the results of Theorems 10.2 and 10.3, we find that the SW-condition is a necessary and sufficient condition for the existence of stable price equilibria, as soon as the money supply satisfies the AB-condition, a stable price equilibrium allocation maximizes social welfare and equilibrium prices are regular price vectors. For *unstable* price equilibria the last two statements need not be true. In Example 10.4 the equilibrium price $p = (15, 40)$ is not regular and the reallocation $\emptyset \to 1$ and $\{\alpha\beta\} \to 2$ does not maximize social welfare. Comparing this result with the results of Bikhchandani and Mamer (1997) we find the following difference. Bikhchandani and Mamer (1997) assume the stronger AB-condition by which every efficient distribution satisfies our AB-condition. Under this assumption they prove the equivalence of the SW-condition and the existence of price equilibria.

10.5 The Nucleolus as Regular Price Vector

For each economy $\mathcal{E}$, we define the partial TU-game $(\bar{N}, \mathcal{B}, \bar{v})$ as follows. The 'player'-set $\bar{N}$ is $Q \cup N$. The collection $\mathcal{B}$ consists of $\bar{N}$ and all coalitions $T \subset \bar{N}$ with $|T \cap N| = 1$. So, if $C \subseteq Q$ and $i \in N$, then $T := C \cup \{i\} \in \mathcal{B}$. The value of T is defined by $\bar{v}(T) := V_i(C)$. The value of the grand coalition $\bar{v}(\bar{N}) := G$ is chosen to be so large that the $\mathcal{B}$-core of $(\bar{N}, \mathcal{B}, \bar{v})$ is nonempty.

In order to prove that the $\mathcal{B}$-nucleolus is a regular price vector, we first have to show that it is a singleton.

Proposition 10.6 *Let $(\bar{N}, \mathcal{B}, \bar{v})$ be a partial TU-game. Then the following two statements are equivalent:*

(i) $\mathcal{B}$ is balanced and complete,

(ii) the $\mathcal{B}$-nucleolus is a singleton.

Proof. Let us call a vector in $\mathbb{R}^{\bar{N}}$ of which the coordinates sum up to zero a *side payment.* A side payment z is called *beneficial* if $z(T) \geq 0$ for all $T \in \mathcal{B}$. Corollary 6 of Derks and Reijnierse (1998) shows that $\mathcal{B}$ is balanced and complete if and only if the zero vector is the only beneficial side payment.

(i)$\Longrightarrow$ (ii): Since the only beneficial side payment is the zero vector, $\min_{T\in\mathcal{B}} z(T) < 0$ for every side payment $z \neq 0$. As the function $z \mapsto \min_{T\in\mathcal{B}} z(T)$ is continuous, there is a number $t > 0$ such that $\min_{T\in\mathcal{B}} z(T) \leq -t$ for all side payments z with $\|z\| = 1$. In words, moving from one pre-imputation to another with unit length distance, will always cost at least t to some coalition T in $\mathcal{B}$.

Let x be any pre-imputation of $\bar{v}$. How far can we move from x without enlarging the maximal excess? An upper bound can be given as follows. Let $x \in I^*(\bar{N}, \bar{v})$ and let exc_{max} and exc_{min} be the largest and smallest excess, respectively, with respect to x for coalitions in $\mathcal{B}\backslash\{\bar{N}\}$. Let y be any other pre-imputation.

If $\|y - x\| > \frac{1}{t}(\text{exc}_{\text{max}} - \text{exc}_{\text{min}})$, then moving from x to y will cost some coalition T more than $(\text{exc}_{\text{max}} - \text{exc}_{\text{min}})$, which gives:

$$\text{exc}(T, y \mid \bar{v}) > \text{exc}(T, x \mid \bar{v}) + \text{exc}_{\text{max}} - \text{exc}_{\text{min}} \geq \text{exc}_{\text{max}}.$$

Define:

$$\Pi := \{y \in I^*(\bar{N}, \bar{v}) : \ \|x - y\| \leq \max_{S,T\in\mathcal{B}} \frac{\text{exc}(S, x \mid \bar{v}) - \text{exc}(T, x \mid \bar{v})}{t}\}.$$

Pre-imputations y outside Π satisfy $\theta \circ E(x \mid \bar{v}) \prec_{\text{lex}} \theta \circ E(y \mid \bar{v})$. So we might as well restrict the set of candidates of the $\mathcal{B}$-nucleolus to Π; if we have a lexicographically best candidate in Π, it is a global best candidate. The set Π is compact and therefore we can apply Theorem 2.3 (P_0) of Maschler *et al.* (1992): the $\mathcal{B}$-nucleolus is nonempty.

Let x and y be elements of the $\mathcal{B}$-nucleolus. Theorem 4.3 (P_{12}) of Maschler *et al.* (1992) gives that the excesses at the nucleolus are constant:

$$\text{exc}(S, x \mid \bar{v}) = \text{exc}(S, y \mid \bar{v}) \qquad \text{for all } S \in \mathcal{B}.$$

Hence:

$$x(S) = y(S) \qquad \text{for all } S \in \mathcal{B}.$$

The collection $\mathcal{B}$ is complete by assumption, so x and y coincide; the nucleolus is a singleton.

(ii) $\Longrightarrow$ (i): Let z be a beneficial side payment. Then adding z to the $\mathcal{B}$-nucleolus leads to a (weakly) lexicographically better pre-imputation.

Hence, z must be the zero vector. We can apply again Corollary 6 of Derks and Reijnierse (1998). □

With the help of the previous proposition, it is not difficult to prove that the $\mathcal{B}$-nucleolus of $(\bar{N}, \mathcal{B}, \bar{v})$ is a singleton:

Lemma 10.7 *Let $(\bar{N}, \mathcal{B}, \bar{v})$ be a partial TU-game arising from an exchange economy. If $n \geq 2$, the $\mathcal{B}$-nucleolus of $(\bar{N}, \mathcal{B}, \bar{v})$ consists of one point.*

Proof. To show the completeness of $\mathcal{B}$, it suffices to show that e_i and e_α are in the span of $\{e_S \mid S \in \mathcal{B}\}$ for all $i \in N$ and $\alpha \in Q$. This is true, because for all $i \in N$, $\alpha \in Q$, we have $\{i\} \in \mathcal{B}$, $\{i, \alpha\} \in \mathcal{B}$ and $e_\alpha = e_{\{i,\alpha\}} - e_i$.

To show the balancedness of $\mathcal{B}$, it is sufficient to prove that every coalition S in $\mathcal{B}$ is a member of a partition that is a subcollection of $\mathcal{B}$. Let $C \subseteq Q$, $i \in N$ and let $S := C \cup \{i\} \in \mathcal{B}$. Then, for $j \neq i$ ($n \geq 2$ is necessary):

$$\{S, (Q \backslash C) \cup \{j\}\} \cup \{\{k\}\}_{k \in N \backslash \{i,j\}} \text{ is a partition in } \mathcal{B}.$$

Now the lemma is a direct consequence of the previous Proposition. □

Theorem 10.8 *Let $\mathcal{E}$ be an exchange economy with indivisible goods and money. If $(x, y) \in I\!R^Q \times I\!R^N$ is the $\mathcal{B}$-nucleolus of the associated partial game $(\bar{N}, \mathcal{B}, \bar{v})$, then x is a regular price vector.*

Proof. Let $-t$ be the maximal excess with respect to (x, y):

$$t := -\max_{S \in \mathcal{B} \backslash \{\bar{N}\}} e(S, (x, y) \mid \bar{v}).$$

This means that $x(C) + (y_i - t) \geq V_i(C)$ for all $i \in N$ and all $C \subseteq Q$. Accordingly, $(p, q) := (x, y - t\, e_N)$ is a feasible point of (LP)*. Suppose, $(\bar{p}, \bar{q})$ satisfies $\bar{p}(C) + \bar{q}_i \geq V_i(C)$ for all (C, i), and moreover $\bar{p}(Q) + \bar{q}(N) < p(Q) + q(N)$. Then $(\bar{x}, \bar{y}) := (\bar{p}, \bar{q} + t\, e_N)$ satisfies:

$$\begin{aligned} \bar{x}(C) + \bar{y}_i &\geq V_i(C) + t \quad \text{for all pairs } (C, i), \text{ and} \\ \bar{x}(Q) + \bar{y}(N) &< x(Q) + y(N) = \bar{v}(\bar{N}). \end{aligned}$$

There is a number $\varepsilon > 0$ such that $(\bar{x}, \bar{y} + \varepsilon\, e_N)$ is a pre-imputation. All coalitions $S \in \mathcal{B} \backslash \{\bar{N}\}$ have an excess strictly lower than $-t$ with respect to this imputation. Hence, (x, y) is not the $\mathcal{B}$-nucleolus. This contradiction gives that (p, q) is an optimal point of (LP)*, so x is a regular price vector. □

Define for any pre-imputation (x', y') the collection $\mathcal{B}_{\max}(x', y')$ as the subcollection of $\mathcal{B}$ with maximal excess:

$$\mathcal{B}_{\max}(x', y') := \operatorname*{argmax}_{S \in \mathcal{B} \setminus \{\bar{N}\}} \operatorname{exc}(S, (x', y') \mid \bar{v}), \quad (x', y') \in I^*(\bar{N}, \bar{v})$$

We abbreviate $\mathcal{B}_{\max}(x, y)$ by $\mathcal{B}_{\max}$. The $\mathcal{B}$-nucleolus can help to check the SW-condition:

Proposition 10.9 *The SW-condition holds if and only if $\mathcal{B}_{\max}$ contains a partition.*

Proof. We have seen that $(x, y - te_N)$ is an optimal solution of (LP)* (x, y and t as in the previous proof). If $\{B_i \cup \{i\}\}_{i \in N}$ is a partition in $\mathcal{B}_{\max}$, the (stochastic) reallocation:

$$\begin{aligned} y_{C,i} &:= 1 \quad \text{if } C = B_i, \\ y_{C,i} &:= 0 \quad \text{else} \end{aligned}$$

satisfies complementary slackness and is therefore optimal in (LP). This implies that the SW-condition holds.

Conversely, if the SW-condition holds, there is an N-reallocation $\{B_i\}_{i \in N}$ that maximizes expected social welfare. As $(x, y - te_N)$ is an optimal solution of (LP)*, complementary slackness holds:

$$x(B_i) + (y_i - t) = V_i(B_i) \text{ for all } i \in N.$$

Then $B_i \cup \{i\} \in \mathcal{B}_{\max}$ for every agent i, so $\mathcal{B}_{\max}$ contains a partition. □

So we are left to answer the question: does $\mathcal{B}_{\max}$ contain a *partition* of $Q \cup N$? If not, we can combine the results of Theorem 10.3 and Proposition 10.9 and conclude that there is no stable price equilibrium.

If $\{B_i \cup \{i\}\}_{i \in N}$ is a partition of $Q \cup N$ in $\mathcal{B}_{\max}$, the proof of Proposition 10.9 shows that this is a distribution of the indivisibilities maximizing expected social welfare. If p is the regular price vector[2], obtained by computing the $\mathcal{B}$-nucleolus of $(\bar{N}, \mathcal{B}, \bar{v})$, one selects the coalitions $C \cup \{i\} \in \mathcal{B}_{\max}$ with $p(C) \leq p(A_i) + m_i$. If this subcollection contains a partition $\{B_i \cup \{i\}\}_{i \in N}$, then the assignment:

$$i \mapsto (B_i, x_i := p(A_i) + m_i - p(B_i)), \quad (i \in N)$$

is a stable price equilibrium. This can be deduced by a reasoning similar to the proof of Theorem 10.2.

[2]The regularity follows by Theorem 10.8.

Note that the condition $p(B_i) \leq p(A_i) + m_i$ for $i \in N$ is weaker than the AB-condition $V_i(B_i) \leq V_i(A_i) + m_i$ for $i \in N$.

References

Aumann, R.J., and M. Maschler (1964): "The bargaining set for cooperative games," in: Dresher, M., Shapley, L.S., Tucker, A.W. (eds.), *Advances in Game Theory.* Princeton: Princeton University Press, 443–476.

Beviá, C., M. Quinzii, and J.A. Silva (1999): "Buying several indivisible goods," *Mathematical Social Sciences,* 37, 1–23.

Bikhchandani, S., and J.W. Mamer (1997): "Competitive equilibrium in an exchange economy with indivisibilities," *Journal of Economic Theory,* 74, 385–413.

Debreu, G. (1959): *Theory of Value.* New York: John Wiley and Son Inc.

Derks, J., and J.H. Reijnierse (1998): "On the core of a collection of coalitions," *International Journal of Game Theory,* 27, 451–459.

Maschler, M., J.A.M. Potters, and S.H. Tijs (1992): "The general nucleolus and the reduced game property," *International Journal of Game Theory,* 21, 85–106.

Potters, J.A.M., and S.H. Tijs (1992): "The nucleolus of matrix games and other nucleoli," *Mathematics of Operations Research,* 17, 164–174.

Reijnierse, J.H., and J.A.M. Potters (1998): "The $\mathcal{B}$-nucleolus of TU-games," *Games and Economic Behavior,* 24, 77–96.

Schmeidler, D. (1969): "The nucleolus of a characteristic function game," *SIAM Journal of Applied Mathematics,* 17, 1163–1170.

Shapley, L.S. (1953): "A value for n-person games," in: Kuhn, H.W., Tucker, A.W. (eds.) *Contribution to the Theory of Games II, Annals of Mathematics Study,* 28. Princeton: Princeton University Press, 307–317.

Snijders, C. (1995): "Axiomatization of the nucleolus," *Mathematics of Operations Research,* 20, 189–196.

Young, H. (1985): "Monotonic solutions of cooperative games," *International Journal of Game Theory,* 14, 65–72.

Chapter 11

Network Formation, Costs, and Potential Games

BY MARCO SLIKKER AND ANNE VAN DEN NOUWELAND

11.1 Introduction

We study the endogenous formation of networks in situations where the values obtainable by coalitions of players can be described by a coalitional game. To do so, we model network formation as a strategic-form game in which an exogenous allocation rule is used to determine the payoffs to the players in various networks. We only consider exogenous allocation rules that divide the value of each group of interacting players among these players. Such allocation rules are called component efficient. In the network-formation game, the players have to weigh the possible advantages of forming links, such as occupying a more central position in a network and therefore maybe increasing their payoff, against the costs of forming links. The starting point of this chapter is the strategic-form network-formation game that was introduced in Dutta *et al.* (1998) and that was extended to include a cost for forming a link by Slikker and van den Nouweland (2000).[1] We show that this strategic-form network-formation game is a potential game if and only if the exogenous allocation rule is the cost-extended Myerson value that was introduced in Slikker and van den Nouweland (2000). Potential games,

[1]This model was actually first mentioned, briefly, in Myerson (1991).

P. Borm and H. Peters (eds.), Chapters in Game Theory, 223–246.

which were introduced by Monderer and Shapley (1996), are easy to analyze because for such a game all the information necessary to compute its Nash equilibria can be captured in a potential function, a function that assigns to each strategy profile a single number. Also, the existence of a potential function gives rise to a refinement of Nash equilibrium, namely the set of strategy profiles that maximize this potential function. We study which networks emerge according to the potential-maximizing strategy profiles. We find for games with three symmetric players, that the pattern of networks supported by potential-maximizing strategy profiles as the costs for forming links increase depends on whether the underlying coalitional game is superadditive and/or convex. In all cases, though, higher costs for forming links result in the formation of fewer links. The results that we obtain for 3-player symmetric games are surprisingly similar to those found for coalition-proof Nash equilibrium in Slikker and van den Nouweland (2000). We conclude the current chapter by extending the result that, according to the potential maximizer, higher costs for forming links result in the formation of fewer links, to games with more than three players who are not necessarily symmetric.

The outline of the chapter is as follows. We start with a review of the literature on network formation in Section 11.2. In Section 11.3 we describe cost-extended communication situations and the cost-extended Myerson value as well as the network-formation game in strategic form. In Section 11.4 we describe potential games and we show that the network-formation game in strategic form is a potential game if and only if the cost-extended Myerson value is used to determine the payoffs of the players. In Section 11.5 we then use the potential maximizer as an equilibrium refinement in these games and we study which networks are formed according to the potential maximizer. We obtain the result that higher costs for forming links result in the formation of fewer links.

11.2 Literature Review

In this section we provide a brief review of the literature on network formation.

The game-theoretical literature on the formation of networks was initiated by Aumann and Myerson (1988). They study situations in which the profits obtainable by coalitions of players can be described by a coalitional game. For such situations, they introduce an extensive-form game of network formation in which links are formed sequentially and in which a link that is formed at some point cannot be broken later in the game. The Myerson value (cf. Myerson, 1977) is used as an exogenous allocation rule to determine the payoffs to the players in various networks. Aumann and Myerson (1988) study which networks are supported by subgame-perfect Nash equilibria of this network-formation game. They show that there exist superadditive games such that only incomplete or even non-connected networks are supported by subgame-perfect Nash equilibria. They also show that in weighted majority games with several small players who each have one vote and who as a group have a majority and one large player who needs at least one small player to form a majority, the subgame-perfect Nash equilibrium predicts the formation of the complete network on a minimal winning coalition of small players. Aumann and Myerson (1988) also provide two examples of weighted majority games with several large players in which complete networks containing one large player and several small players are supported by subgame-perfect Nash equilibria. They pose the question whether there exists a weighted majority game for which a network that is not internally complete can be supported by a subgame-perfect Nash equilibrium. This question is addressed by Feinberg (1988), who provides an example of a weighted majority game and a network that is not internally complete such that no new links will be formed once this network has been formed. Slikker and Norde (2000) use the extensive-form game of Aumann and Myerson (1988) to study network formation in symmetric convex games. They show that for symmetric convex games with up to five players the complete network is always supported by a subgame-perfect Nash equilibrium and that all networks that are supported by a subgame-perfect Nash equilibrium are payoff equivalent to the complete network. Furthermore, Slikker and Norde (2000) show that this result cannot be extended to games with more than five players. They provide an example of a 6-player symmetric convex game for which there are networks supported by subgame-perfect Nash equilibria in which the players have payoffs that are different from those they get in the complete network.

Dutta *et al.* (1998) study a network-formation game in strategic form in which links are formed simultaneously. Like Aumann and Myerson (1988), they study situations in which the profits obtainable by coalitions of players can be described by a coalitional game. They also use an exogenous allocation rule to determine the payoffs to the players in various networks. However, rather than focusing on the Myerson value only, they consider a class of allocation rules that includes the Myerson value. They restrict their attention to superadditive coalitional games. Their focus is on the identification of networks that are supported by various equilibrium concepts. After showing that every network can be supported by a Nash equilibrium of the strategic-form network-formation game, they proceed by studying refinements of Nash equilibrium. Because strong Nash equilibria might not exist, they focus on less demanding refinements such as Nash equilibria in undominated strategies and coalition-proof Nash equilibria. They show that both of these equilibrium refinements predict the formation of the complete network or of some network in which the players get the same payoffs as in the complete network.

Qin (1996) studies the relation between potential games and strategic-form network-formation games. He shows that the Myerson value is the unique component efficient allocation rule that results in the network-formation game being a potential game. He then applies the equilibrium refinement called the potential maximizer, which Monderer and Shapley (1996) defined for potential games, to strategic-form network-formation games that use the Myerson value to determine the payoffs to the players in various networks. He shows that the potential maximizer predicts the formation of the complete network or of some network in which the players get the same payoffs as in the complete network.

In both the extensive-form network-formation game of Aumann and Myerson (1988) and the strategic-form network-formation game of Dutta *et al.* (1998), forming links is free of charge. Slikker and van den Nouweland (2000) introduce costs for establishing links in these two models and study how the level of these costs influences which networks are supported by equilibria. They use the cost-extended Myerson value to determine the payoffs to the players in various networks. For various equilibrium refinements, they identify which networks are supported in equilibrium as the costs for establishing links increase. For the extensive-form network-formation game they obtain the perhaps counterintuitive result that in some cases rising costs for forming links may result in the formation of *more* links in subgame-perfect equilibrium. In the strategic-form network-formation game, they concentrate on Nash equilibria in undominated strategies and coalition-proof Nash equilibria. They show that generally for very low costs these equilibria predict the formation of the complete network, while the number of links formed in equilibrium decreases as the costs increase.

Slikker and van den Nouweland (2001a) introduce link and claim games, strategic-form network-formation games in which players bargain over the division of payoffs while forming links. This makes their model very different from those described before, where bargaining over payoff division occurs *after* a network has been formed. Following previous papers, they study situations in which the profits obtainable by coalitions of players can be described by a coalitional game. They find that Nash equilibrium does, in general, not support networks that contain a cycle. The main focus in Slikker and van den Nouweland (2001a) is on the payoffs to the players that can emerge according to various equilibrium refinements. They show that any payoff vector that is in the core of the underlying coalitional game is supported by a Nash equilibrium of the link and claim game but not necessarily by a strong Nash equilibrium, while any strong Nash equilibrium of the link and claim game results in a payoff vector that is in the core of the underlying coalitional game. They also provide an overview of all coalition-proof Nash equilibria for 3-player games that satisfy a mild form of superadditivity.

All the papers described above study situations in which the profits obtainable by coalitions of players can be described by a coalitional game. In recent years, however, a number of papers have been published that study the formation of networks in situations where the profits obtainable by a coalition of players do not depend solely on whether they are connected or not, but also on exactly how they are connected to each other. In this setting, Jackson and Wolinsky (1996) expose a tension between stability and optimality of networks. Dutta and Mutuswami (1997) further study this issue using the strategic-form network-formation game of Dutta *et al.* (1998). They show that the conflict between stability and optimality of networks can be avoided by taking an implementation approach.

We end this very brief review by pointing the reader to several papers that study dynamic models of network formation in which players are not forward looking. Papers in this area mostly focus on specific parametric models. Without going into any detail, we refer the reader to Bala and Goyal (2000), Goyal and Vega-Redondo (2000), Jackson and Watts (2000), Johnson and Gilles (2000), Watts (2000), and Watts (2001).

For an extensive and up-to-date overview of the game-theoretical literature on networks and network formation we refer the reader to Slikker and van den Nouweland (2001b).

11.3 Network Formation Model in Strategic Form

In this section we describe cost-extended communication situations and the cost-extended Myerson value as defined in Slikker and van den Nouweland (2000). We also describe the network-formation game in strategic form that was studied in Dutta *et al.* (1998) and Slikker and van den Nouweland (2000) and we describe the results that were obtained in those papers.

Let N be a group of players whose cooperative possibilities are described by the characteristic function $v : 2^N \rightarrow \mathbb{R}$ that assigns to every coalition $T \subseteq N$ of players a value $v(T)$ with $v(\emptyset) = 0$. The *coalitional game* (N, v) describes for every coalition the value that its members can obtain *if* they cooperate, but it does not address the issue of which players actually cooperate. In communication situations, cooperation is achieved through bilateral relationships that are called (communication) links. The set of all possible links is $L^N = \{\{i, j\} \mid i, j \in N, i \neq j\}$ and a *(communication) network* is a graph (N, L) in which the players are the nodes, who are connected via the bilateral links in $L \subseteq L^N$. The formation of each link costs $c \geq 0$. A tuple (N, v, L, c) in which (N, v) is a coalitional game, (N, L) is a network, and $c \geq 0$ is the cost of establishing a link, is called a *cost-extended communication situation.*

Let (N, v, L, c) be a cost-extended communication situation. The cost-extended network-restricted game $(N, v^{L,c})$ associated with this situation incorporates three elements, namely the information on the cooperative possibilities of the players as described by the coalitional game (N, v), the restrictions on cooperation as described by the network (N, L), and the costs for establishing links. Let $T \subseteq N$ be a coalition of players. These players can use the links in $L(T) = \{ij \in L \mid i, j \in T\}$ to communicate.[2] This induces a natural partition T/L of T into components, in which each component consists of a subgroup of players in T who are either directly connected or indirectly connected through other players in T, and in which two players in different components are not connected in the network $(T, L(T))$. The value of T in the *cost-extended network-restricted game* $(N, v^{L,c})$ is defined as the sum of the values of its components in the game (N, v) minus the costs for the links between

[2]For notational convenience, we omit brackets and denote a link $\{i, j\}$ by ij. Note that $ij = ji$. We will also omit brackets in other expressions and, for example, write $S\backslash i$ rather than $S\backslash\{i\}$ and $L\backslash ij$ instead of $L\backslash\{ij\}$.

the players in T, i.e.,

$$v^{L,c}(T) = \sum_{C \in T/L} v(C) - c\,|L(T)|.$$

An allocation rule γ for cost-extended communication situations assigns to every cost-extended communication situation (N, v, L, c) a vector $\gamma(N, v, L, c) \in \mathbb{R}^N$ of payoffs to the players. The cost-extended Myerson value is an allocation rule for cost-extended communication situations that is defined using the Shapley value. The Shapley value (cf. Shapley, 1953) is a well-known solution concept for coalitional games and it is easiest described using unanimity games. For a coalition of players $T \in 2^N \backslash \{\emptyset\}$, the *unanimity game* (N, u_T) is defined by $u_T(R) = 1$ if $T \subseteq R$ and $u_T(R) = 0$ otherwise. Shapley (1953) showed that every coalitional game (N, v) can be written as a linear combination of unanimity games $v = \sum_{T \in 2^N \backslash \{\emptyset\}} \lambda_T(v) u_T$ in a unique way. In terms of the unanimity coefficients $\lambda_T(v)$, $T \in 2^N \backslash \{\emptyset\}$, the *Shapley value* Φ of a game (N, v) is given by

$$\Phi_i(N, v) = \sum_{T \subseteq N : i \in T} \frac{\lambda_T(v)}{|T|} \text{ for each } i \in N.$$

The *cost-extended Myerson value* ν of a cost-extended communication situation (N, v, L, c) is the Shapley value Φ of the associated cost-extended network-restricted game, i.e.,

$$\nu(N, v, L, c) = \Phi(N, v^{L,c}).$$

The cost-extended Myerson value can be axiomatically characterized using two of its properties, component efficiency and fairness.

Component Efficiency An allocation rule γ on a class $\mathcal{CS}$ of cost-extended communication situations is *component efficient* if for every cost-extended communication situation $(N, v, L, c) \in \mathcal{CS}$ and every component $C \in N/L$

$$\sum_{i \in C} \gamma_i(N, v, L, c) = v(C) - c\,|L(C)|.$$

Fairness An allocation rule γ on a class $\mathcal{CS}$ of cost-extended communication situations is *fair* if for every cost-extended communication situation $(N, v, L, c) \in \mathcal{CS}$ and every link $ij \in L$ it holds that $(N, v, L \backslash ij, c) \in \mathcal{CS}$ and

$$\gamma_i(N, v, L, c) - \gamma_i(N, v, L \backslash ij, c) = \gamma_j(N, v, L, c) - \gamma_j(N, v, L \backslash ij, c).$$

For any coalitional game (N,v) and $c \geq 0$, we define the class $CS^N_{v,c}$ of cost-extended communication situations with underlying coalitional game (N,v) and a cost c for establishing a link. The cost-extended Myerson value is the unique allocation rule on a class $CS^N_{v,c}$ that satisfies component efficiency and fairness. Theorem 11.1 follows from Theorem 4 in Jackson and Wolinsky (1996) and we omit its proof.[3]

Theorem 11.1 *For any coalitional game (N,v) and $c \geq 0$, the cost-extended Myerson value is the unique allocation rule on $CS^N_{v,c}$ that satisfies component efficiency and fairness.*

We now proceed by describing network-formation games in strategic form. In such a game, the players decide with whom they want to form links, taking into account their possible gains from cooperation as described by an underlying coalitional game and the costs of forming links. A link between two players is then formed if and only if both these players indicate that they want to form it. This results in the formation of a specific network and the payoffs to the players in the network are determined using some exogenously given allocation rule γ for cost-extended communication situations.

Let (N,v) be a coalitional game, $c \geq 0$ the cost for forming a link, and let γ be an allocation rule on the class $CS^N_{v,c}$ of cost-extended communication situations with underlying coalitional game (N,v) and a cost c for establishing a link. In the network-formation game in strategic form $\Gamma^{nf}(N,v,c,\gamma)$, the set of strategies available to player $i \in N$ is $S_i = 2^{N\backslash i}$. By choosing a strategy $s_i \subseteq N\backslash i$, player i indicates that he is willing to form links with the players in s_i. Because a link between two players is formed if and only if both players want to form it, a strategy profile $s \in S = \prod_{i \in N} S_i$ results in the formation of a network with links

$$L(s) = \{ij \mid i \in s_j,\ j \in s_i\}.$$

The payoffs to the players are their payoffs in the induced cost-extended communication situation $(N,v,L(s),c)$ as prescribed by γ, i.e., $\gamma(N,v,L(s),c)$. The *network-formation game in strategic form* $\Gamma^{nf}(N,v,c,\gamma)$ is described by the tuple $(N;(S_i)_{i\in N};(f_i^{\gamma})_{i\in N})$, where

[3]The theorem in Jackson and Wolinsky (1996) is presented in a setting of reward functions. Theorem 8.1 in Slikker and van den Nouweland (2001b) explicitly shows the correspondence between the value of Jackson and Wolinsky (1996) and the cost-extended Myerson value.

$S_i = 2^{N\setminus i}$ for each $i \in N$ and the payoff function $f^\gamma = (f_i^\gamma)_{i\in N}$ is defined by

$$f^\gamma(s) = \gamma(N, v, L(s), c) \text{ for all } s \in S.$$

We illustrate the network-formation game in strategic form in the following example.

Example 11.2 Let (N, v) be the 3-player game with $N = \{1, 2, 3\}$ and v given by

$$v(T) = \begin{cases} 0 & \text{if } |T| \leq 1; \\ 60 & \text{if } |T| = 2; \\ 72 & \text{if } T = N. \end{cases}$$

Suppose that the cost for establishing a link is $c = 10$. We use the cost-extended Myerson value to determine the payoffs to the players for any given network. In the network-formation game $\Gamma^{nf}(N, v, c, \nu)$ every player has 4 strategies, representing whether he wants to form links with none of the other two players, one of them, or both of them. Link ij will be formed only if both players i and j indicate that they want to form it and it will not be formed if at least one of these two players indicates that he does not want to form it. For example, if player 1 plays $s_1 = \{2\}$, player 2 plays $s_2 = \{1, 3\}$, and player 3 plays $s_3 = \{1, 2\}$, then only links 12 and 23 are formed, i.e., $L(s) = \{12, 23\}$. Hence, the players find themselves in cost-extended communication situation $(N, v, \{12, 23\}, c)$ and their payoffs are the cost-extended Myerson value of this situation $\nu(N, v, \{12, 23\}, c)$. To compute this cost-extended Myerson value, we first compute the associated cost-extended network-restricted game $(N, v^{L(s),c})$. Because in network $(N, L(s))$ all coalitions but the coalition consisting of players 1 and 3 are connected, we find

$$v^{L(s),c}(T) = \begin{cases} 0 & \text{if } |T| \leq 1 \text{ or } T = \{1, 3\}; \\ 50 & \text{if } T = \{1, 2\} \text{ or } T = \{2, 3\}; \\ 52 & \text{if } T = N. \end{cases}$$

Expressed in unanimity games, we have $v^{L(s),c} = 50u_{\{1,2\}} + 50u_{\{2,3\}} - 48u_N$. The Shapley value of $(N, v^{L(s),c})$ is easily computed from this as $\Phi(N, v^{L(s),c}) = (9, 34, 9)$. Hence, in the network-formation game $\Gamma^{nf}(N, v, c, \nu)$, we now have $f^\nu(s) = (9, 34, 9)$.

Proceeding like described above, we find that the network-formation game in strategic form $\Gamma^{nf}(N, v, c, \nu)$ is as represented in Figure 11.1, where player 1 chooses a row, player 2 chooses a column, and player 3 chooses one of the four payoff matrices. ◁

	$\emptyset$	$\{1\}$	$\{3\}$	$\{1,3\}$
$\emptyset$	0, 0, 0	0, 0, 0	0, 0, 0	0, 0, 0
$\{2\}$	0, 0, 0	25, 25, 0	0, 0, 0	25, 25, 0
$\{3\}$	0, 0, 0	0, 0, 0	0, 0, 0	0, 0, 0
$\{2,3\}$	0, 0, 0	25, 25, 0	0, 0, 0	25, 25, 0

$\emptyset$

	$\emptyset$	$\{1\}$	$\{3\}$	$\{1,3\}$
$\emptyset$	0, 0, 0	0, 0, 0	0, 0, 0	0, 0, 0
$\{2\}$	0, 0, 0	25, 25, 0	0, 0, 0	25, 25, 0
$\{3\}$	25, 0, 25	25, 0, 25	25, 0, 25	25, 0, 25
$\{2,3\}$	25, 0, 25	34, 9, 9	25, 0, 25	34, 9, 9

$\{1\}$

	$\emptyset$	$\{1\}$	$\{3\}$	$\{1,3\}$
$\emptyset$	0, 0, 0	0, 0, 0	0, 25, 25	0, 25, 25
$\{2\}$	0, 0, 0	25, 25, 0	0, 25, 25	9, 34, 9
$\{3\}$	0, 0, 0	0, 0, 0	0, 25, 25	0, 25, 25
$\{2,3\}$	0, 0, 0	25, 25, 0	0, 25, 25	9, 34, 9

$\{2\}$

	$\emptyset$	$\{1\}$	$\{3\}$	$\{1,3\}$
$\emptyset$	0, 0, 0	0, 0, 0	0, 25, 25	0, 25, 25
$\{2\}$	0, 0, 0	25, 25, 0	0, 25, 25	9, 34, 9
$\{3\}$	25, 0, 25	25, 0, 25	9, 9, 34	9, 9, 34
$\{2,3\}$	25, 0, 25	34, 9, 9	9, 9, 34	14, 14, 14

$\{1,2\}$

Figure 11.1: *The network-formation game* $\Gamma^{nf}(N, v, c, \nu)$

Dutta *et al.* (1998) studied the network-formation game $\Gamma^{nf}(N, v, c, \gamma)$ in the absence of costs, i.e., $c = 0$. They restrict their attention to superadditive coalitional games (N, v), which satisfy $v(T_1)+v(T_2) \leq v(T_1 \cup T_2)$ for all disjoint $T_1, T_2 \subseteq N$. Also, they require the exogenous allocation rules for communication situations to satisfy three appealing properties that are all satisfied by the Myerson value. Their focus is on the identification of networks that are supported by various equilibrium concepts. After showing that every network can be supported by a Nash equilibrium of the network-formation games $\Gamma^{nf}(N, v, 0, \gamma)$, they proceed by studying refinements of Nash equilibrium. Because strong Nash equilibria might not exist, they focus on less demanding refinements such as Nash equilibria in undominated strategies and coalition-proof Nash equilibria. They show that both of these equilibrium refinements predict the formation of the complete network (N, L^N) or of some network in which the players get the same payoffs as in the complete network.

Slikker and van den Nouweland (2000) introduce costs for establishing links into communication situations and study how the level of these costs influences which networks are supported by equilibria. They use the cost-extended Myerson value to determine the payoffs to the players in various cost-extended communication situations. For computational reasons, they limit the scope of their analysis to symmetric 3-player games throughout most of their paper. For various equilibrium refinements, they identify which networks are supported in equilibrium as the costs for establishing links increase. For the network-formation game in strategic form $\Gamma^{nf}(N, v, c, \nu)$ that we studied in Example 11.2, their results imply that every network is supported by Nash equilibrium, while only the complete network is supported by undominated Nash equilibrium and coalition-proof Nash equilibrium. As the cost for forming a link increases, the complete network is no longer supported by a Nash equilibrium and undominated Nash equilibrium and coalition-proof Nash equilibrium support the three networks containing exactly one link. Generally, as the costs increase, networks with fewer links are supported in equilibrium.

11.4 Potential Games

We start out this section by describing potential games and several results obtained for such games by other authors. We then show that the network-formation game in strategic form $\Gamma^{nf}(N, v, c, \gamma)$ is a potential game if and only if γ is the cost-extended Myerson value.

Strategic-form potential games were introduced by Monderer and Shapley (1996). A strategic-form game is a potential game if there exists a real-valued function on the set of strategy profiles that captures for any deviation by a single player the change in payoff of the deviating player. Such a function is called a potential function or simply a potential for the game in strategic form. Formally, a *potential* for a strategic-form game $\Gamma = (N; (S_i)_{i \in N}; (f_i)_{i \in N})$ is a function P on $S = \prod_{i \in N} S_i$ that satisfies the property that for every strategy profile $s \in S$, every $i \in N$, and every $t_i \in S_i$ it holds that

$$f_i(s_i, s_{-i}) - f_i(t_i, s_{-i}) = P(s_i, s_{-i}) - P(t_i, s_{-i}),$$

where s_{-i} denotes the restriction of s to $\prod_{j \in N \setminus i} S_j$. A game Γ that admits a potential is called a *potential game*. Monderer and Shapley (1996) showed that there exist many potential functions for each potential game. Specifically, they showed that if P is a potential function a strategic-form game Γ, then adding a constant (function) to P results in another potential for Γ. Moreover, any two potentials P and Q for a game Γ differ by a constant (function). If a strategic-form game is a potential game, then each of its potential functions contains all the information necessary to determine its Nash equilibria because the change in payoff of a unilaterally deviating player is captured in the potential. Moreover, the existence of a potential function naturally leads to a refinement of Nash equilibrium by selecting the strategy profiles that maximize the potential function.

A relation between cooperation structure formation games and potential games was already established in Monderer and Shapley (1996). They considered a two-stage model, called a participation game. In the first stage, each player chooses whether or not he wants to participate. In the second stage, the players who chose not to participate receive some stand-alone value and the participating players are assumed to form a coalition. The players in the coalition receive payoffs that are determined by applying an exogenously given allocation rule. Monderer and Shapley (1996) show that the Shapley value is the unique efficient allocation rule that results in the participation game being a potential game. Qin (1996) studied the relation between potential games and strategic-form network-formation games (as introduced in Section 11.3) in the absence of costs. He showed that the Myerson value is the unique component efficient allocation rule that results in the network-formation game being a potential game.

The work of Monderer and Shapley (1996) and Qin (1996) indicates that there may be a relation between the existence of potential functions for games in strategic form and Shapley values of coalitional games. This relation is studied by Ui (2000). To describe his result, we need some additional notation. Let N be a set of players and $S = \prod_{i\in N} S_i$ a set of strategy profiles for these players. After choosing a strategy profile $s \in S$, the players play a cooperative game (N, v_s) that depends on the strategy profile chosen. In the cooperative game that is played, the value of a coalition depends only on the strategies of the players in this coalition, i.e., it is independent of the strategies of the players outside this coalition. Formally, $v_s(R) = v_t(R)$ for any coalition $R \subseteq N$ and any two strategy profiles $s, t \in S$ such that $s_R = t_R$, where s_R denotes the restriction of s to $\prod_{j\in R} S_j$. Hence, with every player set N and set of strategy profiles $S = \prod_{i\in N} S_i$, we associate an indexed set of coalitional games in

$$\begin{aligned}\mathcal{G}_{N,S} &= \Big\{\{(N,v_s)\}_{s\in S} \in (\mathrm{TU}^N)^S \mid \text{ for all } s,t\in S \text{ and } R\subseteq N \\ &\qquad \text{it holds that } v_s(R)=v_t(R) \text{ if } s_R = t_R\Big\}.\end{aligned}$$

Here, TU^N denotes the set of coalitional games with player set N.

The following theorem, due to UI (2000), provides a general relation between Shapley values of coalitional games and strategic-form potential games.

Theorem 11.3 *Let $\Gamma = (N; (S_i)_{i\in N}; (f_i)_{i\in N})$ be a game in strategic form. Γ is a potential game if and only if there exists an indexed set of coalitional games $\{(N, v_s)\}_{s\in S} \in \mathcal{G}_{N,S}$ such that*

$$f_i(s) = \Phi_i(N, v_s)$$

for each $i \in N$ and each $s \in S$. Furthermore, if Γ is a potential game and $\{(N, v_s)\}_{s\in S} \in \mathcal{G}_{N,S}$ are as described above, then the function P described by

$$P(s) = \sum_{R\subseteq N: R\neq\emptyset} \frac{\lambda_R(v_s)}{|R|}$$

for all $s \in S$ is a potential for Γ.

We now turn our attention to network formation in a setting in which there are costs for establishing links. We first show that the strategic-form network-formation game is a potential game if the cost-extended

Myerson value is used to determine the payoffs for the players. We point out that the following lemma extends a result by Qin (1996), who proves a similar result in the absence of costs.

Lemma 11.4 *For any coalitional game (N,v) and cost per link c it holds that the network-formation game $\Gamma^{nf}(N,v,c,\nu)$ is a potential game.*

Proof. Let (N,v) be a coalitional game and let c be the cost for establishing a link. For any strategy profile $s \in S$ in the strategic-form game $\Gamma^{nf}(N,v,c,\nu)$, we consider the network-restricted game $(N, v^{L(s),c})$ associated with cost-extended communication situation $(N,v,L(s),c)$. This defines an indexed set of coalitional games $\{(N, v^{L(s),c})\}_{s\in S}$. We will prove that $\{(N, v^{L(s),c})\}_{s\in S} \in \mathcal{G}_{N,S}$. Let $R \subseteq N$ and $s = (s_R, s_{N\backslash R}) \in S$. Since $v^{L(s),c}(R) = \sum_{C\in R/L(s)} v(C) - c|(L(s))(R)|$ and both $(L(s))(R) = \{ij \mid i \in s_j \cap R, j \in s_i \cap R\}$ and $R/L(s)$ do not depend on $s_{N\backslash R}$, it follows that $v^{L(s),c}(R)$ does not depend on $s_{N\backslash R}$. This implies that $\{(N, v^{L(s),c})\}_{s\in S} \in \mathcal{G}_{N,S}$.

Also, by the definition of the payoff functions $(f_i^{\nu})_{i\in N}$ of the network-formation game $\Gamma^{nf}(N,v,c,\nu)$, it holds that

$$f_i^{\nu}(s) = \nu_i(N,v,L(s),c) = \Phi_i(N, v^{L(s),c})$$

for all $s \in S$.

It now follows from Theorem 11.3 that $\Gamma^{nf}(N,v,c,\nu)$ is a potential game. □

Proving that some allocation rule results in the network-formation game being a potential game begs another question, namely whether there exist other allocation rules with this property. We will answer this question in Theorem 11.6, whose proof uses the following lemma. The lemma states that a network-formation game is a potential game only if the exogenous allocation rule used satisfies fairness.

Lemma 11.5 *Let (N,v) be a coalitional game and $c \geq 0$. Let γ be an allocation rule on the class $CS^N_{v,c}$ of cost-extended communication situations with underlying coalitional game (N,v) and cost c for establishing a link. If $\Gamma^{nf}(N,v,c,\gamma)$ is a potential game, then γ satisfies fairness.*

Proof. Suppose $\Gamma^{nf}(N,v,c,\gamma)$ is a potential game and let P be a potential function for this game. Fix a network (N,L). For each $i \in N$, we define the strategy

$$s_i = \{j \mid ij \in L\}.$$

Then, obviously, $L(s) = L$. Choose a link $ij \in L$. We use the notation s_{-ij} to denote the restriction of s to the players in $N\backslash\{i,j\}$. Then it holds that

$$\begin{aligned} P(s_i\backslash j, s_j, s_{-ij}) &= P(s_i\backslash j, s_j\backslash i, s_{-ij}) \\ &= P(s_i, s_j\backslash i, s_{-ij}), \end{aligned}$$

because the three strategy tuples $(s_i\backslash j, s_j, s_{-ij})$, $(s_i\backslash j, s_j\backslash i, s_{-ij})$, and $(s_i, s_j\backslash i, s_{-ij})$ all result in the formation of the same network, namely $(N, L\backslash ij)$, and, hence, in the same payoffs for the players.

Using the definition of the payoff functions $(f_i^\gamma)_{i\in N}$ of the network-formation game $\Gamma^{nf}(N,v,c,\gamma)$, we now find

$$\begin{aligned} \gamma_i(N,v,L,c) - \gamma_i(N,v,L\backslash ij,c) &= f_i^\gamma(s) - f_i^\gamma(s_i\backslash j, s_{-i}) \\ &= P(s) - P(s_i\backslash j, s_{-i}) \\ &= P(s) - P(s_j\backslash i, s_{-j}) \\ &= f_j^\gamma(s) - f_j^\gamma(s_j\backslash i, s_{-j}) \\ &= \gamma_j(N,v,L,c) - \gamma_j(N,v,L\backslash ij,c). \end{aligned}$$

Because $ij \in L$ was chosen arbitrarily, we may now conclude that γ satisfies fairness. □

Combining Lemmas 11.4 and 11.5, we derive the following theorem.

Theorem 11.6 *Let (N,v) be a coalitional game and $c \geq 0$. Let γ be a component efficient allocation rule on the class $CS_{v,c}^N$ of cost-extended communication situations with underlying coalitional game (N,v) and cost c for establishing a link. Then $\Gamma^{nf}(N,v,c,\gamma)$ is a potential game if and only if γ coincides with ν on $CS_{v,c}^N$.*

Proof. The if-part in the theorem follows directly from Lemma 11.4.

To prove the only-if-part, suppose that the network-formation game $\Gamma^{nf}(N,v,c,\gamma)$ is a potential game. Then it follows from Lemma 11.5 that γ satisfies fairness on $CS_{v,c}^N$. Because γ is component efficient by assumption, it now follows from Theorem 11.1 that γ coincides with ν on $CS_{v,c}^N$. □

In the following theorem, we describe a potential for a strategic-form network-formation game $\Gamma^{nf}(N,v,c,\nu)$ in terms of unanimity coefficients of the associated network-restricted game.

Theorem 11.7 *Let (N, v) be a coalitional game and $c \geq 0$ the cost for establishing a link. Then the function P defined by*

$$P(s) = \sum_{T \subseteq N : T \neq \emptyset} \frac{\lambda_T(v^{L(s),c})}{|T|} = \sum_{T \subseteq N : T \neq \emptyset} \left[\frac{\lambda_T(v^{L(s),0})}{|T|} \right] - \frac{1}{2} c \, |L(s)|.$$

for each $s \in S$, is a potential for the network-formation game $\Gamma^{nf}(N, v, c, \nu)$.

Proof. In the proof of Lemma 11.4 we showed that $\{(N, v^{L(s),c})\}_{s \in S} \in \mathcal{G}_{N,S}$ and that for the payoff functions $(f_i^\nu)_{i \in N}$ of the network-formation game $\Gamma^{nf}(N, v, c, \nu)$, it holds that

$$f_i^\nu(s) = \nu_i(N, v, L(s), c) = \Phi_i(N, v^{L(s),c})$$

for all $s \in S$. We can then conclude from the second part of Theorem 11.3 that the function P given by

$$P(s) = \sum_{T \subseteq N : T \neq \emptyset} \frac{\lambda_T(v^{L(s),c})}{|T|}$$

for all $s \in S$, is a potential function for $\Gamma^{nf}(N, v, c, \nu)$. The second equality in the statement of the theorem now follows easily by noting that for all $s \in S$ and any $T \subseteq N$ with $T \neq \emptyset$ it holds that[4]

$$\lambda_T(v^{L(s),c}) = \begin{cases} \lambda_T(v^{L(s),0}) - c & \text{if } T \in L(s); \\ \lambda_T(v^{L(s),0}) & \text{if } T \notin L(s). \end{cases}$$

□

11.5 Potential Maximizer

In the previous section we showed that strategic-form network-formation games with costs for establishing links are potential games if the cost-extended Myerson value is used as the exogenous allocation rule. This paves the path for us to use the potential maximizer as an equilibrium refinement in these games. In the current section, we study the networks that are formed according to the potential maximizer.

For a potential game Γ, the potential maximizer selects the strategy profiles that maximize a potential function. This equilibrium refinement

[4]Note that $T \in L(s)$ implies that $|T| = 2$.

was introduced by Monderer and Shapley (1996), who also prove that it is well defined because for every potential game Γ the set of strategy profiles that maximize a potential function is independent of the particular potential function used. As a motivation for this equilibrium refinement, they remark that in the so-called stag-hunt game that was described by Crawford (1991), potential maximization selects strategy profiles that are supported by the experimental results of van Huyck *et al.* (1990). Additional motivation for the potential maximizer as an equilibrium refinement is provided by Ui (2001), who showed that Nash equilibria that maximize a potential function are generically robust.

In a setting in which establishing links is free, Qin (1996) analyzed strategic-form network-formation games using the Myerson value to determine the players' payoffs. He showed that for any superadditive game the complete network is supported by a potential-maximizing strategy profile. Furthermore, he showed that any potential-maximizing strategy profile gives rise to the formation of a network that results in the same payoffs to the players as the complete network. We extend the work of Qin (1996) and investigate which networks are supported by potential-maximizing strategy profiles in the presence of costs for establishing links.

In the following example, we consider the coalitional game of Example 11.2 and analyze the networks that are supported by potential-maximizing strategy profiles for varying levels of the cost for establishing a link.

Example 11.8 Consider the 3-player coalitional game (N, v) with characteristic function v defined by

$$v(T) = \begin{cases} 0 & \text{if } |T| \leq 1; \\ 60 & \text{if } |T| = 2; \\ 72 & \text{if } T = N. \end{cases}$$

We established in Lemma 11.4 that the network-formation game $\Gamma^{nf}(N, v, c, \nu)$ is a potential game. To find the potential-maximizing strategy profiles in this game, we start by describing a potential function P. It follows from Theorem 11.7 that the value that a potential function assigns to a strategy profile s only depends on the network $(N, L(s))$. Hence, we can describe a potential function by the values it assigns to strategy profiles resulting in the formation of various networks. Because the players in the game (N, v) are symmetric, we can restrict attention to nonisomorphic networks only.[5] Consider, for example, a network (N, L) with two links, say $L = \{12, 23\}$. Denoting the costs for establishing links by c, the associated cost-extended network-restricted game $(N, v^{L,c})$ is described by

$$v^{L,c}(T) = \begin{cases} 0 & \text{if } |T| \leq 1 \text{ or } T = \{1,3\}; \\ 60 - c & \text{if } |T| = 2 \text{ and } T \neq \{1,3\}; \\ 72 - 2c & \text{if } T = N. \end{cases}$$

In terms of unanimity games, the network-restricted game is given by

$$v^{L,c} = (60 - c)u_{\{1,2\}} + (60 - c)u_{\{2,3\}} - 48u_N.$$

Hence, it follows for the potential P described in Theorem 11.7 that for any strategy profile s that results in the formation of links 12 and 23

$$P(s) = \sum_{T \subseteq N: T \neq \emptyset} \frac{\lambda_T(v^{L(s),c})}{|T|} = \frac{60 - c}{2} + \frac{60 - c}{2} - \frac{48}{3} = 44 - c.$$

It is easily seen that the potential P takes the same value for every strategy profile that results in the formation of a network with two links. The values that P assigns to strategy profiles that result in the formation of networks with 0, 1, or 3 links are determined in a similar manner. We provide the results in Table 11.1. It readily follows using

Number of links in $L(s)$	Potential $P(s)$
0	0
1	$30 - \frac{1}{2}c$
2	$44 - c$
3	$54 - 1\frac{1}{2}c$

Table 11.1: *Potentials for various strategy profiles*

Table 11.1, that in the absence of costs ($c = 0$) the potential maximizer predicts the formation of the complete network (N, L^N). This is in line with the results of Qin (1996). For positive costs, we derive that the potential maximizer predicts the formation of fewer links as the costs for establishing links rise. The results are represented in Figure 11.2.

[5]Two networks (N_1, L_1) and (N_2, L_2) are isomorphic if there is a one-to-one correspondence between the vertices in N_1 and those in N_2 with the additional property that a link between two vertices in N_1 is included in L_1 if and only if the link between the corresponding two vertices in N_2 is included in L_2.

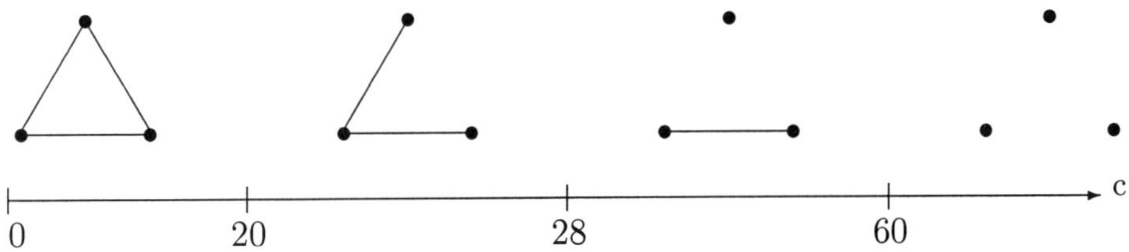

Figure 11.2: *Networks according to the potential maximizer in example 11.8.*

Figure 11.2 schematically represents the networks that can result according to the potential maximizer for different levels of the cost c. The way to read this figure, as well as the figures to come, is as follows. For $c < 20$ (and $c \geq 0$) the complete network is the only network that results according to the potential maximizer, for c with $20 < c < 28$ all three networks with two links are supported by the potential maximizer, and so on. On the boundaries between these intervals all the networks that appear on either side of this boundary are supported by the potential maximizer. So, for example, if $c = 60$, then four networks are supported by the potential maximizer; the empty network and three networks with one link each.

We conclude this example with the observation that, for the coalitional game (N, v) in this example, the cost-network pattern in Figure 11.2 also results if we use coalition-proof Nash equilibrium instead of the potential maximizer. That pattern can be found in Slikker and van den Nouweland (2000). $\triangleleft$

We now turn our attention to the class of symmetric 3-player games. In such a game, the value of a coalition of players does not depend on the identities of its members, but solely on how many players it contains. Hence, a 3-player symmetric game (N, v) can be described by the values that it assigns to coalitions of various sizes. To keep notations to a minimum, we assume (without loss of generality) that 1-player coalitions have a value of zero, and we denote the values of 2-player coalitions and 3-player coalitions by v_2 and v_3, respectively. In addition to this, we restrict our analysis to non-negative games and assume that $v_2 \geq 0$ and $v_3 \geq 0$. In the setting of 3-player symmetric games, Slikker and van den Nouweland (2000) find that for various equilibrium refinements,

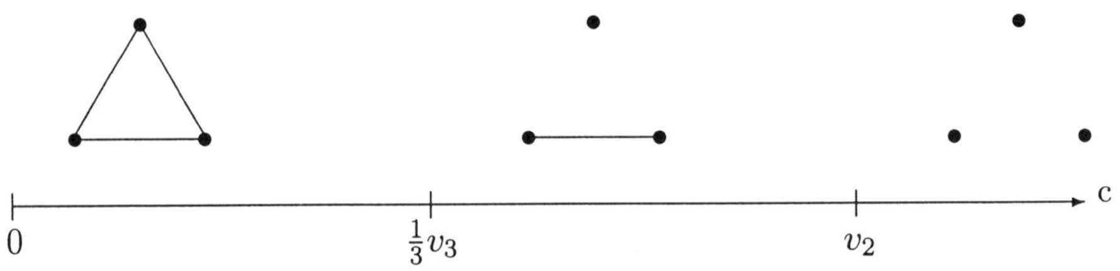

Figure 11.3: *Networks according to the potential maximizer if* $v_2 > v_3$

the cost-network patterns for network-formation games with costs for establishing links depend on whether the underlying coalitional game is superadditive and/or convex. We find that for the structures that are supported by the potential maximizer a similar distinction holds. To derive these patterns, we use the values according to the potential function P described in Theorem 11.7, which we provide in Table 11.2. The cost-network patterns for the classes of games that contain only non-

Number of links in $L(s)$	Potential $P(s)$
0	0
1	$\frac{1}{2}v_2 - \frac{1}{2}c$
2	$\frac{1}{3}v_3 + \frac{1}{3}v_2 - c$
3	$\frac{1}{3}v_3 + \frac{1}{2}v_2 - 1\frac{1}{2}c$

Table 11.2: *Potentials for various strategy profiles*

superadditive games, superadditive but non-convex games, and convex games can be found in Figures 11.3, 11.4, and 11.5, respectively.

We notice that the number of links formed if the players play a potential-maximizing strategy profile declines as the cost for forming a link increases. For non-superadditive games, networks with two links are never formed according to the potential maximizer and for convex games, networks with 1 link are not supported by the potential maximizer for any cost. For any coalitional game, we find that if the cost is very low, then all three links are formed, and if the cost is very high, then no links are formed.

The predictions according to the potential maximizer are remarkably

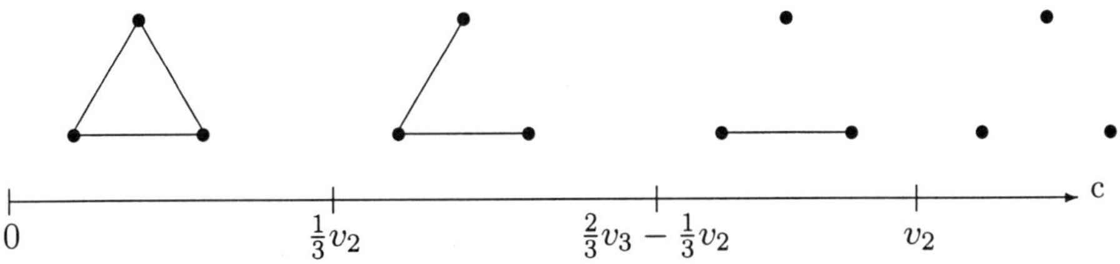

Figure 11.4: *Networks according to the potential maximizer if* $2v_2 > v_3 > v_2$

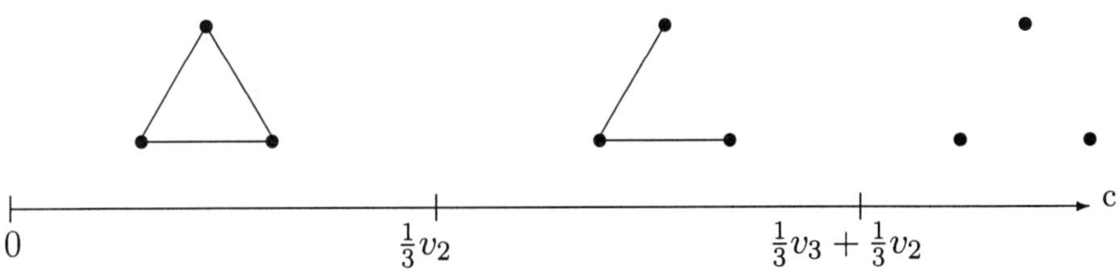

Figure 11.5: *Networks according to the potential maximizer if* $v_3 > 2v_2$

similar to the predictions according to coalition-proof Nash equilibrium (see Figures 11-13 in Slikker and vanden Nouweland, 2000). The only difference is the transition point from networks with 3 links to networks with 1 link for the class with nonsuperadditive games only, i.e., for games with $v_2 > v_3$. This point is $\frac{1}{3}v_3$ for the potential maximizer (see Figure 11.3) and $\frac{2}{3}v_3 - \frac{1}{3}v_2$ for coalition-proof Nash equilibrium.

We are able to extend the result that the potential maximizer predicts the formation of fewer links if the costs increase, to games with an arbitrary number of players that are not necessarily symmetric.[6]

Theorem 11.9 *Let (N, v) be a coalitional game and let $c \geq 0$ and $d \geq 0$ denote two levels of costs for establishing links such that $c < d$.*

[6]This result may seem straightforward from an intuitive point of view. However, we stress that a similar result cannot be obtained for subgame-perfect Nash equilibria of extensive-form network-formation games. Slikker and van den Nouweland (2000) show that in this game increasing costs can lead to more links being formed in equilibrium.

Let (N, L^c) be a network that is supported by a potential-maximizing strategy profile in $\Gamma^{nf}(N, v, c, \nu)$ and (N, L^d) a network that is supported by a potential-maximizing strategy profile in $\Gamma^{nf}(N, v, d, \nu)$. Then $|L^c| \geq |L^d|$.

Proof. Let s^c be a potential-maximizing strategy profile in $\Gamma^{nf}(N, v, c, \nu)$ such that $L(s^c) = L^c$ and let s^d be defined analogously. We denote by P^c (P^d) the potential for the game $\Gamma^{nf}(N, v, c, \nu)$ ($\Gamma^{nf}(N, v, d, \nu)$) described in Theorem 11.7. Note that both network-formation games $\Gamma^{nf}(N, v, c, \nu)$ and $\Gamma^{nf}(N, v, d, \nu)$ have the same set of strategy profiles, which we denote by S. Because the potential function P^c takes a maximum value for strategy profile s^c, it holds for all $t \in S$ that

$$P^c(s^c) \geq P^c(t).$$

Now, let $t \in S$ be such that $|L(t)| > |L^c|$. Then we find using the second equality sign in the expression in Theorem 11.7 for P^c and P^d, that

$$P^d(s^c) - P^d(t) = P^c(s^c) - P^c(t) + \frac{1}{2}(d - c)\left(|L(t)| - |L^c|\right) > 0.$$

Because $P^d(s^c) > P^d(t)$ for every $t \in S$ with $|L(t)| > |L^c|$, the strategy profile s^d, which maximizes the potential P^d, results in the formation of at most $|L^c|$ links. Hence, $|L^d| \leq |L^c|$. □

References

Aumann, R., and R. Myerson (1988): "Endogenous formation of links between players and coalitions: an application of the Shapley value," in Roth, A. (ed.) *The Shapley Value*. Cambridge, UK: Cambridge University Press, 175–191.

Bala, V., and S. Goyal (2000): "A noncooperative model of network formation," *Econometrica*, 68, 1181–1229.

Crawford, V. (1991): "An evolutionary interpretation of van Huyck, Battalio, and Beil's experimental results on coordination," *Games and Economic Behavior*, 3, 25–59.

Dutta, B. and S. Mutuswami (1997): "Stable networks," *Journal of Economic Theory*, 76, 322–344.

Dutta, B., A. van den Nouweland, and S. Tijs (1998): "Link formation in cooperative situations," *International Journal of Game Theory*, 27, 245–256.

Feinberg, Y. (1998): "An incomplete cooperation structure for a voting game can be strategically stable," *Games and Economic Behavior*, 24, 2–9.

Goyal, S., and F. Vega-Redondo (2000): "Learning, network formation and coordination," Mimeo.

Jackson, M., and A. Watts (2000): "On the formation of interaction networks in social coordination games," Mimeo.

Jackson, M., and A. Watts (2001): "The evolution of social and economic networks," *Journal of Economic Theory* (to appear).

Jackson, M., and A. Wolinsky (1996): "A strategic model of social and economic networks," *Journal of Economic Theory*, 71, 44–74.

Johnson, C., and R. Gilles (2000): "Spatial social networks," *Review of Economic Design*, 5, 273–299.

Monderer, D., and L. Shapley (1996): "Potential games," *Games and Economic Behavior*, 14, 124–143.

Myerson, R. (1977): "Graphs and cooperation in games," *Mathematics of Operations Research*, 2, 225–229.

Myerson, R. (1991): *Game Theory: Analysis of Conflict.* Cambridge, Mass.: Harvard University Press.

Qin, C. (1996): "Endogenous formation of cooperation structures," *Journal of Economic Theory*, 69, 218–226.

Shapley, L. (1953): "A value for n-person games," in: Tucker, A. and Kuhn, H. (eds.), *Contributions to the Theory of Games II.* Princeton: Princeton University Press, 307–317.

Slikker, M., and H. Norde (2000): "Incomplete stable structures in symmetric convex games," CentER Discussion Paper 2000-97, Tilburg University, Tilburg, The Netherlands.

Slikker, M., and A. van den Nouweland (2000): "Network formation with costs for establishing links," *Review of Economic Design*, 5, 333–362.

Slikker, M., and A. van den Nouweland (2001a): "A one-stage model of link formation and payoff division," *Games and Economic Behavior*, 34, 153–175.

Slikker, M., and A. van den Nouweland (2001b): *Social and Economic Networks in Cooperative Game Theory.* Boston: Kluwer Academic Publishers.

Ui, T. (2000): "A Shapley value representation of potential games," *Games and Economic Behavior*, 31, 121–135.

Ui, T. (2001): "Robust equilibria of potential games," *Econometrica*, to appear.

van Huyck, J., R. Battalio, and R. Beil (1990): "Tactic coordination games, strategic uncertainty, and coordination failure," *American Economic Review*, 35, 347–359.

Watts, A. (2000): "Non-myopic formation of circle networks," Mimeo.

Watts, A. (2001): "A dynamic model of network formation," *Games and Economic Behavior*, 34, 331–341.

Chapter 12

Contributions to the Theory of Stochastic Games

BY FRANK THUIJSMAN AND KOOS VRIEZE

12.1 The Stochastic Game Model

In this introductory section we give the necessary definitions and notations for the two-person case of the stochastic game model and we briefly present some basic results. In Section 12.2 we discuss the main existence results for zero-sum stochastic games, while in Section 12.3 we focus on general-sum stochastic games. In each section we discuss several examples to illustrate the most important phenomena. Unless mentioned otherwise, we shall assume the state space and the action spaces to be finite. In our discussion we shall address in particular the contributions by Dutch researchers to the field. For a more general and more detailed discussion we refer to Neyman and Sorin (2001).

It all started with the fundamental paper by von Neumann (1928) in which he proves the minimax theorem which says that for each finite matrix of reals $A = [a_{ij}]_{i=1,j=1}^{m\ \ n}$ there exist probability vectors $\bar{x} = (\bar{x}_1, \bar{x}_2, \ldots, \bar{x}_m)$ and $\bar{y} = (\bar{y}_1, \bar{y}_2, \ldots, \bar{y}_n)$ such that for all x and y it holds that $xA\bar{y} \leq \bar{x}A\bar{y} \leq \bar{x}Ay$. (Note that we do not distinguish row vectors from column vectors. In the matrix products this should be clear from the context.) In other words: $\max_x \min_y xAy = \min_y \max_x xAy$. This theorem can be interpreted to say that each matrix game has a

P. Borm and H. Peters (eds.), Chapters in Game Theory, 247–265.

value. A matrix game A is played as follows. Simultaneously, and independent from each other, player 1 chooses a row i and player 2 chooses a column j of A. Then player 2 has to pay the amount a_{ij} to player 1. Each player is allowed to randomize over his available actions and we assume that player 1 wants to maximize his expected payoff, while player 2 wants to minimize the expected payoff to player 1. The minimax theorem tells us that, for each matrix A there is a unique amount, the value denoted by $\text{val}(A)$, which player 1 can guarantee as his minimal expected payoff, while at the same time player 2 can guarantee that the expected payoff to player 1 will be at most this amount.

Later Nash (1951) considered the N-person extension of matrix games, in the sense that all N players, simultaneously and independently choose actions that determine a payoff for each and every one of them. Nash (1951) showed that in such games there always exists at least one (Nash-)equilibrium: a tuple of strategies such that each player is playing a best reply against the joint strategy of his opponents. For the two-player case this boils down to a "bimatrix game" where players 1 and 2 receive a_{ij} and b_{ij} respectively in case their choices determine entry (i, j). The result of Nash says that there exist $\bar{x}$ and $\bar{y}$ such that for all x and y it holds that $\bar{x}A\bar{y} \geq xA\bar{y}$ and $\bar{x}B\bar{y} \geq \bar{x}By$, where $A = [a_{ij}]$ and $B = [b_{ij}]$ are finite matrices of the same size.

Shapley (1953) introduced dynamics into game theory by considering the situation that at discrete stages in $\mathbb{N}$ the players play one of finitely many matrix games, where the choices of the players determine a payoff to player 1 (by player 2) as well as a stochastic transition to go to a next matrix game. He called these games "stochastic games", which brings us to the topic of this chapter. Formally, a two-person stochastic game with finite state and action spaces can be represented by a finite set of matrices $A^1, A^2, \ldots, A^z$ corresponding to the set of states $S = \{1, 2, \ldots, z\}$. For $s \in S$ matrix A^s has size $m_s \times n_s \in \mathbb{N} \times N$ and entry (i, j) of A^s contains:

a) a payoff $r^k(s, i, j) \in \mathbb{R}$ for each player $k \in \{1, 2\}$

b) a transition probability vector $p(s, i, j) = (p(1|s, i, j), p(2|s, i, j), \ldots, p(z|s, i, j))$ where $p(s'|s, i, j)$ is the probability of a transition from state s to state s' whenever entry (i, j) of A^s is selected.

Play can start in any state of S and evolves by players independently choosing actions i_t and j_t of A^{s_t}, where s_t denotes the state visited at stage t. In case $r^1(s,i,j)+r^2(s,i,j)=0$ for all s,i and j, then the game is called zero-sum, otherwise it is called general-sum. In zero-sum games players have strictly opposite interests, since they are paying each other.

At any stage t each player knows the history $h_t=(s_1,i_1,j_1,s_2,i_2,j_2,\ldots,s_{t-1},i_{t-1},j_{t-1},s_t)$ up to stage t, so the players know the sequence of states visited and the actions that were actually chosen in any of these. The players do not know how their opponents choose those actions, i.e. they do not know their opponent's strategy. A strategy is a plan that tells a player what mixed action to use in state s_t at stage $t\in \mathbb{N}$, given the full history h_t. Such behavior strategies will be denoted by π for player 1 and by σ for player 2.

For initial state s and any pair of strategies (π,σ) the limiting average reward and the β-discounted reward, $\beta\in(0,1)$, to player $k\in\{1,2\}$ are respectively given by

$$\gamma^k(s,\pi,\sigma)=E_{s\pi\sigma}\left(\liminf_{T\to\infty}\tfrac{1}{T}\textstyle\sum_{t=1}^{T}r^k(S_t,I_t,J_t)\right)$$

$$\gamma^k_\beta(s,\pi,\sigma)=E_{s\pi\sigma}\left((1-\beta)\textstyle\sum_{t=1}^{\infty}\beta^{t-1}r^k(S_t,I_t,J_t)\right)$$

where S_t,I_t,J_t are random variables for the state and actions at stage t. Let $\gamma^k(\pi,\sigma)$ and $\gamma^k_\beta(\pi,\sigma)$ denote vectors of rewards with coordinates corresponding to the initial states.

A stationary strategy for a player consists of a mixed action for each state, to be used whenever that state is being visited, regardless of the history. Stationary strategies for player 1 are denoted by $x=(x_1,x_2,\ldots,x_z)$, where $x_s=(x_s(1),x_s(2),\ldots,x_s(m_s))$ is the mixed action used by player 1 in state s. For player 2's strategies we write y. A pair (x,y) of stationary strategies determines a Markov-chain (with transition matrix) $P(x,y)$ on S, where entry (s,s') of $P(x,y)$ is $p(s'|s,x_s,y_s)=\sum_{i=1}^{m_s}\sum_{j=1}^{n_s}x_s(i)p(s'|s,i,j)y_s(j)$. If we use the notation $r^k(x,y)=(r^k(1,x_1,y_1),r^k(2,x_2,y_2),\ldots,r^k(z,x_z,y_z))$ with $r^k(s,x_s,y_s)=\sum_{i=1}^{m_s}\sum_{j=1}^{n_s}x_s(i)r^k(s,i,j)y_s(j)$ then

$$\gamma^k_\beta(x,y)=(1-\beta)(I-\beta P(x,y))^{-1}r^k(x,y) \qquad (12.1)$$

where I is the identity matrix, and

$$\gamma^k(x,y)=Q(x,y)r^k(x,y) \qquad (12.2)$$

with

$$Q(x,y)=\lim_{T\to\infty}\tfrac{1}{T}\textstyle\sum_{t=1}^{T}(P(x,y))^t. \qquad (12.3)$$

It is well-known (cf. Blackwell (1962)) that

$$Q(x,y)P(x,y) = Q(x,y) \tag{12.4}$$

$$Q(x,y) = \lim_{\beta\uparrow 1}\ (1-\beta)(I-\beta P(x,y))^{-1} \tag{12.5}$$

and hence (12.1), (12.2) and (12.5) give

$$\gamma^k(x,y) = \lim_{\beta\uparrow 1}\ \gamma^k_\beta(x,y).$$

Notice that (12.3) and (12.4) imply that row s of $Q(x,y)$ is the unique stationary distribution for the Markov chain $P(x,y)$ starting in state s. A stationary strategy x is called pure if $\#\{i : x_s(i) > 0\} = 1$ for all s. Pure stationary strategies shall be denoted by f and g for players 1 and 2 respectively. The following lemma is due to Hordijk *et al.* (1983). It says that, when playing against a fixed stationary strategy, a player always has a pure stationary best reply:

Lemma 12.1 *For all $\beta \in (0,1)$ and for all stationary strategies y for player 2, there exist pure stationary strategies f^1 and f^1_β for player 1, such that for all strategies π:*

$$\gamma^1(f^1,y) \geq \gamma^1(\pi,y) \quad \textit{and} \quad \gamma^1_\beta(f^1_\beta,y) \geq \gamma^1_\beta(\pi,y).$$

A similar result applies for stationary strategies x for player 1.

Finally, we wish to mention one more type of strategies, namely Markov strategies. These are strategies that, at any stage of play, prescribe actions that only depend on the current state and stage. Thus, the past actions of the opponent are not being taken into account. Strategies for which these choices do depend on those past actions shall be called history dependent.

12.2 Zero-Sum Stochastic Games

In zero-sum stochastic games it is customary to consider only the payoffs to player 1, which player 1 wishes to maximize and which player 2 wants to minimize. In his ancestral paper on stochastic games Shapley (1953) shows

Theorem 12.2 *For each stochastic game and for all $\beta \in (0,1)$ there exists $v^1_\beta \in I\!R^z$ and there exist stationary strategies x^1_β and y^1_β such that for all strategies π and σ*

$$\gamma^1_\beta(x^1_\beta, \sigma) \geq v^1_\beta \geq \gamma^1_\beta(\pi, y^1_\beta).$$

The vector v^1_β is called the β-discounted value and the strategies x^1_β, y^1_β are called stationary β-discounted optimal strategies.

Thus we have that v^1_β is the highest reward that player 1 can guarantee:

$$v^1_\beta = \max_\pi \min_\sigma \ \gamma^1_\beta(\pi,\sigma) = \min_\sigma \max_\pi \ \gamma^1_\beta(\pi,\sigma),$$

while player 2 can make sure that player 1's reward will not exceed v^1_β, and each player can do so by some specific stationary strategy. Shapley's proof is based on the observation that v^1_β is the unique solution of the following system of equations:

$$\alpha_s = \text{val}\ [(1-\beta)r^1(s,i,j) + \beta \textstyle\sum_{s'} p(s'|s,i,j)\alpha_{s'}]_{i=1,j=1}^{m_s \ \ n_s} \quad (s \in S), \tag{12.6}$$

where 'val' denotes the matrix game value operator.

Everett (1957) and Gillette (1957) were the first to consider undiscounted rewards. Everett (1957) examined recursive games, which can be defined as stochastic games where the only non-zero payoffs can be obtained in absorbing states, i.e. states that have the property that once play gets there, it remains there forever. Although optimal strategies need not exist for such games, Everett (1957) showed the following:

Theorem 12.3 *For each recursive game the limiting average value exists, and it can be achieved by using stationary ε-optimal strategies, i.e. there exists v^1 and for each $\varepsilon > 0$ there exist strategies $x^1_\varepsilon, y^1_\varepsilon$ such that for all π and σ:*

$$\gamma^1(x^1_\varepsilon, \sigma) + \varepsilon 1_z \geq v^1 \geq \gamma^1(\pi, y^1_\varepsilon) - \varepsilon 1_z.$$

Here 1_z denotes the vector $(1,1,\ldots,1)$ in $I\!R^z$.

Example 12.4 Consider the following recursive game.

State 1:

0 1	0 3
0 3	0 2

State 2:

0 1	0 4
0 4	0 2

State 3:

1 3

State 4:

−1 4

To explain this notation: Player 1 chooses rows; player 2 chooses columns; for each entry the above diagonal number is the payoff to player 1; in case of a general-sum game the payoff tuple is written at this place. The below diagonal number is the state at which play is to proceed; in case of a stochastic transition we write the transition probability vector at this place.

States 3 and 4 are absorbing and obviously states 1 and 2 are the only interesting initial states. For this game the limiting average value is $v^1 = (1, -1, 1, -1)$. For player 1 a stationary limiting average ε-optimal strategy is given by $((1-\varepsilon, \varepsilon), (1, 0))$ for states 1 and 2 respectively (clearly, in states 3 and 4 he can only choose the one available action). As can be verified using (6), the β-discounted value is $v_\beta^1 = (\frac{1-\sqrt{1-\beta^2}}{\beta}, \frac{-1+\sqrt{1-\beta^2}}{\beta}, 1, -1)$ and for player 1 the unique stationary β-discounted optimal strategies are given by playing Top, the first row, with probability $\frac{1-\beta^2-\sqrt{1-\beta^2}}{\beta-\beta^2-\beta\sqrt{1-\beta^2}}$ in state 1 as well as in state 2. $\triangleleft$

An elementary proof for Everett's (1957) result is given by Thuijsman and Vrieze (1992), where for the recursive game situation a stationary limiting average ε-optimal strategy is constructed from an arbitrary sequence of stationary β_t-discounted optimal strategies, with $\lim_{t\to\infty} \beta_t = 1$.

Example 12.5 This famous game is the so called big match introduced by Gillette (1957).

State 1:

	1 / 1	0 / 1
	0 / 2	1 / 3

State 2:

0 / 2

State 3:

1 / 3

For this game the unique stationary β-discounted optimal strategies are given by $x_\beta^1 = (\frac{1}{2-\beta}, \frac{1-\beta}{2-\beta})$ and $y_\beta^1 = (\frac{1}{2}, \frac{1}{2})$ for players 1 and 2 respectively, and $v_\beta^1 = \frac{1}{2}$ for initial state 1. However, it was not clear for a long time, whether or not the limiting average value would exist. The problem was that against any Markov strategy for player 1 and for any $\varepsilon > 0$ player 2 has a Markov strategy such that player 1's limiting average reward is less than ε. On the other hand, player 2 can guarantee that he has to pay a limiting average reward of at most $\frac{1}{2}$, but he cannot guarantee anything less than $\frac{1}{2}$. Hence there is an apparent gap between the amounts the players can guarantee using only Markov strategies. The matter was settled by Blackwell and Ferguson (1968), who formulated, for arbitrary $\varepsilon > 0$, a history dependent strategy for player 1 which guarantees a limiting average reward of at least $\frac{1}{2} - \varepsilon$ against any strategy of player 2. This history dependent limiting average ε-optimal strategy is of the following type. At stage t suppose that play is still in state 1 where player 2 has chosen Left $l(t)$ times, while he has chosen Right $r(t)$ times. Then, player 1 should play Bottom (his second row) with probability $\varepsilon^2(1-\varepsilon)^{d(t)}$, where $d(t) = \max\{0, l(t) - r(t)\}$. $\lhd$

This result on the big match was generalized by Kohlberg (1974), who showed that every repeated game with absorbing states has a limiting average value. A repeated game with absorbing states is a stochastic game in which, just like in the big match, all states but one are absorbing. Finally, by an ingenious proof Mertens and Neyman (1981) showed:

Theorem 12.6 *For every stochastic game there exists $v^1 \in R^z$ and, for each $\varepsilon > 0$ there exist strategies π_ε and σ_ε such that for all strategies π and σ:*

$$\gamma^1(\pi^1_\varepsilon, \sigma) + \varepsilon 1_z \geq v^1 \geq \gamma^1(\pi, \sigma^1_\varepsilon) - \varepsilon 1_z.$$

Their proof exploits the remarkable observation by Bewley and Kohlberg (1976) that the β-discounted value as well as the stationary β-discounted optimal strategies can be expanded as Puiseux series in powers of $1 - \beta$. For example, for the above big match we have that $x^1_\beta = (1,0) + (-1,1)(1-\beta) + (1,-1)(1-\beta)^2 + (-1,1)(1-\beta)^3 + \ldots$

Before deriving this breakthrough on the easy initial states (Tijs and Vrieze, 1986), the same authors have studied structural properties of stochastic games in a number of papers. In Tijs and Vrieze (1980) the effect that perturbations in the game parameters have on the value and on the (ε-)optimal strategies is being examined. In Vrieze and Tijs (1980) the results of Bohnenblust *et al.* (1950) and of Shapley and Snow (1950) have been extended to the case of β-discounted stochastic games. Besides, it is shown how games with a given solution can be constructed. At about the same time, Tijs and Vrieze (1981) also generalized the results of Vilkas (1963) and Tijs (1980), who characterized the value for matrix games, to the case of β-discounted stochastic games. Slightly earlier, Tijs (1979) examined N-person β-discounted stochastic games with finite state spaces and metric action spaces. He showed that, under continuity assumptions with respect to reward and transition functions, as well as some assumptions on the topological size of the action spaces, the existence of ε-equilibria can be established. As far as games with non-finite action spaces are concerned, we should also mention the work by Sinha *et al.* (1991) who examined semi-infinite stochastic games, extending earlier work by Tijs (1979) on semi-infinite matrix games.

As far as structural properties are concerned, we would like to mention the very important result by Tijs and Vrieze (1986), which says that for every stochastic game there is for each player a non-empty set of initial states for which a stationary limiting average optimal strategy exists. Their proof relies on the Puiseux series work by Bewley and Kohlberg (1976). A new and direct proof for the same result is given in Thuijsman and Vrieze (1991) and Thuijsman (1992). A detailed study of the possibilities for limiting average optimality by means of stationary strategies can be found in Thuijsman and Vrieze (1993), while in Flesch *et al.* (1996b) it is proved that the existence of a limiting average optimal strategy implies the existence of stationary limiting average ε-optimal strategies. The idea of easy initial states also plays a key-role in the existence proof for ε-equilibria in general-sum stochastic games by Vieille (2000a,b).

Apart from these general results, specially structured stochastic games have been examined. We already discussed recursive games and repeated games with absorbing states, but we should also mention the following classes: irreducible/unichain stochastic games (cf. Rogers, 1969; Sobel, 1971; or Federgruen, 1978), i.e. stochastic games for which for any pair of stationary strategies the related Markov chain is irreducible/unichain; single controller stochastic games (cf. Parthasarathy and Raghavan, 1981), i.e. games in which the transitions only depend on the actions of one and the same player for all states; switching control stochastic games (cf. Filar, 1981; Vrieze *et al.*, 1983), i.e. games with transitions for each state depending on the action of only one player; perfect information stochastic games (cf. Liggett and Lippman, 1969), where in each state one of the players has only one action available; stochastic games with additive rewards and additive transitions ARAT (cf. Raghavan *et al.*, 1985), i.e. there are r_1^k, r_2^k, p_1, p_2 such that $r^k(s,i,j) = r_1^k(s,i) + r_2^k(s,j)$ and $p(s,i,j) = p_1(s,i) + p_2(s,j)$ for all s,i,j, and, finally, stochastic games with separable rewards and state independent transitions (cf. Parthasarathy *et al.*, 1984), i.e. there are r_a^k, r_b^k, p_a such that $r^k(s,i,j) = r_a^k(s) + r_b^k(i,j)$ and $p(s,i,j) = p_a(i,j)$ for all s,i,j. All these classes admit stationary limiting average optimal strategies. Later, in Thuijsman and Vrieze (1991, 1992) and in Thuijsman (1992) new (and more simple) proofs were provided for the existence of stationary solutions in several of these classes. Characterizations, in terms of game properties, for the existence of stationary limiting average optimal strategies are provided in Vrieze and Thuijsman (1987), Filar *et al.* (1991) and Thuijsman (1992).

12.3 General-Sum Stochastic Games

The first persons to examine general-sum stochastic games were Fink (1964) and Takahashi (1964), who, independent from each other, showed the existence of stationary β-discounted equilibria for stochastic games:

Theorem 12.7 *For each stochastic game and for all $\beta \in (0,1)$ there exist stationary strategies x and y, for players 1 and 2 respectively, such that for all strategies π and σ:*

$$\gamma_\beta^1(\pi, y) \leq \gamma_\beta^1(x, y) \quad \text{and} \quad \gamma_\beta^2(x, \sigma) \leq \gamma_\beta^2(x, y).$$

Since, by its definition, for the zero-sum situation an equilibrium can only consist of a pair of optimal strategies, the big match (cf. Example 12.5) immediately shows that limiting average equilibria do not always exist. Where we introduced ε-optimal strategies for the zero-sum case, we now have to introduce ε-equilibria for the general-sum case.

Definition 12.8 *A pair of strategies $(\pi_\varepsilon, \sigma_\varepsilon)$ is called a limiting average ε-equilibrium ($\varepsilon > 0$) if neither player 1 nor player 2 can gain more than ε by a unilateral deviation, i.e. for all strategies π and σ:*

$$\gamma^1(\pi, \sigma_\varepsilon) \leq \gamma^1(\pi_\varepsilon, \sigma_\varepsilon) + \varepsilon 1_z \quad \text{and} \quad \gamma^2(\pi_\varepsilon, \sigma) \leq \gamma^2(\pi_\varepsilon, \sigma_\varepsilon) + \varepsilon 1_z.$$

◁

The existence of limiting average ε-equilibria for arbitrary general-sum two-person stochastic games has recently been established by Vieille (2000a,b):

Theorem 12.9 *For each stochastic game and for all $\varepsilon > 0$ there exists a limiting average ε-equilibrium $(\pi_\varepsilon, \sigma_\varepsilon)$.*

Preceding, and leading to this breakthrough, are the following results. First, the result by Tijs and Vrieze (1986) on easy initial states was generalized to the case of general-sum stochastic games, i.e., it was shown that in every stochastic game there is a non-empty set of initial states for which ε-equilibria exist (cf. Thuijsman and Vrieze, 1991; Thuijsman, 1992; or Vieille, 1993). Our proof of this result was based on ergodicity properties of a converging sequence of stationary β_t-discounted equilibria, with $\lim_{t\to\infty}\beta_t = 1$ (please note that, here t is just a counter and is not related to the stage parameter). However, the equilibrium strategies are of a behavioral type: at all stages players must take into account the history of past moves of their opponent. Nevertheless, a side-result of this approach was a simple and straightforward proof for the existence of stationary limiting average equilibria for irreducible/unichain stochastic games (which was earlier derived by Rogers, 1969; Sobel, 1971; and Federgruen, 1978).

Concerning the existence of limiting average ε-equilibria for all initial states (simultaneously), sufficient conditions have been formulated in Thuijsman (1992), which are based on properties of a converging sequence of stationary β_t-discounted equilibria, with $\lim_{t\to\infty}\beta_t = 1$, while in Thuijsman and Vrieze (1997) quite general sufficient conditions are formulated in terms of stationary strategies, and of observability and punishability of deviations. This punishability principle is based on the observation that in any equilibrium each player should get at least as much as he can guarantee himself in the worst case. To be more precise, we have seen in the previous section that player can guarantee himself an amount

$$v^1 = \sup_{\pi}\inf_{\sigma} \;\; \gamma^1(\pi,\sigma) = \inf_{\sigma}\sup_{\pi} \;\; \gamma^1(\pi,\sigma),$$

and, by its definition, player 1 can not guarantee any higher reward. For player 2 we have v^2 with similar properties:

$$v^2 = \sup_{\sigma}\inf_{\pi} \;\; \gamma^2(\pi,\sigma) = \inf_{\pi}\sup_{\sigma} \;\; \gamma^2(\pi,\sigma).$$

Thus player 1 has the power to restrict player 2's reward to be at most v^2, while, at the same time, in any equilibrium player 2 should always get at least v^2, for otherwise he would have a profitable deviation. Therefore we call this approach the threat approach, since the players are constantly checking after each other, and any "wrong" move of the opponent will immediately trigger a punishment that will push the reward down to v. Thus the threats are the stabilizing force in the limiting average

(ε-)equilibria. Using this threat approach, the existence of ε-equilibria is proved for repeated games with absorbing states (cf. Vrieze and Thuijsman, 1989), where a prototype threat approach is being used), as well as for stochastic games with state independent transitions (cf. Thuijsman, 1992), as well as for stochastic games with three states (cf. Vieille, 1993), as well as for stochastic games with switching control (cf. Thuijsman and Raghavan, 1997), and existence of pure 0-equilibria has been shown for stochastic games with additive rewards and additive transitions (ARAT, cf. Thuijsman and Raghavan, 1997). The latter class includes the class of perfect information games; a perfect information game has the property that in each state one of the players has only one action available. The use of threats is also indispensable in the existence proof given by Vieille (2000a,b).

We remark that prior to our threat approach for none of these classes, the existence of limiting average ε-equilibria was known, even though the zero-sum solutions have been derived a long time ago. Also note that even for perfect information stochastic games stationary limiting average equilibria generally do not exist, although for the zero-sum case pure stationary limiting average optimal strategies are available (cf. Liggett and Lippman, 1969). Example 12.11 below will illustrate this point.

For recursive repeated games with absorbing states (cf. Flesch *et al.*, 1996) and for ARAT repeated games with absorbing states (cf. Evangelista *et al.*, 1996) stationary limiting average ε-equilibria do exist (without threats).

We conclude this chapter with three very special examples. In Example 12.10 we examine a repeated game with absorbing states for which there is a gap between the β-discounted equilibrium rewards and the limiting average equilibrium rewards. In Example 12.11 we discuss a perfect information stochastic game which does not have stationary limiting average ε-equilibria, but where the only equilibria known to us, are of the threat type. In Example 12.12 we discuss a three person recursive repeated game with absorbing states for which the only limiting average equilibria consist of cyclic Markov strategies. This is very remarkable since, in that game stationary limiting average ε-equilibria do not exist.

Example 12.10 Consider the following example with three states:

State 1	
$1,0$ 1	$0,1$ 1
$0,2$ 2	$1,0$ 3

State 2
$0,2$ 2

State 3
$1,0$ 3

This is an example of a repeated game with absorbing states, where play remains in the initial state 1 as long as player 1 chooses Top, but play reaches an absorbing state as soon as player 1 ever chooses Bottom. Sorin (1986) examined this example in great detail. The (sup-inf) limiting average values (for initial state 1) are given by $(v^1, v^2) = (\frac{1}{2}, \frac{2}{3})$. Clearly then, there can be no stationary limiting average ε-equilibrium, because against any stationary strategy of player 1, player 2 can get at least 1, and by doing so player 1 would get $0 < v^1$, which he can always achieve by playing limiting average ε-optimal in the 1-zero-sum game. However, for each pair in $\text{Conv}\{(\frac{1}{2}, 1), (\frac{2}{3}, \frac{2}{3})\}$, where Conv stands for convex hull, Sorin (1986) gives history dependent limiting average ε-equilibria that yield this pair as an equilibrium reward. Besides, he shows that any limiting average ε-equilibrium corresponds to a reward in $\text{Conv}\{(\frac{1}{2}, 1), (\frac{2}{3}, \frac{2}{3})\}$, while all β-discounted equilibria yield $(\frac{1}{2}, \frac{2}{3})$. Although this observation suggests that the limiting average general-sum case can not be approached from the β-discounted general-sum case, by studying this example Vrieze and Thuijsman (1989) discovered a general principle to construct, starting from any arbitrary sequence of stationary β_t-discounted equilibria with $\lim_{t\to\infty} \beta_t = 1$, a limiting average ε-equilibrium. $\triangleleft$

Example 12.11 The next example has four states:

State 1
$0,0$ 4
$0,0$ $(0, \frac{1}{2}, \frac{1}{2}, 0)$

State 2	
$0,0$ 1	$0,0$ 2

State 3
$3,-1$ 3

State 4
$2,1$ 4

This game is a recursive perfect information game for which there is no stationary limiting average ε-equilibrium. One can prove this as follows. Suppose player 2 puts positive weight on Left in state 2, then player 1's only stationary limiting average ε-best replies are those that put weight at most $\frac{\varepsilon}{2-\varepsilon}$ on Top in state 1; against any of these strategies, player

2's only stationary limiting average ε-best replies are those that put weight 0 on Left in state 2. So there is no stationary limiting average ε-equilibrium where player 2 puts positive weight on Left in state 2. But there is neither a stationary limiting average ε-equilibrium where player 2 puts weight 0 on Left in state 2, since then player 1 should put at most 2ε weight on Bottom in state 1, which would in turn contradict player 2's putting weight 0 on Left. Following the construction of Thuijsman and Raghavan (1997), where existence of limiting average 0-equilibria is proved for arbitrary N-person games with perfect information, we can find an equilibrium by the following procedure. Take a pure stationary limiting average optimal strategy f^1 for player 1 (this exists by Liggett and Lippman, 1969); let g^1 be pure stationary limiting average optimal strategy for player 2 minimizing player 1's reward; let g^2 be a pure stationary limiting average best reply for player 2 maximizing his own reward against f^1(which exists by Lemma 1). Now define g^* for player 2 by: play g^2 unless at some stage player 1 has ever deviated from playing f^1, then play g^1. Here, $f^1 = (1,0) = g^2$ and $g^1 = (0,1)$. Now it can be verified that (f^1, g^*) is a limiting average equilibrium. ◁

Example 12.12 Our final example is described by the following payoff matrices:

N

	L	*R*
T	$0,0,0$	$0,1,3$ *
B	$1,3,0$ *	$1,0,1$ *

F

	$3,0,1$ *	$1,1,0$ *
	$0,1,1$ *	$0,0,0$ *

This is a three-person recursive repeated game with absorbing states, where an asterisk in any particular entry denotes a transition to an absorbing state with the same payoff as in this particular entry. There is only one entry for which play will remain in the non-trivial initial state. One should picture the game as a $2 \times 2 \times 2$ cube, where the layers belonging to the actions of player 3 (Near and Far) are represented separately. As before, player 1 chooses Top or Bottom and player 2 chooses Left or Right. The entry (T, L, N) is the only non-absorbing entry for the initial state. Hence, as long as play is in the initial state the only possible history is the one where entry (T, L, N) was played at all previous stages. This rules out the use of any non-trivial history dependent strategy for this game. Therefore, the players only have Markov strategies at their disposal. In Flesch *et al.* (1997) it is shown that, although (cyclic) Markov limiting average 0-equilibria exist for this game, there are no stationary limiting average ε-equilibria in this game. Moreover, the set of all limiting average equilibria is being characterized completely. An example of a Markov equilibrium for this game is (π, σ, τ), where π is defined by: at stages $1, 4, 7, 10, \ldots$ play T with probability $\frac{1}{2}$ and at all other stages play T with probability 1. Similarly, σ is defined by: at stages $2, 5, 8, 11, \ldots$ play L with probability $\frac{1}{2}$ and at all other stages play L with probability 1. Likewise, τ is defined by: at stages $3, 6, 9, 12, \ldots$ play N with probability $\frac{1}{2}$ and at all other stages play N with probability 1. The limiting average reward corresponding to this equilibrium is $(1, 2, 1)$. $\triangleleft$

So far, existence of ε-equilibria has been the main issue in the theory stochastic games, whereas in other areas of non-cooperative game theory refinements to the Nash-equilibrium concept have been introduced. Only few of these refinements have been generalized to the area of stochastic games. One such extension is that of (trembling hand) perfect equilibria. A perfect equilibrium is one where the strategies played are not only best replies to eachother, but to small perturbations of the opponent's strategy as well. Thuijsman *et al.* (1991) showed the existence of such perfect stationary β-discounted equilibria for arbitrary stochastic games, and also the existence of perfect stationary limiting average equilibria for irreducible stochastic games.

Finally we would like to mention three recent contributions to the field of stochastic games. The first one is the work of Potters *et al.* (1999) who examined β-discounted stochastic games with a potential function. For the classes of additive reward and additive transition stochastic games (ARAT), as well as for the class of stochastic games with separable rewards and state independent transitions (SER-SIT), the potential function is used to derive the existence of pure stationary optimal strategies in the zero-sum case and pure stationary equilibria in the general-sum case.

The second paper to mention is the one by Herings and Peeters (2000) who introduce an algorithm, a tracing procedure, to compute stationary equilibria in discounted stochastic games. Moreover, convergence of the algorithm for almost all such games is proved and the issue of equilibrium selection is addressed.

The third one is by Schoenmakers *et al.* (2001) who introduce a new approach for extending the method of fictitious play developed by Brown (1951) and Robinson (1950) to the situation of stochastic games. A different approach on applying fictitious play to stochastic games was studied by Vrieze and Tijs (1982).

References

Bewley, T., and E. Kohlberg (1976): "The asymptotic theory of stochastic games," *Math. Oper. Res.*, 1, 197–208.

Blackwell, D. (1962): "Discrete dynamic programming," *Ann. Math. Statist.*, 33, 719–726.

Blackwell, D., and T.S. Ferguson (1968): "The big match," *Ann. Math. Statist.*, 39, 159–163.

Bohnenblust, H.F., S. Karlin, and L.S. Shapley (1950): "Solutions of discrete two-person games," *Annals of Mathematics Studies*, 24. Princeton: Princeton University Press, 51–72.

Brown, G.W. (1951): "Iterative solution of games by fictitious play," in: Koopmans, T.C. (ed.), *Activity Analysis of Production and Allocation.* New York: Wiley, 374–376.

Evangelista, F.S., T.E.S. Raghavan, and O.J. Vrieze (1996): "Repeated ARAT games," in: Ferguson, T.S. *et al.* (eds.), *Statistics, Probability and Game Theory; Papers in honor of David Blackwell*, IMS Lecture Notes Monograph Series, 30, pp 13–28.

Everett, H. (1957): "Recursive games," in: Dresher, M., *et al.* (eds.), *Contributions to the Theory of Games*, III, *Annals of Mathematical Studies*, 39. Princeton: Princeton University Press, 47–78.

Federgruen, A. (1978): "On N-person stochastic games with denumerable state space," *Adv. Appl. Prob.*, 10, 452–471.

Filar, J.A. (1981): "Ordered field property for stochastic games when the player who controls transitions changes from state to state," *J. Opt. Theory Appl.*, 34, 503–515.

Filar, J.A., T.A. Schultz, F. Thuijsman, and O.J. Vrieze (1991): "Nonlinear programming and stationary equilibria in stochastic games," *Math. Progr.*, 50, 227–237.

Fink, A.M. (1964): "Equilibrium in a stochastic n-person game," *J. Sci. Hiroshima Univ.*, Series A-I, 28, 89–93.

Flesch, J., F. Thuijsman, and O.J. Vrieze (1996): "Recursive repeated games with absorbing states," *Math. Oper. Res.*, 21, 1016–1022.

Flesch, J., F. Thuijsman, and O.J. Vrieze (1997): "Cyclic Markov equilibria in stochastic games," *Int. J. Game Theory*, 26, 303–314.

Flesch, J., F. Thuijsman, and O.J. Vrieze (1998): "Simplifying optimal strategies in stochastic games," *SIAM Journal of Control and Optimization*, 36, 1331–1347.

Gillette, D. (1957): "Stochastic games with zero stop probabilities," in: Dresher, M., *et al.* (eds.), *Contributions to the Theory of Games*, III, *Annals of Mathematical Studies*, 39. Princeton: Princeton University Press, 179–187.

Herings, P.J.J., R.J.A.P. Peeters (2000): "Stationary equilibria in stochastic games: structure, selection and computation," Report RM/00/031, Meteor, Maastricht University.

Hordijk, A., O.J. Vrieze, and G.L. Wanrooij (1983): "Semi-Markov strategies in stochastic games," *Int. J. Game Theory*, 12, 81–89.

Kohlberg, E. (1974): "Repeated games with absorbing states," *Annals of Statistics*, 2, 724–738.

Liggett, T.M., and S.A. Lippman (1969): "Stochastic games with perfect information and time average payoff," *SIAM Review*, 11, 604–607.

Mertens, J.F., and A. Neyman (1981): "Stochastic games," *Int. J. Game Theory*, 10, 53–66.

Nash, J. (1951): "Non-cooperative games," *Annals of Mathematics*, 54, 286–295.

Neyman, A., and S. Sorin (2001): *Stochastic Games.* Proceedings of he 1999 NATO Summer Institute on Stochastic Games held at Stony Brook (forthcoming).

Parthasarathy, T., and T.E.S. Raghavan (1981): "An orderfield property for tochastic games when one player controls transition probabilities," *J. Opt. Theory Appl.*, 33, 375–392.

Parthasarathy, T., S.H. Tijs, and O.J. Vrieze (1984): "Stochastic games with atate independent transitions and separable rewards," in: Hammer, G., and D. Pallaschke (eds.), *Selected Topics in Operations Research and Mathematical Economics*. Berlin: Springer Verlag, 262–271.

Potters, J.A.M., T.E.S. Raghavan, and S.H. Tijs (1999): "Pure equilibrium strategies for stochastic games via potential functions. Report 9910, Department of Mathematics, University of Nijmegen.

Raghavan, T.E.S., S.H. Tijs, and O.J. Vrieze (1985): "On stochastic games with additive reward and transition structure," *J. Opt. Theory Appl.*, 47, 451–464.

Robinson, J. (1950): "An iterative method of solving a game," *Annals of Mathematics*, 54, 296–301.

Rogers, P.D. (1969): *Non-zerosum stochastic games.* PhD thesis, report ORC 69-8, Operations Research Center, University of California, Berkeley.

Schoenmakers, G., J. Flesch, and F. Thuijsman (2001): "Fictitious play in stochastic games," Report M01-02, Department of Mathematics, Maastricht University.

Shapley, L.S. (1953): "Stochastic games," *Proc Nat Acad Sci USA*, 39, 1095–1100.

Shapley, L.S., and R.N. Snow (1950): "Basic solutions of discrete games," *Annals of Mathematics Studies*, 24. Princeton: Princeton University Press, 27–35.

Sinha, S., F. Thuijsman, and S.H. Tijs (1991): "Semi-infinite stochastic games," in: Raghavan, T.E.S., *et al.* (eds.), *Stochastic Games and Related Topics.* Dordrecht: Kluwer Academic Publishers, 71–83.

Sobel, M.J. (1971): "Noncooperative stochastic games," *Ann. Math. Statist.*, 42, 1930–1935.

Sorin, S. (1986): "Asymptotic properties of a non-zerosum stochastic game," *Int. J. Game Theory*, 15, 101–107.

Takahashi, M. (1964): "Equilibrium points of stochastic noncooperative n-person games," *J. Sci. Hiroshima Univ.*, Series A-I, 28, 95-99.

Thuijsman, F. (1992): *Optimality and Equilibria in Stochastic Games.* CWI-tract 82, Center for Mathematics and Computer Science, Amsterdam.

Thuijsman, F., and T.E.S. Raghavan (1997): "Perfect information stochastic games and related classes," *Int. J. Game Theory*, 26, 403–408.

Thuijsman, F., S.H. Tijs, and O.J. Vrieze (1991): "Perfect equilibria in stochastic games," *J. Opt. Theory Appl.*, 69, 311–324.

Thuijsman, F., and O.J. Vrieze (1991): "Easy initial states in stochastic games," in: Raghavan, T.E.S., *et al.* (eds.), *Stochastic Games and Related Topics*. Dordrecht: Kluwer Academic Publishers, 85–100.

Thuijsman, F., and O.J. Vrieze (1992): "Note on recursive games," in: Dutta, B., *et al.* (eds.), *Game Theory and Economic Applications,*

Lecture Notes in Economics and Mathematical Systems, 389. Berlin: Springer, 133–145.

Thuijsman, F., and O.J. Vrieze (1993): "Stationary ε-optimal strategies in stochastic games," *OR Spektrum*, 15, 9–15.

Thuijsman, F., and O.J. Vrieze (1998): "The power of threats in stochastic games," in: Bardi *et al.* (eds.), *Stochastic Games and Numerical Methods for Dynamic Games*. Boston: Birkhauser, 339–353.

Tijs, S.H. (1979): "Semi-infinite linear programs and semi-infinite matrix games," *Nieuw Archief voor de Wiskunde*, 27, 197–214.

Tijs, S.H. (1980): "Stochastic games with one big action space in each state," *Methods of Operations Research*, 38, 161–173.

Tijs, S.H. (1981): "A characterization of the value of zero-sum two-person games," *Naval Research Logistics Quarterly*, 28, 153–156.

Tijs, S.H., and O.J. Vrieze (1980): "Perturbation theory for games in normal form and stochastic games," *J. Opt. Theory Appl.*, 30, 549–567.

Tijs, S.H., and O.J. Vrieze (1981): "Characterizing properties of the value function of stochastic games," *J. Opt. Theory Appl.*, 33, 145–150.

Tijs, S.H., and O.J. Vrieze (1986): "On the existence of easy initial states for undiscounted stochastic games," *Math. Oper. Res.*, 11, 506–513.

Vieille, N. (1993): "Solvable states in stochastic games," *Int. J. Game Theory*, 21, 395–404.

Vieille, N. (2000a): "2-person stochastic games I: a reduction," *Israel Journal of Mathematics*, 119, 55–91.

Vieille, N. (2000b): "2-person stochastic games II: the case of recursive games," *Israel Journal of Mathematics*, 119, 93–126.

Vilkas, E.I. (1963): "Axiomatic definition of the value of a matrix game," *Theory of Probability and its Applications*, 8, 304–307.

Von Neumann, J. (1928): "Zur Theorie der Gesellschafsspiele," *Mathematische Annalen*, 100, 295–320.

Vrieze, O.J. (1987): *Stochastic Games with Finite State and Action Spaces*. CWI-tract 33, Center for Mathematics and Computer Science, Amsterdam.

Vrieze, O.J., and F. Thuijsman (1987): "Stochastic games and optimal tationary strategies, a survey," in: Domschke, W., *et al.* (eds.), *Methods of Operations Research*, 57, 513–529.

Vrieze, O.J., and F. Thuijsman (1989): "On equilibria in repeated games with absorbing states," *Int J Game Theory*, 18, 293–310.

Vrieze, O.J., and S.H. Tijs (1980): "Relations between the game parameters, value and optimal strategy spaces in stochastic games and construction of games with given solution," *J. Opt. Theory Appl.*, 31, 501–513.

Vrieze, O.J., and S.H. Tijs (1982): "Fictitious play applied to sequences of games and discounted stochastic games," *Int. J. Game Theory*, 11, 71–85.

Vrieze, O.J., S.H. Tijs, T.E.S. Raghavan, and J.A. Filar (1983): "A finite algorithm for the switching control stochastic game," *OR Spektrum*, 5, 15–24.

Chapter 13

Linear (Semi-) Infinite Programs and Cooperative Games

BY JUDITH TIMMER AND NATIVIDAD LLORCA

13.1 Introduction

In 1975 Stef Tijs defended his Ph.D. thesis entitled "Semi-infinite and infinite matrix games and bimatrix games". Following this, his paper "Semi-infinite linear programs and semi-infinite matrix games" was published in 1979. Both these works deal with programs and noncooperative games in a (semi-)infinite setting. Several decades later these works and Stef Tijs himself inspired some researchers from Italy, Spain and The Netherlands to study cooperative games arising from linear (semi) infinite programs. These studies were performed under the inspiring supervision of Stef Tijs.

While studying these games it turned out that results from Tijs (1975, 1979) were very useful. For example, the critical number that is introduced in Tijs (1975) shows up again in the study of semi-infinite assignment problems (see Section 13.3.1), and some results about semi-infinite linear programs in Tijs (1979) are useful when studying semi-infinite linear production problems, as in Section 13.2.2. Hence, the early work of Stef provided a basis for studying cooperative games in a semi-infinite setting.

The aim of this work is to provide the reader with an overview of

P. Borm and H. Peters (eds.), Chapters in Game Theory, 267–285.

cooperative games arising from linear (semi-)infinite programs. In Section 13.2 semi-infinite programs and their corresponding games are presented, like flow games (Section 13.2.1), linear production games (Section 13.2.2) and games involving the linear transformation of products (Section 13.2.2). Section 13.3 concentrates on games arising from infinite programs like assignment games (Section 13.3.1) and transportation games (Section 13.3.2). For transportation games, a distinction is made between the transportation of an indivisible good (Section 13.3.2) or a divisible good (Section 13.3.2).

13.2 Semi-infinite Programs and Games

In this section we discuss three types of cooperative games that arise from semi-infinite problems. These are flow games, linear production games and games involving the linear transformation of products. All these games and their underlying problems have in common that they deal with a finite number of agents while another component is available in a countable infinite amount. For example, we consider linear production problems with a countable infinite number of production techniques. The main result is that each of these games has a non-empty core, just like its finite counterpart.

As far as we know, next to these three types a few other problems and corresponding cooperative games have been studied in a semi-infinite setting. Connection problems and games are studied in a semi-infinite setting in Fragnelli *et al.* (1999) and recently Fragnelli (2001) obtained some results for semi-infinite sequencing problems.

13.2.1 Flow games

Flow games in an infinite setting are introduced in Fragnelli *et al.* (1999). The authors consider a network with an infinite number of arcs that connect the source to the sink. These arcs are owned by a finite number of players. Each arc has one owner. A group of players can pool their privately owned arcs and thus obtains a subnetwork of the original network. Their goal is to maximize the flow on this subnetwork given the capacities of the arcs.

Formally, a network with privately owned arcs is described by a tuple

$$H = (M, A, f, N, p, S_0, P_0)$$

where M is a countable set of nodes and A is an infinite collection of arcs, which are elements $(i,j) \in M \times M$, $i \neq j$. If there is an arc $(i,j) \in A$ then there can be a flow from node i to node j but not from node j to node i. Multiple arcs between two nodes are also allowed. The map $f : A \to (0,\infty)$ assigns to each arc $a \in A$ its capacity $f(a)$. N is a finite set of players and the map $p : A \to N$ assigns to each arc a the player who owns it. Finally, S_0 and P_0 are special nodes in M that are called the source and the sink, respectively.

Given a network H, define

$$\omega(m)_- = \{a \in A | \exists i \in M : a = (i,m)\}$$

and

$$\omega(m)_+ = \{a \in A | \exists j \in M : a = (m,j)\}.$$

The sets $\omega(m)_-$ and $\omega(m)_+$ denote the set of arcs entering and leaving node m, respectively. A flow on network H is a map $\phi : A \to [0,\infty)$ such that

$$\phi(a) \leq f(a) \text{ for all arcs } a,$$

that is, a flow on an arc is restricted by its capacity, and

$$\sum_{a \in \omega(m)_-} \phi(a) = \sum_{a \in \omega(m)_+} \phi(a) \text{ for all } m \in M \setminus \{S_0, P_0\},$$

at each node the incoming flow is as large as the outgoing flow. The value $w(\phi)$ of a flow ϕ is defined as the outgoing flow at the source,

$$w(\phi) = \sum_{a \in \omega(S_0)_+} \phi(a).$$

In order to achieve results like those for finite flows, the authors assume that the total capacity of the arcs is finite:

$$\sum_{a \in A} f(a) < \infty. \tag{13.1}$$

Given this assumption they show that each flow has a finite value and that there exists a flow ϕ_0 that attains the maximal value on this network, that is, $w(\phi_0) \geq w(\phi)$ for all flows ϕ. Denote this maximal value by $w^*(H)$.

The flow game (N, v_H) corresponding to the network H is defined as follows. Let $S \subset N$ be a coalition of players. Let H_S be the subnetwork of H obtained by removing all arcs not in $p^{-1}(S)$. The value $v_H(S)$ of coalition S is the maximal value of a flow on its subnetwork H_S, $v_H(S) = w^*(H_S)$. The main result in Fragnelli *et al.* (1999) for flow games is that the core $C(v_H)$,

$$C(v_H) = \left\{ x \in \mathbb{R}^N \,\middle|\, \sum_{i \in N} x_i = v_H(N), \sum_{i \in S} x_i \geq v_H(S), S \subset N \right\},$$

of the flow game is non-empty. Hence, there exists an allocation of $v_H(N)$ to the players in N such that no coalition S has an incentive to deviate because they receive at least as much as they can obtain on their own.

Theorem 13.1 (Fragnelli et al., 1999) *Given a network H satisfying (13.1), the game (N, v_H) has a non-empty core.*

13.2.2 Linear Production Games

A semi-infinite linear production (LP) problem describes a production situation with a countable infinite number of linear production techniques and a finite number of resources. Each producer owns a bundle of resources. He may use these to produce on his own or to cooperate with other producers. In the latter case the cooperating producers pool their resources and act like one large producer. All produced goods can be sold on the market at exogenous market prices. This means that the producers cannot influence the market prices. It is assumed that there are no production costs. The goal of each producer is to maximize the total revenue of the products given the amount of resources that are available.

Such a problem is described by a tuple (N, A, B, c) where N is the finite set of producers. Let R be the finite set of resources and $Q = \mathbb{N} = \{1, 2, 3, \ldots\}$ the countable infinite set of linear production techniques. The matrix $A \in \mathbb{R}_+^{R \times Q}$ is the technology matrix where element A_{jk} describes how much of resource j is needed to produce 1 unit of product k. Because production techniques are linear one needs, for example, $5A_{jk}$ units of resource j to produce five units of product k, and so on. The resource matrix $B \in \mathbb{R}_+^{R \times N}$ tells us that producer i has B_{ji} units of resource j. The vector of market prices of the produced goods is denoted by $c \in \mathbb{R}_+^Q \setminus \{0\}$.

For the moment assume that for any resource there is at least one producer who owns a positive quantity of it. Furthermore, if product

k has a positive market price c_k then at least one resource is needed to produce k. In other words, there is a resource j such that $A_{jk} > 0$. Finally, all the producers take the market prices as given and all products can be sold on the market.

The cooperative LP game corresponding to such a semi-infinite LP problem is denoted by the pair (N, v) with the function v defined by

$$\begin{aligned} v(S) = \sup \quad & x^T c \\ \text{s.t.} \quad & Ax \leq Be_S \\ & x \geq 0 \end{aligned}$$

for all coalitions S of producers where

$$e_{S,i} = \begin{cases} 1, & i \in S, \\ 0, & i \in N \setminus S, \end{cases}$$

and $v(\emptyset) = 0$. Hence, the value of coalition S is equal to the maximal revenue it can achieve from selling the products that are produced from its resources. These LP games are studied by Fragnelli *et al.* (1999) and Tijs *et al.* (2001). Specific attention is paid to the question when a core-element can be constructed via a related dual linear program. Owen (1975) shows that this is always possible for LP games corresponding to finite LP problems. H is argument goes along the following lines. The linear program that determines the value of coalition N is

$$\begin{aligned} \max \quad & x^T c \\ \text{s.t.} \quad & Ax \leq Be_N \\ & x \geq 0 \end{aligned}$$

and its related dual linear program equals

$$\begin{aligned} \min \quad & y^T Be_N \\ \text{s.t.} \quad & A^T y \geq c \\ & y \geq 0. \end{aligned}$$

The assumptions on A, B and c, and the finiteness of these programs imply that both programs have the same finite value. Owen (1975) shows that for any optimal solution $\bar{y}$ of the dual program, the vector $z \in \mathbb{R}^N$ defined by $z_i = \bar{y}^T Be_{\{i\}}$ is a core-element of the corresponding LP game. Thus one can find a core-element with little effort since only one linear program has to be solved instead of determining the values $v(S)$ for all coalitions S in order to calculate the core $C(v)$.

For games corresponding to semi-infinite LP problems, this construction of core-elements need not always work as the example below shows.

Example 13.2 (Tijs, 1979) Let (N, A, B, c) be the semi-infinite LP problem where $N = \{1\}$,

$$A = \begin{bmatrix} 1 & 1 & 1 & 1 & \cdots \\ 0 & 1 & \frac{1}{2} & \frac{1}{3} & \cdots \end{bmatrix},$$

$Be_N = (1,0)^T$ and $c^T = (1,2,2,2,\ldots)$. Then $v(N) = 1$ and this is unequal to the value of the dual program for coalition N, which equals 2. We are confronted with a so-called duality gap, that is, the linear program for N and its dual program do not have the same value. Consequently, we cannot construct a core-element in the same fashion as Owen did, because such a core-element z would not satisfy $\sum_{i \in N} z_i = v(N)$.$\triangleleft$

Fragnelli *et al.* (1999) give two conditions on semi-infinite LP problems such that there is no duality gap for coalition N and a core-element can be constructed via the dual program.

Theorem 13.3 (Fragnelli et al., 1999) *Let (N, A, B, c) be a semi-infinite LP problem such that*

- $\sup_k c_k = \gamma < \infty$,
- $\sup_j A_{jk} \geq \alpha > 0$ *for all* $k \in Q$.

Then the corresponding LP game (N, v) has a non-empty core.

The first condition says that all market prices c_k have a finite upper bound and according to the second condition there is a minimal amount of resources, α, that is useful for production.

A more general analysis of semi-infinite LP problems and games can be found in Tijs *et al.* (2001). They study semi-infinite LP problems (N, A, B, c) with no other assumptions than $A_{jk} \geq 0$, $B_{ji} \geq 0$ and $c_k \geq 0$ for all $i \in N$, $j \in R$ and $k \in Q$. Let

$$v_d(N) = \inf\{y^T Be_N | A^T y \geq c,\ y \geq 0\}$$

be the value of the dual program for coalition N and $v_p(N) = v(N)$. Further, let

$$\text{Owen}(A, B, c) = \{(y^T Be_{\{i\}})_{i \in N} | y \text{ is optimal for the dual of } N\}$$

be the Owen set, which is the set of all vectors that can be constructed along the same lines as Owen did for finite LP problems, and let $\text{Core}(A, b, c) = C(v)$. The relations between the Owen set and the core are summarized in the following theorem.

Theorem 13.4 (Tijs et al., 2001) *Let (N, A, B, c) be a semi-infinite LP problem. If $v_p(N) = v_d(N)$ then $Owen(A, B, c) \subset Core(A, B, c)$. Otherwise, $Owen(A, B, c) \cap Core(A, B, c) = \emptyset$.*

Hence, if there is no duality gap, that is, $v_p(N) = v_d(N)$, then an element of the Owen set is a core-element of the game. Otherwise, one cannot use the Owen set to find core-elements. Finally, if coalition N has a finite value in the corresponding game then the core is non-empty as is shown below.

Theorem 13.5 (Tijs et al., 2001) *Let (N, A, B, c) be a semi-infinite LP problem with corresponding LP game (N, v). If $v(N) < \infty$ then the game has a non-empty core.*

13.2.3 Games Involving Linear Transformation of Products

Problems involving the linear transformation of products (LTP) are introduced by Timmer *et al.* (2000a) as generalizations of LP problems. Two assumptions for LP problems are that all producers have the same production techniques (represented by the production matrix A) and any production technique has only one output good. These assumptions do not hold for LTP problems. Hence, in an LTP problem different producers may have different production (transformation) techniques and such a transformation technique may have by-products.

A semi-infinite extension of LTP problems, where the number of transformation techniques is countable infinite, is studied in Timmer *et al.* (2000b) and *Tijs et al.* (2001). A semi-infinite LTP problem is denoted by a tuple (N, A, D, ω, p) where N is the finite set of producers. Let $D(i)$ be the set of all transformation techniques of producer i. Then $D = \cup_{i \in N} D(i)$ is the infinite set of available techniques. Let M be the finite set of goods. Then $A \in I\!R^{M \times D}$ denote the transformation matrix where a^k is the column corresponding to transformation technique $k \in D = \{1, 2, 3, \ldots\}$. Each row in A corresponds to a good in M. The resource bundle of producer $i \in N$ is $\omega(i) \in I\!R^M_+$, $\omega = (\omega(i))_{i \in N}$ and $p \in I\!R^M_+ \setminus \{0\}$ denotes the exogenous market prices.

Because positive elements in A indicate output goods and negative elements input goods, the resource matrix $G \in I\!R^{M \times D}_+$ can be defined by $G_{jk} = \max\{0, -A_{jk}\}$ for all $j \in M$, $k \in D$, and the kth column of G is denoted by g^k. The resources owned by a coalition $S \subset N$

of producers are $\omega(S) = \sum_{i \in S} \omega(i)$. Denote by $D(S) = \cup_{i \in S} D(i)$ the set of all techniques available for coalition S; thus $D = D(N)$. Let $y_k \geq 0$ be the activity level, or productivity factor, of technique $k \in D$ and $y = (y_k)_{k \in D}$. For the moment assume that each transformation technique uses at least one input good to produce at least one output good.

In the corresponding LTP game (N, v) the value of coalition S is the smallest upper bound of its profit,

$$\begin{aligned} v(S) = \sup \quad & p^T(\omega(S) + Ay) \\ \text{s.t.} \quad & Gy \leq \omega(S) \\ & y_k = 0 \text{ if } k \notin D(S) \\ & y \geq 0. \end{aligned} \tag{13.2}$$

In Timmer *et al.* (2000b) the authors are interested in finding a core-element of the LTP game via a related dual program, as is done in the previous subsection. If one considers coalition N then (13.2) reduces to

$$\begin{aligned} v(N) = \sup \quad & p^T(\omega(N) + Ay) \\ \text{s.t.} \quad & Gy \leq \omega(N) \\ & y \geq 0. \end{aligned}$$

because all techniques belong to $D = D(N)$. The dual program related to this problem is

$$\begin{aligned} \inf \quad & (z + p)^T \omega(N) \\ \text{s.t.} \quad & G^T z \geq A^T p \\ & z \geq 0. \end{aligned}$$

If $\underline{z}$ is an optimal solution of this program and if there is no duality gap, that is, $v(N) = (\underline{z}+p)^T \omega(N)$, then $x \in \mathbb{R}^N$ defined by $x_i = (\underline{z}+p)^T \omega(i)$ for all $i \in N$ is a core-element of the LTP game. Unfortunately, there need not always exist an optimal solution of the dual program, and also the absence of a duality gap is not guaranteed.

Example 13.6 (Timmer et al., 2000b) Consider the semi-infinite LTP problem with a single producer, two goods, $\omega(N) = (0, 1)^T$, $p = (1, 1)^T$ and

$$A = \begin{bmatrix} -1 & -\frac{1}{2} & -\frac{1}{3} & -\frac{1}{4} & \cdots & -\frac{1}{k} & \cdots \\ 1 & 1 & 1 & 1 & \cdots & 1 & \cdots \end{bmatrix}.$$

Then $v(N) = 1$ with optimal activity vector $y = 0$. The value of the dual program is ∞ since there exists no feasible solution z. Hence, there is no optimal solution to the dual program and there is a duality gap.$\triangleleft$

This example indicates that we need conditions on a semi-infinite LTP problem if we want to find a core-element of the corresponding game via the dual program. In Timmer *et al.* (2000b) two such sets of conditions are presented. For the first set, let 0_M be the zero-vector in $\mathbb{R}^M$ and e^j the jth unit vector in $\mathbb{R}^M$ with $e^j_m = 1$ if $m = j$ and $e^j_m = 0$ otherwise. Denote by $CC(B)$ the convex cone generated by the (infinite) set B of vectors in $\mathbb{R}^M \times \mathbb{R}$. Define

$$K = CC\left(\left\{\begin{pmatrix} g^k \\ p^T a^k \end{pmatrix}_{k\in D}, \begin{pmatrix} e^j \\ 0 \end{pmatrix}_{j\in M}\right\}\right)$$

and let $cl(K)$ be the closure of the set K. Now one can show the following result.

Theorem 13.7 (Timmer et al., 2000b) *Let (N, A, D, ω, p) be a semi-infinite LTP problem. If*

- $\omega(N) \in \mathbb{R}^M_{++}$,
- $\begin{pmatrix} 0_M \\ 1 \end{pmatrix} \notin cl(K)$

then the corresponding LTP game has a non-empty core.

The first condition says that for any good there is a producer who owns a positive amount of it and according to the second condition, the producers cannot earn a positive profit when using no inputs. The proof of this theorem shows that these conditions are sufficient to allow us to construct a core-element via the dual program.

A second set of conditions is similar to the conditions in Theorem 13.3 for semi-infinite LP problems.

Theorem 13.8 (Timmer et al., 2000b) *Let (N, A, D, ω, p) be a semi-infinite LTP problem such that*

- $\sup_{k\in D} p^T a^k = \gamma < \infty$,
- $\max_{j\in M} G_{jk} \geq \alpha > 0$ *for all* $k \in D$.

Then the corresponding LTP game and all its subgames have a non-empty core.

Tijs *et al.* (2001) also study semi-infinite LTP problems and related games but they only require that N is a finite set of agents, $A \in \mathbb{R}^{M\times D}$, $D = \{1, 2, 3, \ldots\}$, $\omega(i) \in \mathbb{R}^M_+$ and $p \in \mathbb{R}^M_+$. These conditions are sufficient to show the following result.

Theorem 13.9 (Tijs et al. (2001)) *Let (N, A, D, ω, p) be a semi-infinite LTP problem and (N, v) its corresponding game. If $v(N) < \infty$ then the game has a non-empty core.*

13.3 Infinite Programs and Games

This section is devoted to semi-infinite assignment and transportation problems and corresponding games. These problems involve linear programs with an infinite number of variables and of constraints. Nevertheless the corresponding games are referred to as being semi-infinite because their player sets are partitioned into two disjoint subsets: one set is finite while the other is countable infinite. Also in these problems duality gaps may arise. Hence, in all cases considered below one of the first results will be about the absence of such a gap. Further, the emphasis lies on showing the existence of (ε-)core-elements.

13.3.1 Assignment Games

In finite assignment problems two finite sets of agents have to be matched in such a way that the reward obtained from these matchings is as large as possible. Shapley and Shubik (1972) study these problems in a game-theoretic setting, resulting in cooperative assignment games. A semi-infinite extension of assignment problems and games is given in Llorca *et al.* (1999). An example of such a semi-infinite assignment problem is the following. Consider a textile firm whose marketing policy is to produce unique pieces of textile. The firm owns a finite number of printing machines that can be programmed to print a piece of fabric. There are an infinite number of patterns available. The machines can print all of these patterns, but with different (bounded) rewards. The firm wan ts to maximize the total reward from matching machines with patterns. Therefore, it has to tackle an assignment problem in which there are a finite number of one type (machines) and an infinite number of the other type (possible designs).

A semi-infinite (bounded) assignment problem is denoted by a tuple (M, W, A) where $M = \{1, 2, ..., m\}$ is a finite set of agents of one type and $W = \mathbb{N}$ is the countable infinite set of agents of the other type. Matching agent $i \in M$ to agent $j \in W$ results in the nonnegative reward a_{ij} which is bounded from above. All these rewards are gathered in the matrix A. In the sequel we write $\mathcal{A}$ to denote the assignment problem (M, W, A).

An *assignment plan* $Y = [y_{ij}]_{i \in M, j \in W}$ is a matrix with 0,1-entries where $y_{ij} = 1$ if i is assigned to j and $y_{ij} = 0$ otherwise. Each agent $i \in M$ will be assigned to at most one agent $j \in W$ and vice versa, therefore $\sum_{j \in W} y_{ij} \leq 1$ and $\sum_{i \in M} y_{ij} \leq 1$. Then

$$v_p(\mathcal{A}) = \sup \left\{ \sum_{(i,j) \in M \times W} a_{ij} y_{ij} \,\middle|\, \sum_{j \in W} y_{ij} \leq 1, \ \sum_{i \in M} y_{ij} \leq 1, \ y_{ij} \in \{0,1\} \right\}$$

is the smallest upper bound of the benefit that the agents in M and W together can achieve.

Given an assignment problem $\mathcal{A} = (M, W, A)$, the corresponding semi-infinite bounded assignment game (N, w) is a cooperative game with countable infinite player set $N =$ '$M \cup W$', that is, each player corresponds to an agent in M or to an agent in W. Let S be a coalition of players in N and define $M_S = M \cap S$ and $W_S = W \cap S$. Then the worth of coalition S is $w(S) = 0$ if $S = M_S$ or $S = W_S$. If there is only one type of agents present then no matchings can be made. Otherwise, $w(S) = v_p(\mathcal{A}_S)$ where $\mathcal{A}_S$ denotes the (semi-infinite) assignment problem $(M_S, W_S, [a_{ij}]_{i \in M_S, j \in W_S})$.

The value $w(N) = v_p(\mathcal{A})$ of the grand coalition N is determined by a linear program, the so-called primal program. According to Sánchez-Soriano *et al.* (2001) the condition $y_{ij} \in \{0,1\}$ may be replaced by $y_{ij} \geq 0$. When doing so, the corresponding dual program is

$$\begin{aligned} v_d(\mathcal{A}) = \inf \quad & \textstyle\sum_{i \in M} u_i + \sum_{j \in W} v_j \\ \text{s.t.} \quad & u_i + v_j \geq a_{ij}, \\ & u_i, v_j \geq 0, \ \text{for all i} \in \text{M}, \ \text{j} \in \text{W}. \end{aligned}$$

Both the primal and the dual program have an infinite number of variables and an infinite number of constraints. Hence, they are infinite programs, for which a gap between the optimal values can appear. Therefore, one would like to know if the primal and the dual program in semi-infinite assignment problems have the same value and if there exists an optimal solution of the dual problem. If so, then one can construct a core-element like Owen did for LP problems.

Theorem 13.10 (Llorca et al., 1999) *Let $\mathcal{A}$ be a semi-infinite bounded assignment problem. Then $v_p(\mathcal{A}) = v_d(\mathcal{A})$, there is no duality gap, and there exists an optimal solution for the dual program.*

A corollary of this theorem is that semi-infinite assignment games have a non-empty core.

To continue the analysis of semi-infinite bounded assignment problems, the so-called critical number is introduced. The origin of this number is based on a similar concept introduced by Tijs (1975) for (semi-) infinite matrix games and bi-matrix games. For assignment problems it is defined as follows. If there exists an $n \in \mathbb{N}$ with $v_p(\mathcal{A}_n) = v_p(\mathcal{A})$ where $\mathcal{A}_n$ denotes the finite assignment problem $(M, \{1, \ldots, n\}, [a_{ij}]_{i \in M, j=1,\ldots,n})$ then

$$c(\mathcal{A}) = \min \{n \in \mathbb{N} \mid v_p(\mathcal{A}_n) = v_p(\mathcal{A})\}.$$

Otherwise, $c(\mathcal{A}) = \infty$. The critical number tells whether there exists a finite subproblem with the same value as $\mathcal{A}$ or not. In terms of semi-infinite assignment games, $c(\mathcal{A}) < \infty$ says that there exists an optimal assignment plan. If $c(\mathcal{A}) = \infty$, then there exists no optimal assignment plan but we can use a finite auxiliary matrix H corresponding to the matrix A to approach the value $v_p(\mathcal{A})$. For this we need a new concept, namely the hard-choice number. The next example explains these new terms.

Example 13.11 Let $M = \{1, 2\}$, $W = \mathbb{N}$ and

$$A = \begin{bmatrix} 1 & \frac{1}{2} & 1 & \frac{2}{3} & \frac{3}{4} & \frac{4}{5} & \cdots \\ 1\frac{1}{2} & 1\frac{2}{3} & 1\frac{3}{4} & 1\frac{4}{5} & 1\frac{5}{6} & 1\frac{6}{7} & \cdots \end{bmatrix}.$$

Agent $1 \in M$ attains a maximal value of 1 if she is assigned to agent 1 or 3 in W. These agents in W are the best two ($= |M|$) choices for agent $1 \in M$ and this is denoted by $C_1 = \{1, 3\}$.

However, there is no largest value that agent $2 \in M$ can attain because the reward a_{2n} reaches the value 2 from below when n goes to infinity. Hence, $C_2 = \emptyset$: agent $2 \in M$ has no best choice. Now the hard-choice number $n^*(\mathcal{A})$ is the smallest number in $\mathbb{N} \cup \{0\}$ such that

$$\cup_{i \in M} C_i \subset \{1, 2, \ldots, n^*(\mathcal{A})\},$$

in this example $n^*(\mathcal{A}) = 3$. Further, $c(\mathcal{A}) = \infty$ which means that there exists no optimal assignment plan. Therefore we construct a finite auxiliary problem $\mathcal{H}$ to approach the value $v_p(\mathcal{A})$. $\mathcal{H}$ is the finite assignment problem $(M, \{1, \ldots, n^*(\mathcal{A}), \eta\}, H)$ where η is an artificial agent and the matrix $H \in I\!R_+^{M \times (n^*(\mathcal{A})+1)}$ is defined by $h_{ij} = a_{ij}$ if $j \leq n^*(\mathcal{A})$ and $h_{i\eta} = \sup_{k > n^*(\mathcal{A})} a_{ik}$. In this example

$$H = \left[\begin{array}{ccc|c} \mathbf{1} & \frac{1}{2} & 1 & 0 \\ 1\frac{1}{2} & 1\frac{2}{3} & 1\frac{3}{4} & \mathbf{2} \end{array} \right]$$

where the vertical line seperates the artificial agent, agent 4, from the others. Now $v_p(\mathcal{H}) = 3 = v_p(\mathcal{A})$. An optimal assignment plan in $\mathcal{H}$ is Y' with $y'_{11} = 1$, $y'_{24} = 1$ and $y'_{ij} = 0$ otherwise. From this, it follows that the assignment plan Y for $\mathcal{A}$ defined by $y_{11} = 1$, $y_{2n} = 1$ for one $n > 3$ and $y_{ij} = 0$ otherwise, is a $\frac{1}{n}$-optimal assignment, which means that the total reward from the assignment plan Y equals $v_p(\mathcal{A}) - \frac{1}{n}$. $\triangleleft$

In general, if $c(\mathcal{A}) = \infty$ then $v_p(\mathcal{A})$ and an ε-optimal assignment plan can be obtained with the aid of the corresponding finite assignment problem $\mathcal{H}$.

Theorem 13.12 (Llorca et al., 1999) *Let $\mathcal{A}$ be a semi-infinite bounded assignment problem with $c(\mathcal{A}) = \infty$ and let $\mathcal{H}$ be the corresponding finite problem. Then*

- $v_p(\mathcal{A}) = v_p(\mathcal{H})$;
- *From each optimal assignment plan Y' for $\mathcal{H}$ and for each $\varepsilon > 0$ one can determine an ε-optimal assignment plan for $\mathcal{A}$.*

13.3.2 Transportation Games

Sánchez-Soriano *et al.* (2001b) introduce finite transportation games corresponding to transportation problems where a good has to be transported from suppliers to demanders. A semi-infinite extension of these games with indivisible goods is studied by Sánchez-Soriano *et al.* (2001a). Sánchez-Soriano *et al.* (2000) deals with semi-infinite transportation situations with divisible goods.

In a semi-infinite transportation problem the demand for a single good at a countable infinite number of places has to be covered from a finite number of supply points. The transportation of one unit of the good from a supply location to a demand point generates a certain (bounded) profit and the goal of the suppliers and demanders is to maximize the total profit derived from transportation.

Let P be the finite set of supply points and $Q = \mathbb{N}$ the countable infinite set of demand points. Supply point $i \in P$ has s_i units of the good available for transport and the demand at point $j \in Q$ equals d_j units. Both s_i and d_j are positive numbers for all $i \in P$ and $j \in Q$. The profit of transporting one unit of the good from supplier i to demander j is a nonnegative real number t_{ij}, which is bounded from above. Thus, a semi-infinite bounded transportation problem an be described by the 5-tuple (P, Q, T, s, d) where $T = [t_{ij}]_{i\in P, j\in Q}$ is the matrix of profits, and $s = (s_i)_{i\in P}$ and $d = (d_j)_{j\in Q}$ are the supply and demand vectors, respectively. Denote by $\mathcal{T}$ the transportation problem (P, Q, T, s, d).

Indivisible goods

In this subsection the good to be transported is indivisible. Therefore the supply and demand vectors s and d will only consist of positive integer numbers. A transportation plan $X = [x_{ij}]_{i\in P, j\in Q}$ is a matrix with integer entries where x_{ij} is the number of units of the good that will be transported from supply point i to demand point j. Each supply point $i \in P$ cannot supply more than s_i units of the good, $\sum_{j\in Q} x_{ij} \leq s_i$. Similarly, each demand point $j \in Q$ wants to receive at most d_j units, $\sum_{i\in P} x_{ij} \leq d_j$. Thus the maximal profit that the supply and demand points can achieve is

$$v_p(\mathcal{T}) = \sup\left\{ \sum_{(i,j)\in P\times Q} t_{ij}x_{ij} \,\middle|\, \sum_{j\in Q} x_{ij} \leq s_i,\ \sum_{i\in P} x_{ij} \leq d_j,\ x_{ij} \in \{0\} \cup \mathbb{N} \right\}.$$

The semi-infinite transportation game (N, w) corresponding to a semi-infinite transportation problem $\mathcal{T}$ is a cooperative game with countable infinite player set $N = P \cup Q$. Let $S \subset N$ be a coalition of players and define $P_S = P \cap S$ and $Q_S = Q \cap S$. If $S = P_S$ or $S = Q_S$ then there are no demand or supply points present in coalition S and therefore no transportation plans can be made. In this case, $w(S) = 0$. Otherwise, the worth of coalition S equals

$$\begin{aligned} w(S) &= \sup\{ \sum_{(i,j)\in P_S\times Q_S} t_{ij}x_{ij} | \sum_{j\in Q_S} x_{ij} \leq s_i, \\ &\qquad \sum_{i\in P_S} x_{ij} \leq d_j,\ x_{ij} \in \{0\} \cup \mathbb{N}\} \\ &= v_p(\mathcal{T}_S) \end{aligned}$$

where $\mathcal{T}_S = (P_S, Q_S, [t_{ij}]_{i\in P_S, j\in Q_S}, \{s_i\}_{i\in P_S}, \{d_j\}_{j\in Q_S})$ is the problem $\mathcal{T}$ restricted to agents in coalition S.

It is shown in Sánchez-Soriano *et al.* (2001) that

$$\inf\left\{\sum_{i\in P} s_i u_i + \sum_{j\in Q} d_j v_j \,\middle|\, u_i + v_j \geq t_{ij},\; u_i, v_j \geq 0 \text{ for all i} \in \text{P},\; \text{j} \in \text{Q}\right\}$$

is the dual program corresponding to the program that determines $v_p(\mathcal{T})$. Let $v_d(\mathcal{T})$ be the value and $O_d(\mathcal{T})$ the set of optimal solutions of this dual program. As was done for assignment problems, also here one would like to show the existence of an optimal solution of the dual program and the absence of a duality gap. This can be done using the results from semi-infinite assignment problems because such an assignment problem is a semi-infinite transportation problem and vice versa.

Theorem 13.13 (Sánchez-Soriano et al., 2001a) *Let $\mathcal{T}$ be a semi-infinite bounded transportation problem. Then $v_p(\mathcal{T}) = v_d(\mathcal{T})$ and $O_d(\mathcal{T})$ is non-empty.*

Due to the absence of a duality gap, one can find a core-element of the semi-infinite transportation game via the Owen set, which for semi-infinite transportation problems is defined by

$$\text{Owen}(\mathcal{T}) = \left\{ z \in I\!R^N \,\middle|\, \begin{array}{l} \exists (u,v) \in O_d(\mathcal{T}) \text{ such that } z_k = s_k u_k \\ \text{if } k \in P \text{ and } z_k = d_k v_k \text{ if } k \in Q \end{array} \right\}.$$

Notice that each element of a vector $(u, v) \in O_d(\mathcal{T})$ is the mean profit that a player will receive per unit of his supply or demand. Hence, an Owen vector $(s_i u_i, d_j v_j)_{i\in P, j\in Q}$ is a vector of profits that agents obtain from their supply or demand.

Theorem 13.14 (Sánchez-Soriano et al., 2001a) *Let $\mathcal{T}$ be a semi-infinite transportation problem and (N, w) the corresponding game. Then* Owen$(\mathcal{T}) \subset$ C(w).

Combining the Theorems 13.13 and 13.14, one can conclude that a transportation game corresponding to a transportation problem with an indivisible good has a non-empty core.

Perfectly Divisible Goods

After having studied semi-infinite transportation problems with indivisible goods, attention will be paid to problems with perfectly divisible goods like e.g. gas, electricity or sand. These goods need not be supplied in integer units and therefore the elements of the supply and demand vectors s and d are (positive) real numbers and a transportation plan X is a matrix with entries $x_{ij} \geq 0$. A transportation problem with a perfectly divisible good is called a continuous transportation problem to distinguish it from problems with indivisible goods.

Using the absence of a duality gap for semi-infinite transportation problems with indivisible goods, one can establish that also transportation problems with perfectly divisible goods have no duality gap.

Theorem 13.15 (Sánchez-Soriano et al., 2000) *Let $\mathcal{T}$ be a semi-infinite continuous transportation problem. Then $v_p(\mathcal{T}) = v_d(\mathcal{T})$.*

Showing the existence of a core-element for these problems turned out to be not that easy. An intermediate result is that so-called ε-core-elements exist. Given an arbitrary cooperative game (N, w) a vector $z \in I\!R^N$ is said to be an ε-core-element of this game if

$$\sum_{i \in N} z_i = w(N)$$

and

$$\sum_{i \in S} z_i \geq w(S) - \varepsilon$$

for all $S \subset N$. Thus, an ε-core-element shares $w(N)$ among the players in such a way that a coalition S can gain at most ε by splitting off.

Theorem 13.16 (Sánchez-Soriano et al. 2000) *Let $\varepsilon > 0$ and let (N, w) be the cooperative game corresponding to the semi-infinite continuous transportation problem $\mathcal{T}$. Then there exists an ε-core-element of the game (N, w).*

Sánchez-Soriano *et al.* (2000) show for two types of semi-infinite continuous transportation problems that the corresponding games have a non-empty core. The first type are problems $\mathcal{T}$ with a finite total demand.

Theorem 13.17 (Sánchez-Soriano et al., 2000) *Let $\mathcal{T}$ be a semi-infinite continuous transportation problem with $\sum_{j \in Q} d_j < \infty$. Then the corresponding transportation game (N, w) has a non-empty core.*

If the total demand is not finite, an extra condition is needed to ensure the existence of core-elements. This condition is the following: there exists a positive number η such that

$$d_j \geq \eta > 0 \text{ for all } j \in Q. \tag{13.3}$$

The number η may be interpreted as the minimal amount of the good that is useful. Sánchez-Soriano *et al.* (2000) show that because of (13.3) one can find a specific finite transportation problem $\bar{\mathcal{T}}$ with the same value as the original problem $\mathcal{T}$, $v_p(\bar{\mathcal{T}}) = v_p(\mathcal{T})$. Now the following result holds.

Theorem 13.18 (Sánchez-Soriano et al., 2000) *Let $\mathcal{T}$ be a semi-infinite continuous transportation problem with $\sum_{j\in Q} d_j = \infty$ that satisfies (13.3). Then $O_d(\mathcal{T}) \neq \emptyset$.*

Consequently, the Owen set, Owen($\mathcal{T}$), is non-empty. Finally, the absence of a duality gap for semi-infinite continuous transportation problems implies that the Owen set lies in the core of the corresponding game.

Theorem 13.19 (Sánchez-Soriano et al., 2000) *Let $\mathcal{T}$ be a semi-infinite continuous transportation problem with $\sum_{j\in Q} d_j = \infty$ that satisfies (13.3) and let (N, w) be the corresponding game. Then the core $C(w)$ of this game is non-empty.*

Theorems 13.17 and 13.19 present the two types of semi-infinite continuous transportation problems for which the non-emptiness of the core of the corresponding game has been shown.

13.4 Concluding remarks

In this chapter we presented several cooperative games arising from linear semi-infinite and infinite programs. Starting point was the Ph.D. thesis of Stef Tijs (1975) on (semi-)infinite matrix games and bimatrix games, and a subsequent paper, Tijs (1979). Although both these works deal with noncooperative games, they turned out to be inspiring and useful in the study of semi-infinite cooperative games.

When extending a problem to a (semi-) infinite setting, the existence of core-elements of the corresponding game is not that obvious anymore. One can try to prove the non-emptiness of the core in a direct way as in the theorems 13.5 and 13.9, or via the dual program and the Owen set. This latter approach requires the absence of a duality gap; another result that had to be shown using tools from linear (semi-) infinite programming.

These two approaches do not always work, as is shown in the previous section. There Sánchez-Soriano *et al.* (2000) were not able to show that the game corresponding to a semi-infinite continuous transportation problem with infinite total demand and no positive lower bound for the demands, has a non-empty core. Future research should try to solve this.

References

Fragnelli, V. (2001): "On the balancedness of semi-infinite sequencing games," *Preprint*, Dipartimento di Matematica dell'Universià di Genova, N. 442 (2001).

Fragnelli, V., F. Patrone, E. Sideri, and S.H. Tijs (1999): "Balanced games arising from infinite linear models," *Mathematical Methods of Operations Research*, 50, 385–397.

Llorca, N., S. Tijs, and J. Timmer (1999): "Semi-infinite assignment problems and related games," *CentER Discussion Paper* 9974, Tilburg University, The Netherlands.

Owen, G. (1975): "On the core of linear production games," *Mathematical Programming*, 9, 358–370.

Sánchez-Soriano, J., N. Llorca, S.H. Tijs, and J. Timmer (2000): "On the core of semi-infinite transportation games with divisible goods," to appear in *European Journal of Operational Research.*

Sánchez-Soriano, J., N. Llorca, S.H. Tijs, and J. Timmer (2001a): "Semi-infinite assignment and transportation games," in: M.A. Goberna and M.A. López (eds.), *Semi-Infinite Programming: Recent Advances.* Dordrecht: Kluwer Academic Publishers, 349–362.

Sánchez-Soriano, J., M.A. López, and I. García-Jurado (2001b): "On the core of transportation games," *Mathematical Social Sciences* , 41, 215–225.

Shapley, L.S., and S. Shubik (1972): "The assignment game I: the core," *International Journal of Game Theory*, 1, 111–130.

Tijs, S.H. (1975) : *Semi-Infinite and Infinite Matrix Games and Bimatrix Games.* Ph.D. dissertation, University of Nijmegen, The Netherlands.

Tijs, S.H. (1979): "Semi-infinite linear programs and semi-infinite matrix games," *Nieuw Archief voor Wiskunde*, XXVII, 197–214.

Tijs, S.H., J. Timmer, N. Llorca, and J. Sánchez-Soriano (2001): "The Owen set and the core of semi-infinite linear production situations," in: M.A. Goberna and M.A. López (eds.), *Semi-Infinite Programming: Recent Advances.* Dordrecht: Kluwer Academic Publishers, 365–386.

Timmer, J., P. Borm, and J. Suijs (2000a): "Linear transformation of products: games and economies," *Journal of Optimization Theory and Applications*, 105, 677–706.

Timmer, J., N. Llorca, and S. Tijs (2000b): "Games arising from infinite production situations," *International Game Theory Review*, 2, 97–106.

Chapter 14

Population Uncertainty and Equilibrium Selection: a Maximum Likelihood Approach

BY MARK VOORNEVELD AND HENK NORDE

14.1 Introduction

In games with incomplete information (Harsanyi, 1967–1968) as usually studied by game theorists, the characteristics or types of the participating players are possibly subject to uncertainty, but the number of players is common knowledge. Recently, however, Myerson (1998a, 1998b, 1998c, 2000) and Milchtaich (1997) proposed models for situations—like elections and auctions—in which it may be inappropriate to assume common knowledge of the player set. In such games with population uncertainty, the set of actual players and their preferences are determined by chance according to a commonly known probability measure (a Poisson distribution in Myerson's work, a point process in Milchtaich's paper) and players have to choose their strategies before the player set is revealed.

After the introduction of the maximum likelihood principle by R.A. Fisher in the early 1920's (see Aldrich, 1997, for an interesting historical account), the method of selection on the basis of what is *most likely* to

P. Borm and H. Peters (eds.), Chapters in Game Theory, 287–314.

be right has gained tremendous popularity in the field of science dealing with uncertainty. Gilboa and Schmeidler (1999) recently provided an axiomatic foundation for rankings according to the likelihood function.

A first major topic of this chapter is the introduction of a general class of games with population uncertainty, including the Poisson games of Myerson (1998c) and the random-player games of Milchtaich (1997). In line with the maximum likelihood principle, the present chapter stresses those strategy profiles in a game with population uncertainty that are most likely to yield an equilibrium in the game selected by chance. Maximum likelihood equilibria were introduced in Borm *et al.* (1995a) in a class of Bayesian games.

The σ-algebra underlying the chance event that selects the actual game to be played may be too coarse to make the event in which a specific strategy profile yields an equilibrium measurable. A common mathematical approach (also used in a decision theoretic framework; cf. Fagin and Halpern, 1991) to assign probabilities to such events is to use the inner measure induced by the probability measure. Roughly, the inner measure of an event E is the probability of the largest measurable event included in E.

Under mild topological restrictions, an existence result for maximum likelihood equilibria is derived. Since the result establishes the existence of a maximum of the likelihood function, it differs significantly from standard equilibrium existence results that usually rely on a fixed point argument.

The use of inner measures is intuitively appealing and avoids measurability conditions. Still, the measurability issue is briefly addressed and it is shown that under reasonable restrictions the inner measure can be replaced by the probability measure underlying the chance event that selects the actual game. Moreover, it is shown that games with population uncertainty can be used to model situations in which, in addition to the set of players and their preferences, also the set of feasible actions of each player is subject to uncertainty. This captures a common problem in decision making, namely the situation in which decision makers have to plan or decide on their course of action, while still uncertain about contingencies that may make their choices impossible to implement.

A second major topic of this chapter is the use of maximum likelihood equilibria as an equilibrium selection device for finite strategic games. A finite strategic game can be perturbed by assuming that with a commonly known probability distribution there are trembles in the payoff functions of the players. We take two different approaches to equilibrium selection. In the first approach we search for strategy profiles that are most likely to be a Nash equilibrium if the players take into account that according to a certain probability distribution the real *payoffs* of the game differ slightly from those provided by the payoff functions. In the second approach, players take into account that according to a certain probability distribution their real *actions* differ slightly from those actions which they intend to play. We search for strategy profiles that are equilibria of the original game and give rise to nearby equilibria if the game is perturbed.

14.2 Preliminaries

For easy reference, this section summarizes results and definitions from topology, measure theory, and game theory that are used in the rest of the chapter. See Aliprantis and Border (1994) for additional information.

14.2.1 Topology

Let X and Y be topological spaces. A function $f : X \to Y$ is *sequentially continuous* if for every $x \in X$ and every sequence $(x^n)_{n \in \mathbb{N}}$ in X converging to x, it holds that $\lim_n f(x^n) = f(x)$. Sequential continuity is implied by continuity of functions; the converse is not true (Aliprantis and Border, 1994, Theorem 2.25). A function $f : X \to \mathbb{R}$ is

- *sequentially upper semicontinuous* if for every $x \in X$ and every sequence $(x^n)_{n \in \mathbb{N}}$ in X converging to x, it holds that $\limsup_n f(x^n) \leq f(x)$.
- *sequentially lower semicontinuous* if for every $x \in X$ and every sequence $(x^n)_{n \in \mathbb{N}}$ in X converging to x, it holds that $f(x) \leq \liminf_n f(x^n)$.

Sequential upper (lower) semicontinuity is implied by upper (lower) semicontinuity of functions; the converse is not true (Aliprantis and Border, 1994, Lemma 2.40). The topological space X is *separable* if it includes a countable dense subset. A set $A \subseteq X$ is

- *sequentially closed* if for every $x \in X$ and every sequence $(x^n)_{n \in \mathbb{N}}$ in A converging to x, it holds that $x \in A$. Every closed set is sequentially closed; the converse is not true (Aliprantis and Border, 1994, Example 2.10).

- *sequentially compact* if every sequence in A has a subsequence converging to an element of A.

The closure of a set A is denoted by $\overline{A}$, the cardinality of A is denoted by $|A|$. Weak and strict set inclusion are denoted by $\subseteq$ and $\subset$, respectively.

14.2.2 Measure Theory

Let (Ω, Σ, P) be a probability space, where Ω is a nonempty set, Σ is a σ-algebra on Ω, and P a probability measure on Σ. The *inner measure* $P_*(E)$ of a set $E \subseteq \Omega$ is defined as

$$P_*(E) := \sup \{P(F) \mid F \in \Sigma, F \subseteq E\}.$$

Roughly speaking, the inner measure of an event E is the probability of the largest measurable event contained in E. $P_*(E)$ is well-defined, since the set $\{P(F) \mid F \in \Sigma, F \subseteq E\}$ is nonempty ($\emptyset \in \Sigma, \emptyset \subseteq E$) and bounded above by one (P is a probability measure). Moreover, $P_*(E) = P(E)$ if $E \in \Sigma$. The *lower integral* of a bounded function $f : \Omega \to I\!R$ is defined as

$$\begin{aligned} \int_* f(\omega)\, P(d\omega) \quad := \quad & \sup \{ \int g(\omega)\, P(d\omega) \mid \\ & g \text{ Lebesgue integrable}, g \leq f \}. \end{aligned} \tag{14.1}$$

Boundedness of f implies that $\int_* f(\omega)\, P(d\omega) < \infty$. Inner measures and lower integrals are related via the following equality:

$$\forall E \subseteq \Omega : \quad P_*(E) = \int_* 1_E(\omega)\, P(d\omega), \tag{14.2}$$

where 1_E is the indicator function for the set E. Clearly, if f is itself Lebesgue integrable, then

$$\int_* f(\omega)\, P(d\omega) = \int f(\omega)\, P(d\omega). \tag{14.3}$$

Below, a version of Fatou's Lemma is shown to hold for lower integrals. First, a lemma is needed.

Lemma 14.1 *Let $f : \Omega \to I\!R$ be bounded. Then there exists a Lebesgue integrable function $h : \Omega \to I\!R$ such that $h \leq f$, and $\int_* f(\omega)\, P(d\omega) = \int h(\omega)\, P(d\omega)$.*

Proof. By (14.1) there is a sequence $(h^n)_{n\in\mathbb{N}}$ of Lebesgue integrable functions such that $h^n \leq f$ and $\int h^n(\omega)\ P(d\omega) \geq \int_* f(\omega)\ P(d\omega) - \frac{1}{n}$ for each $n \in \mathbb{N}$. The Lebesgue integrable function $h := \sup_n h^n$ clearly satisfies $h \leq f$. Consequently, $\int h(\omega)\ P(d\omega) \leq \int_* f(\omega)\ P(d\omega)$. Moreover, $\int h(\omega)\ P(d\omega) \geq \int h^n(\omega)\ P(d\omega) \geq \int_* f(\omega)\ P(d\omega) - \frac{1}{n}$ for all $n \in \mathbb{N}$, so $\int h(\omega)\ P(d\omega) \geq \int_* f(\omega)\ P(d\omega)$. Hence $\int h(\omega)\ P(d\omega) = \int_* f(\omega)\ P(d\omega)$. □

Proposition 14.2 *Let $(f^n)_{n\in\mathbb{N}}$ be a sequence of bounded functions $f^n : \Omega \to \mathbb{R}$. Then*

$$\limsup_n \int_* f^n(\omega)\ P(d\omega) \leq \int_* \limsup_n f^n(\omega)\ P(d\omega).$$

Proof. Lemma 14.1 implies that for each $n \in \mathbb{N}$ there exists a Lebesgue integrable function h^n with $h^n \leq f^n$ such that $\int_* f^n(\omega)\ P(d\omega) = \int h^n(\omega)\ P(d\omega)$. To this sequence $(h^n)_{n\in\mathbb{N}}$, the classical Fatou Lemma applies:

$$\begin{aligned}\limsup_n \int_* f^n(\omega)\ P(d\omega) &= \limsup_n \int h^n(\omega)\ P(d\omega) \\ &\leq \int \limsup_n h^n(\omega)\ P(d\omega). \qquad (14.4)\end{aligned}$$

Since $\limsup_n f^n \geq \limsup_n h^n$ and $\limsup_n h^n$ is Lebesgue integrable, it follows from (14.1) and (14.3) that

$$\begin{aligned}\int_* \limsup_n f^n(\omega)\ P(d\omega) &\geq \int_* \limsup_n h^n(\omega)\ P(d\omega) \\ &= \int \limsup_n h^n(\omega)\ P(d\omega). \qquad (14.5)\end{aligned}$$

Combining (14.4) and (14.5) yields the desired result. □

14.2.3 Game Theory

A *finite strategic game* is a tuple $G = \langle N, (A_i)_{i\in N}, (\succeq_i)_{i\in N}\rangle$, where N is a finite, nonempty set of players, each player $i \in N$ has a finite, nonempty set A_i of pure strategies, and a preference relation $\succeq_i$ over $\times_{j\in N} A_j$. A strategy profile $a = (a_i)_{i\in N} \in \times_{i\in N} A_i$ is a Nash equilibrium of G if $a \succeq_i (b_i, a_{-i})$ for each $i \in N$ and $b_i \in A_i$. Often the preference relations of the players are represented by payoff functions $u_i : \times_{j\in N} A_j \to \mathbb{R}$, which enables us to define the mixed extension of G. The set of mixed strategies of player $i \in N$ is denoted by

$$\Delta(A_i) = \{\alpha_i : A_i \to [0,1] \mid \sum_{a_i \in A_i} \alpha_i(a_i) = 1\}.$$

Payoffs are extended to mixed strategy profiles in the usual way. A strategy profile $\alpha = (\alpha_i)_{i \in N} \in \times_{i \in N} \Delta(A_i)$ is a *Nash equilibrium* of G if $u_i(\alpha) \geq u_i(\beta_i, \alpha_{-i})$ for each $i \in N$ and $\beta_i \in \Delta(A_i)$. The strategy profile α is a *strict (Nash) equilibrium* of G if α is a Nash equilibrium of G and for each player $i \in N$, the strategy α_i is the *unique* best response to the strategy profile α_{-i} of the other players. This implies that strict equilibria are equilibria in pure strategies.

14.3 Games with Population Uncertainty

In this section, games with population uncertainty are formally defined. Subsequently, games with population uncertainty are briefly compared with the random-player games of Milchtaich (1997) and the Poisson games of Myerson (1998c).

The set of potential players is a nonempty set N. Each potential player $i \in N$ has a strategy set A_i. The actual player set is determined by chance according to a probability space (Ω, Σ, P). To each state $\omega \in \Omega$ is associated a strategic game $G_\omega = \langle N_\omega, A_\omega, (\succeq_{i,\omega})_{i \in N_\omega} \rangle$ with a nonempty set of actual players $N_\omega \subseteq N$ having strategy space $A_\omega := \times_{i \in N_\omega} A_i$ and each player $i \in N_\omega$ having a preference relation $\succeq_{i,\omega}$ over A_ω. The tuple $\langle N, (A_i)_{i \in N}, \Omega, \Sigma, P, (G_\omega)_{\omega \in \Omega} \rangle$ is a *game with population uncertainty.*

In a game with population uncertainty there is uncertainty about the exact state of nature $\omega \in \Omega$, and consequently about the game $G_\omega = \langle N_\omega, A_\omega, (\succeq_{i,\omega})_{i \in N_\omega} \rangle$ that will be played. The probability measure P, according to which the state of nature is determined, is assumed to be common knowledge among the potential players.

Games with population uncertainty as defined above generalize the Poisson games of Myerson (1998c) and the random-player games of Milchtaich (1997). Milchtaich (1997, p.5) introduces random-player games, consisting of:

- a compact metric space X of potential players;
- a simple point process (cf. Daley and Vere–Jones, 1988) on X that determines the actual set of players;

- strategy sets defined by means of a continuous function ξ from a compact metric space Y to X. The strategy set of player $i \in X$ equals $\xi^{-1}(\{i\})$;
- bounded and measurable payoff functions giving a payoff $u(s, S)$ to an actual player who plays s when the strategies of the other players are S.

Every random-player game is easily seen to be a game $\langle N, (A_i)_{i \in N}, \Omega, \Sigma, P, (G_\omega)_{\omega \in \Omega} \rangle$ with population uncertainty: set N equal to X, A_i equal to $\xi^{-1}(\{i\})$, identify (Ω, Σ, P) with the distribution of the simple point process, and the preferences with the utility functions u. Milchtaich (1997, p.6, Example 3) indicates that the Poisson games of Myerson (1998c) are random-player games and consequently games with population uncertainty.

Some additional notation: $A := \times_{i \in N} A_i$ denotes the collection of strategy profiles of the potential players. Assume the potential players have fixed a strategy profile $a = (a_i)_{i \in N}$. For notational convenience, denote by $a_\omega := (a_i)_{i \in N_\omega}$ the strategy profile of the players engaged in the game $G_\omega = \langle N_\omega, A_\omega, (\succeq_{i,\omega})_{i \in N_\omega} \rangle$ that is played if state $\omega \in \Omega$ is realized. The *best response correspondence* of G_ω is denoted by $BR_\omega : A_\omega \twoheadrightarrow A_\omega$, i.e.,

$$BR_\omega(a_\omega) := \times_{i \in N_\omega} \{b_i \in A_i \mid (b_i, (a_\omega)_{-i}) \succeq_{i,\omega} (c_i, (a_\omega)_{-i}) \text{ for all } c_i \in A_i\},$$

for all $a_\omega \in A_\omega$, where $(a_\omega)_{-i} = (a_j)_{j \in N_\omega \setminus \{i\}}$ denotes the strategy profile of the players in $N_\omega \setminus \{i\}$.

14.4 Maximum Likelihood Equilibria

The equilibrium concepts introduced by Myerson (1998c) and Milchtaich (1997) for their classes of games with population uncertainty are variants of the Nash equilibrium concept based on a suitably defined expected utility function for the players. This section presents an alternative approach by stressing those strategy profiles that are *most likely* to yield a Nash equilibrium in the game selected by chance. Maximum likelihood equilibria were introduced in Borm *et al.* (1995a) for a class of Bayesian games and were considered more recently in Voorneveld (1999, 2000). In this section we define maximum likelihood equilibria for games with population uncertainty and provide an existence result.

Consider a game $\langle N, (A_i)_{i\in N}, \Omega, \Sigma, P, (G_\omega)_{\omega\in\Omega}\rangle$ with population uncertainty. The players in N must plan their strategies in ignorance of the stochastic state of nature ω that is realized. A strategy profile $a = (a_i)_{i\in N} \in A$ gives rise to a Nash equilibrium if the realized state of nature is an element of the set

$$\{\omega \in \Omega \mid a_\omega \text{ is a Nash equilibrium of } G_\omega\} = \{\omega \in \Omega \mid a_\omega \in BR_\omega(a_\omega)\}.$$

How likely is this event? Although this set need not be measurable (i.e., an element of the σ-algebra Σ), a common mathematical approach in such cases is to define its likelihood via its inner measure

$$P_*(\{\omega \in \Omega \mid a_\omega \in BR_\omega(a_\omega)\}),$$

the probability of the largest measurable set of states of nature in which the strategy profile $a = (a_i)_{i\in N}$ gives rise to a Nash equilibrium. See Fagin and Halpern (1991) for another paper using inner measures in a decision theoretic framework. Formally, define the *Nash likelihood function* $L : A \to [0,1]$ for each $a = (a_i)_{i\in N} \in A$ as

$$L(a) := P_*(\{\omega \in \Omega \mid a_\omega \in BR_\omega(a_\omega)\}), \tag{14.6}$$

and define $a = (a_i)_{i\in N}$ to be a *maximum likelihood equilibrium* if

$$L(a) = \sup_{b\in A} L(b).$$

In a recent paper, Gilboa and Schmeidler (1999) provided an axiomatic foundation for rankings according to the likelihood function. The following theorem provides an existence result for maximum likelihood equilibria.

Theorem 14.3 *Consider a game with population uncertainty $\langle N, (A_i)_{i\in N}, \Omega, \Sigma, P, (G_\omega)_{\omega\in\Omega}\rangle$ with nonempty $A := \times_{i\in N} A_i$. If there are topologies on A and the sets A_ω for each $\omega \in \Omega$ such that*

(α) *A is sequentially compact;*

(β) *for every $\omega \in \Omega$ the graph* $\operatorname{gph} BR_\omega := \{(a,b) \in A_\omega \times A_\omega \mid b \in BR_\omega(a)\}$ *is sequentially closed in $A_\omega \times A_\omega$;*

(γ) *for every $\omega \in \Omega$ the function from A to A_ω defined by $a \mapsto a_\omega$ is sequentially continuous,* *then the set of maximum likelihood equilibria is nonempty.*

Proof. The set $\{L(b) \mid b \in A\}$ is nonempty and bounded above by one. Hence its supremum exists. Let $(a^n)_{n\in\mathbb{N}}$ be a sequence in A such that $\lim_n L(a^n) = \sup_{b\in A} L(b)$. Since A is sequentially compact by (α), the sequence $(a^n)_{n\in\mathbb{N}}$ has a subsequence converging to an element $a \in A$. Without loss of generality, this subsequence is taken to be $(a^n)_{n\in\mathbb{N}}$ itself: $\lim_n a^n = a$. This $a \in A$ is shown to be a maximum likelihood equilibrium.

For each $\omega \in \Omega$ and $b_\omega \in A_\omega$ it holds by definition that $b_\omega \in BR_\omega(b_\omega)$ if and only if $(b_\omega, b_\omega) \in \text{gph } BR_\omega$. Hence

$$\forall b \in A, \forall \omega \in \Omega: \quad 1_{\{\omega' \in \Omega | b_{\omega'} \in BR_{\omega'}(b_{\omega'})\}}(\omega) = 1_{\text{gph } BR_\omega}(b_\omega, b_\omega). \tag{14.7}$$

We show that for every $\omega \in \Omega$, the function from A_ω to $\{0,1\}$ defined by $b_\omega \mapsto 1_{\text{gph } BR_\omega}(b_\omega, b_\omega)$ is sequentially upper semicontinuous. Fix $\omega \in \Omega$ and a sequence $(b^n_\omega)_{n\in\mathbb{N}}$ in A_ω converging to $b_\omega \in A_\omega$. To show: $\limsup_n 1_{\text{gph } BR_\omega}(b^n_\omega, b^n_\omega) \le 1_{\text{gph } BR_\omega}(b_\omega, b_\omega)$. Since $(1_{\text{gph } BR_\omega}(b^n_\omega, b^n_\omega))_{n\in\mathbb{N}}$ is a sequence in $\{0,1\}$, the inequality trivially holds if $1_{\text{gph } BR_\omega}(b_\omega, b_\omega) = 1$. So assume that $1_{\text{gph } BR_\omega}(b_\omega, b_\omega) = 0$. It remains to prove that

$$\limsup_n 1_{\text{gph } BR_\omega}(b^n_\omega, b^n_\omega) = \lim_n (\sup\{1_{\text{gph } BR_\omega}(b^m_\omega, b^m_\omega) \mid m \ge n\}) = 0,$$

i.e., that there exists an $n \in \mathbb{N}$ such that $1_{\text{gph } BR_\omega}(b^m_\omega, b^m_\omega) = 0$ for each $m \ge n$. Suppose, to the contrary, that such an n does not exist. Then there is a subsequence $(1_{\text{gph } BR_\omega}(b^{n(k)}_\omega, b^{n(k)}_\omega))_{k\in\mathbb{N}}$ such that $1_{\text{gph } BR_\omega}(b^{n(k)}_\omega, b^{n(k)}_\omega) = 1$ for each $k \in \mathbb{N}$, i.e., $(b^{n(k)}_\omega, b^{n(k)}_\omega) \in \text{gph } BR_\omega$ for each $k \in \mathbb{N}$. Since gph BR_ω is sequentially closed by (β), this implies that $\lim_k (b^{n(k)}_\omega, b^{n(k)}_\omega) = (b_\omega, b_\omega) \in \text{gph } BR_\omega$, contradicting the assumption that $1_{\text{gph } BR_\omega}(b_\omega, b_\omega) = 0$. This settles the preliminary work. In the following sequence of (in)equalities,

- the first equality is (14.6),
- the second equality follows from (14.2) and (14.7),
- the first inequality follows from sequential upper semicontinuity of $b_\omega \mapsto 1_{\text{gph } BR_\omega}(b_\omega, b_\omega)$ and the fact that $a^n_\omega \to a_\omega$, since $a^n \to a$ and $a \mapsto a_\omega$ is sequentially continuous by (γ),
- the second inequality follows from Fatou's Lemma for lower integrals (Proposition 14.2),

- the third equality follows from (14.2) and (14.7),
- the fourth equality follows from (14.6),
- the final equality follows from $\sup_{b \in A} L(b) = \lim_n L(a^n)$ $= \limsup_n L(a^n)$.

The following (in)equalities hold:

$$\begin{aligned}
L(a) &= P_*(\{\omega \in \Omega \mid a_\omega \in BR_\omega(a_\omega)\}) \\
&= \int_* 1_{\text{gph } BR_\omega}(a_\omega, a_\omega)\, P(d\omega) \\
&\geq \int_* \limsup_n 1_{\text{gph } BR_\omega}(a_\omega^n, a_\omega^n)\, P(d\omega) \\
&\geq \limsup_n \int_* 1_{\text{gph } BR_\omega}(a_\omega^n, a_\omega^n)\, P(d\omega) \\
&= \limsup_n P_*(\{\omega \in \Omega \mid a_\omega^n \in BR_\omega(a_\omega^n)\}) \\
&= \limsup_n L(a^n) \\
&= \sup_{b \in A} L(b).
\end{aligned}$$

But then $a \in A$ is a maximum likelihood equilibrium of $\langle N, (A_i)_{i \in N}, \Omega, \Sigma, P, (G_\omega)_{\omega \in \Omega} \rangle$. □

A compactness condition like (α) is standard in equilibrium existence results. The sequential continuity condition (γ) guarantees that a convergent sequence of strategy profiles in A is projected to a convergent sequence of strategy profiles in the games $(G_\omega)_{\omega \in \Omega}$ that are realized in the different states of nature. This condition is automatically fulfilled if for instance the topologies on A and $(A_\omega)_{\omega \in \Omega}$ are taken to be the product topologies of those on the strategy spaces A_i of the players $i \in N$. The closedness condition (β) on the graphs of best response correspondences $(BR_\omega)_{\omega \in \Omega}$ is closely related to the upper semicontinuity conditions imposed on best response correspondences in equilibrium existence proofs using the Kakutani fixed point theorem. As a consequence, even though the existence proof of maximum likelihood equilibria significantly differs from existence proofs involving a fixed point argument, the basic conditions driving the result are the same.

14.5 Measurability

Let $\langle N, (A_i)_{i\in N}, \Omega, \Sigma, P, (G_\omega)_{\omega\in\Omega}\rangle$ be a game with population uncertainty and $a \in A = \times_{i\in N} A_i$. Theorem 14.3 relies on the use of inner measures in case the set

$$E(a) := \{\omega \in \Omega \mid a_\omega \in BR_\omega(a_\omega)\}$$

of states in which a gives rise to an equilibrium turns out to lie outside the σ-algebra Σ. In this section, assumptions are provided that guarantee measurability of this set. Observe that

$$\begin{aligned} E(a) \;&=\; \cap_{i\in N}(\{\omega \in \Omega \mid i \notin N_\omega\} \cup \{\omega \in \Omega \mid i \in N_\omega \\ &\qquad \text{and } a_i \in BR_{i,\omega}(a_\omega)\}) \qquad (14.8)\end{aligned}$$

where $BR_{i,\omega}$ is the best response correspondence of player i in the game G_ω. Hence, if

(a) the set N is countable,

(b) for each $i \in N$, the set $\{\omega \in \Omega \mid i \notin N_\omega\}$ of states in which player i does not participate in the game is measurable, and

(c) for each $i \in N$, the set $\{\omega \in \Omega \mid i \in N_\omega \text{ and } a_i \in BR_{i,\omega}(a_\omega)\}$ of states in which the present action a_i of participant i is a best response against the profile $(a_\omega)_{-i}$ of his opponents, is measurable,

then $E(a)$ is the countable intersection of the finite union of measurable sets and consequently measurable itself. Condition (a) appears innocent in many practical situations. Condition (b) is equivalent with the natural assumption that each player $i \in N$ participates in a measurable set of states. Condition (c) seems conceptually more difficult, but basically amounts to a measurability condition on the preference order of each player as a function of the states.

Assume, for instance, that in state $\omega \in \Omega$ the preferences of player $i \in N_\omega$ are represented by a utility function $u_{i,\omega} : A_\omega \to I\!R$. Let $i \in N$ and define $P_i := \{\omega \in \Omega \mid i \in N_\omega\}$, which is measurable by condition (b). Consider the σ-algebra $\Sigma(P_i) := \{S \cap P_i \mid S \in \Sigma\}$ generated by P_i and let $\mathcal{B}$ denote the Borel σ-algebra on $I\!R$. Then

$$\begin{aligned} & \{\omega \in \Omega \mid i \in N_\omega \text{ and } a_i \in BR_{i,\omega}(a_\omega)\} \\ =\; & \{\omega \in P_i \mid u_{i,\omega}(a_\omega) \geq \sup_{b_i\in A_i} u_{i,\omega}(b_i, (a_\omega)_{-i})\} \\ =\; & \{\omega \in P_i \mid u_{i,\omega}(a_\omega) - \sup_{b_i\in A_i} u_{i,\omega}(b_i, (a_\omega)_{-i}) \geq 0\},\end{aligned}$$

so it suffices to impose measurability conditions such that for each $a \in A$ the functions $\omega \mapsto \sup_{b_i \in A_i} u_{i,\omega}(b_i, (a_\omega)_{-i})$ and $\omega \mapsto u_{i,\omega}(a_\omega)$ are $(\Sigma(P_i), \mathcal{B})$-measurable. This is clearly the case if

1. for each $a \in A$ the function $\omega \mapsto u_{i,\omega}(a_\omega)$ is $(\Sigma(P_i), \mathcal{B})$-measurable,
2. A_i is countable,

since the supremum of countably many measurable functions is measurable; cf. Aliprantis and Border (1994), Theorem 8.17. Below we provide a less obvious result.

Proposition 14.4 *Let $i \in N$. Assume that the following conditions hold:*

(a) for each $a \in A$ the function $\omega \mapsto u_{i,\omega}(a_\omega)$ is $(\Sigma(P_i), \mathcal{B})$-measurable,

(b) the function $u_{i,\omega}$ is sequentially lower semicontinuous in its i-th coordinate,

(c) A_i is separable.

Then the set $\{\omega \in \Omega \mid i \in N_\omega$ and $a_i \in BR_{i,\omega}(a_\omega)\}$ is measurable.

Proof. It suffices to prove that the function $\omega \mapsto \sup_{b_i \in A_i} u_{i,\omega}(b_i, (a_\omega)_{-i})$ is $(\Sigma(P_i), \mathcal{B})$-measurable. Let $\alpha \in \mathbb{R}$ and $a \in A$ and assume that $C \subseteq A_i$ is a countably dense subset of A_i. Then

$$\begin{aligned} & \{\omega \in P_i \mid \sup_{b_i \in A_i} u_{i,\omega}(b_i, (a_\omega)_{-i}) \leq \alpha\} \\ = & \{\omega \in P_i \mid \forall b_i \in A_i : u_{i,\omega}(b_i, (a_\omega)_{-i}) \leq \alpha\} \\ = & \cap_{b_i \in A_i} \{\omega \in P_i \mid u_{i,\omega}(b_i, (a_\omega)_{-i}) \leq \alpha\} \\ = & \cap_{b_i \in C} \{\omega \in P_i \mid u_{i,\omega}(b_i, (a_\omega)_{-i}) \leq \alpha\}, \end{aligned} \tag{14.9}$$

where the last equality is a consequence of assumptions (b) and (c): Let $\omega \in P_i$ be such that $u_{i,\omega}(b_i, (a_\omega)_{-i}) \leq \alpha$ for all $b_i \in C$. Let $b_i \in A_i$. Since $\overline{C} = A_i$, there is a sequence $(b_i^n)_{n \in \mathbb{N}}$ in C converging to b_i. Since $u_{i,\omega}$ is sequentially lower semicontinuous in its i-th coordinate, it follows that $u_{i,\omega}(b_i, (a_\omega)_{-i}) \leq \liminf_n u_{i,\omega}(b_i^n, (a_\omega)_{-i}) \leq \alpha$. The set in (14.9) is a countable intersection of measurable sets and hence measurable, as was to be shown. □

Separability is only a weak condition. Typical examples of action spaces that come to mind are strategy simplices (probability distributions over finitely many pure strategies), an interval $[0, \infty)$ of prices, or a subset of $\mathbb{N}$ denoting possible quantities (like production levels). All such sets are separable.

14.6 Random Action Sets

Voorneveld (1999) proposes a model in which, in addition to the set of players and their preferences, also the set of feasible actions of each player is subject to uncertainty. In the set-up of Voorneveld (2000), this would imply ω-dependence of the set of actions of each participating player. There are two attitudes to this. On one hand, it is possible to model such games explicitly and to derive equilibrium existence results in this more general context. This is the approach taken in Voorneveld (1999). On the other hand, it is possible to show that a simple mathematical trick translates such a more general problem into a game with population uncertainty. This is explained below.

Suppose that—in addition to the randomness incorporated in a game with population uncertainty—also the action set $A_{i,\omega}$ of each player $i \in N_\omega$ is allowed to depend on the realized state of nature $\omega \in \Omega$. This captures a common problem in decision making, namely the situation in which players or decision makers have to plan or decide on their course of action, while still uncertain about contingencies that may make their choices impossible to implement: a planned action can turn out to be infeasible in the realized state of nature. One can translate this into a game $\langle N, (A_i)_{i \in N}, \Omega, \Sigma, P, (G_\omega)_{\omega \in \Omega}\rangle$ with population uncertainty by

1. defining every action to be feasible, i.e., take $A_i \supseteq \cup_{\omega \in \Omega : i \in N_\omega} A_{i,\omega}$;

2. making sure in every realized game G_ω that an action profile a_ω in which any of the players plays an infeasible action $a_i \in A_i \setminus A_{i,\omega}$ is *not* a Nash equilibrium of G_ω, for instance by extending the original preferences $\succeq_{i,\omega}$ of each player $i \in N_\omega$ over the feasible action set $\times_{j \in N_\omega} A_{j,\omega}$ to the larger action set $\times_{j \in N_\omega} A_j$ as follows.

   (a) Any action profile involving feasible action choices is strictly preferred over any action profile in which a nonempty set of players chooses an infeasible action: for each $a_\omega, b_\omega \in \times_{j \in N_\omega} A_j$, if $a_j \in A_{j,\omega}$ for every $j \in N_\omega$, but $b_j \in A_j \setminus A_{j,\omega}$ for some $j \in N_\omega$, then $a_\omega \succ_{i,\omega} b_\omega$.

   (b) Any action profile in which a nonempty subset $S \subseteq N_\omega$ of players chooses an infeasible action is strictly preferred to any action profile in which a superset of S chooses an infeasible action: for each $a_\omega, b_\omega \in \times_{j \in N_\omega} A_j$, if $\emptyset \neq \{j \in N_\omega \mid a_j \in A_j \setminus A_{j,\omega}\} \subset \{j \in N_\omega \mid b_j \in A_j \setminus A_{j,\omega}\}$, then $a_\omega \succ_{i,\omega} b_\omega$.

If the preferences of the players over feasible actions are represented by utility functions $u_{i,\omega} : \times_{j \in N_\omega} A_{j,\omega} \to \mathbb{R}$, then we may assume w.l.o.g. that its range is a subset of $[-1,1]$ by taking $\arctan u_{i,\omega}$ if necessary. One way to extend $u_{i,\omega}$ from $\times_{j \in N_\omega} A_{j,\omega}$ to $\times_{j \in N_\omega} A_j$ would be to take

$$u_{i,\omega}(a_\omega) = -2 - |\{j \in N_\omega \mid a_j \in A_j \setminus A_{j,\omega}\}| \tag{14.10}$$

for every $i \in N_\omega$ and $a_\omega \in \times_{j \in N_\omega} A_j \setminus \times_{j \in N_\omega} A_{j,\omega}$. Notice that (14.10) is well-defined, since N_ω is finite, and indeed satisfies 2(a) and 2(b) above. A simple example will illustrate this procedure.

Example 14.5 Suppose that there is only one state of nature, in which two players each have one good strategy (G), which is feasible, and one bad strategy (B), which is not. Clearly, (G,G) should be the unique equilibrium recommendation. Suppose that (G,G) gives payoff zero to both players. Following (14.10) means that in (B,G) and (G,B), one of the players makes an infeasible choice, giving rise to payoff $-2-1=-3$ to both players, while in (B,B) both players make an infeasible choice, giving rise to payoff $-2-2=-4$ to both players. Hence, the corresponding game would be:

$$\begin{array}{c} \\ G \\ B \end{array} \begin{array}{c} \begin{array}{cc} G & \quad B \end{array} \\ \begin{pmatrix} 0,0 & -3,-3 \\ -3,-3 & -4,-4 \end{pmatrix} \end{array}$$

The unique (maximum likelihood) equilibrium of this game is (G,G). This example illustrates the need for condition 2(b): it is not sufficient to assign, in accordance with condition 2(a), a low payoff, say -3, to any action profile involving infeasible action choices, for this would imply that also (B,B) is an equilibrium recommendation, which is nonsensical. Condition 2(b) takes care of distinctions between action profiles involving infeasible choices: the fewer infeasibilities, the better. $\triangleleft$

14.7 Random Games

In line with Borm *et al.* (1995a), we define a random game to be a game with population uncertainty in which the stochastic variable always selects the same player set (one might say a game with population uncertainty *without* population uncertainty) and preferences are represented by means of payoff functions. Such games are incomplete information games (or Bayesian games) in which the players have no private information. Random games will be of central importance in the remainder of this chapter, which mainly focuses on the likelihood principle as an equilibrium selection device for finite strategic games.

Definition 14.6 A *random game* is a game $\langle N, (A_i)_{i \in N}, \Omega, \Sigma, P, (G_\omega)_{\omega \in \Omega} \rangle$ with population uncertainty such that
(i) $N_\omega = N$ for all $\omega \in \Omega$, with N finite, and
(ii) the preferences $\succeq_{i,\omega}$ in G_ω are represented by functions $u_{i,\omega} : \times_{j \in N} A_j \to I\!R$. ◁

Borm *et al.* (1995a) impose a number of conditions on the random game: *(i)* each A_i is a separable topological space, *(ii)* $A = \times_{i \in N} A_i$ is compact, *(iii)* for each $i \in N$ and $a \in A$ the function $\omega \mapsto u_{i,\omega}(a)$ is measurable, *(iv)* for each $i \in N$ and $\omega \in \Omega$ the function $a \mapsto u_{i,\omega}(a)$ is lower semicontinuous. Under these conditions, the authors provide two results (their Lemma 1 and Theorem 1) leading to the existence of maximum likelihood equilibria in random games. Unfortunately, both results are incorrect.

Recall that a real-valued function $f : X \to I\!R$ on a topological space X is lower semicontinuous if $\{x \in X \mid f(x) > r\}$ is open for every $r \in I\!R$.

Example 14.7 Suppose there is only one state of nature and one player with action space $[0, 1]$, endowed with its standard topology, and payoff $u_1(x) = 1$ for all $x \in (0, 1)$ and payoff zero otherwise. Then

$$\{x \in [0,1] \mid u_1(x) > r\} = \begin{cases} [0,1] & \text{if } r < 0, \\ (0,1) & \text{if } 0 \leq r < 1, \\ \emptyset & \text{if } 1 \leq r. \end{cases}$$

The sets $[0, 1]$ and $\emptyset$ are open by definition in every topology on $[0, 1]$ and $(0, 1)$ is open as well. Hence the payoff function is lower semicontinuous. Measurability is trivial. However, the set of maximizers of u_1, i.e., the set of Nash equilibria of the one-player game, is the interval $(0, 1)$, which is open. The sequence $x^n := \frac{1}{n+1}$ approaches zero. The sequence $L(x^n)$ is the sequence of ones, since x^n is always a Nash equilibrium. But $L(0) = 0$, since 0 is not a Nash equilibrium. This provides a counterexample to Lemma 1 of Borm *et al.* (1995), which erroneously claims that $\lim_n L(x^n) \leq L(0)$. ◁

Example 14.8 Suppose there is only one state of nature and one player with action space $[0, 1]$, endowed with its standard topology, and payoff $u_1(x) = x$ for all $x \in [0, 1)$ and payoff zero for $x = 1$. Then

$$\{x \in [0,1] \mid u_1(x) > r\} = \begin{cases} [0,1] & \text{if } r < 0, \\ (r,1) & \text{if } 0 \leq r < 1, \\ \emptyset & \text{if } 1 \leq r. \end{cases}$$

The sets $[0, 1]$ and $\emptyset$ are open by definition in every topology on $[0, 1]$ and $(r, 1)$ with $r \in [0, 1)$ is open as well. Hence the payoff function is lower semicontinuous. Measurability is trivial. The action set $[0, 1]$ is compact. Still, there is no maximum likelihood equilibrium, contradicting Theorem 1 of Borm *et al.* (1995). $\triangleleft$

Fortunately, our general existence result (Theorem 14.3) applies to random games as well.

14.8 Robustness Against Randomization

A finite strategic game $\langle N, (A_i)_{i\in N}, (u_i)_{i\in N}\rangle$ can be perturbed by assuming that with certain probability there are trembles in the payoff functions.

Definition 14.9 Let $G = \langle N, (A_i)_{i\in N}, (u_i)_{i\in N}\rangle$ be a finite strategic game and $\varepsilon > 0$. An *ε–perturbation* of G is a random game $G(\varepsilon) = \langle N, (A_i)_{i\in N}, \Omega, \Sigma, P, (G_\omega(\varepsilon))_{\omega\in\Omega}\rangle$ with:
(i) the payoffs of each player $i \in N$ are perturbed in a set of states of nature with positive measure: $P_*(u^\varepsilon_{i,\omega} \neq u_i) > 0$ for each $i \in N$ (where $u^\varepsilon_{i,\omega}$ is player i's payoff function in $G_\omega(\varepsilon)$);
(ii) the largest perturbation is ε: $\sup_{i,\omega,a} |u^\varepsilon_{i,\omega}(a) - u_i(a)| = \varepsilon$. $\triangleleft$

An intuitive approach would be to search for strategy profiles which are most likely to be a Nash equilibrium if players take into account that with certain probability the real payoffs of the game may differ slightly from those provided by the payoff function.

Definition 14.10 Let $G = \langle N, (A_i)_{i\in N}, (u_i)_{i\in N}\rangle$ be a finite strategic game and $\mathbb{P}$ a set of perturbations of G. A strategy profile $\alpha \in \times_{i\in N}\Delta(A_i)$ is *robust against randomization* if there exist

1. a sequence $(\varepsilon^n)_{n\in\mathbb{N}}$ of positive numbers with $\varepsilon^n \to 0$;

2. a sequence $(G(\varepsilon^n))_{n\in I\!N}$ of ε^n–perturbed games in $I\!P$;

3. a sequence $(\alpha^n)_{n\in I\!N}$ of strategy profiles converging to α,

such that $\alpha^n \in MLE(G(\varepsilon^n))$ and $L^{\varepsilon^n}(\alpha^n) > 0$ for every $n \in I\!N$, where $L^{\varepsilon^n}(\cdot)$ denotes the likelihood function for the random game $G(\varepsilon^n)$. The set of strategy profiles in G that is robust against randomization is denoted by $RR(G)$. ◁

A strategy profile is robust against randomization if it is the limit of a sequence of maximum likelihood equilibria in perturbed games, each having a strictly positive likelihood. This last restriction essentially means that even though the actual payoffs are subject to chance, the state spaces are such that at least some strategy profiles are Nash equilibria in a set of realized games with positive measure; otherwise, the MLE concept has no cutting power.

We prove that under some conditions on the set $I\!P$ of permissible perturbations the set of strategy profiles that is robust against randomization is nonempty, and that the concept is a refinement of the Nash equilibrium concept.

Theorem 14.11 *Let $G = \langle N, (A_i)_{i\in N}, (u_i)_{i\in N}\rangle$ be a finite strategic game and $I\!P$ a set of perturbations of G.*

(a) If there exist

- *a sequence $(\varepsilon^n)_{n\in I\!N}$ of positive real numbers converging to zero and*
- *a sequence $(G(\varepsilon^n))_{n\in I\!N}$ of ε^n–perturbations in $I\!P$ with a finite state space Ω,*

then $RR(G) \neq \emptyset$;

(b) $RR(G) \subseteq NE(G)$, where $NE(G)$ denotes the set of (mixed) Nash equilibria of G.

Proof. (a): Let $(\varepsilon^n)_{n\in I\!N}$ and $(G(\varepsilon^n))_{n\in I\!N}$ be as in Theorem 14.11(a). Choose $(\alpha^n)_{n\in I\!N}$ such that $\alpha^n \in MLE(G(\varepsilon^n))$ for each $n \in I\!N$. The state space Ω is finite, so $L^{\varepsilon^n}(\alpha^n) > 0$ for each $n \in I\!N$. Since $(\alpha^n)_{n\in I\!N}$ is a sequence in the compact set $\times_{i\in N}\Delta(A_i)$, it has a convergent subsequence; its limit is robust against randomization.
(b): Let $(\varepsilon^n, G(\varepsilon^n), \alpha^n)_{n\in I\!N}$ as in Definition 14.10 support $\alpha \in RR(G)$ as a randomization robust strategy profile. Suppose $\alpha \notin NE(G)$. Then

there is a player $i \in N$ and an action $a_i \in A_i$ such that $u_i(a_i, \alpha_{-i}) > u_i(\alpha)$. Take $\delta = \frac{1}{2}(u_i(a_i, \alpha_{-i}) - u_i(\alpha)) > 0$. Since $\varepsilon^n \to 0$, there is an $N_\delta \in \mathbb{N}$ such that $\varepsilon^n \leq \frac{1}{2}\delta$ for each $n \geq N_\delta$. Definition 14.9*(ii)* implies

$$u_{i,\omega}^{\varepsilon^n}(a_i, \alpha_{-i}) - u_i(a_i, \alpha_{-i}) \geq -\frac{1}{2}\delta \text{ and } u_i(\alpha) - u_{i,\omega}^{\varepsilon^n}(\alpha) \geq -\frac{1}{2}\delta,$$

for all $n \geq N_\delta$ and $\omega \in \Omega$, so that for each $n \geq N_\delta$ and $\omega \in \Omega$:

$$u_{i,\omega}^{\varepsilon^n}(a_i, \alpha_{-i}) - u_{i,\omega}^{\varepsilon^n}(\alpha) \geq -\delta + u_i(a_i, \alpha_{-i}) - u_i(\alpha) = -\delta + 2\delta = \delta. \tag{14.11}$$

Since G is a finite strategic game, u_i is bounded by a certain $M > 0$. Definition 14.9*(ii)* implies that for each $n \geq N_\delta$ and each $\omega \in \Omega$, $u_{i,\omega}^{\varepsilon^n}$ is bounded by $K := M + \frac{1}{2}\delta$. Consequently, for all $\alpha, \beta \in \times_{i \in N}\Delta(A_i), n \geq N_\delta$, and $\omega \in \Omega$:

$$\begin{aligned} |u_{i,\omega}^{\varepsilon^n}(\alpha) - u_{i,\omega}^{\varepsilon^n}(\beta)| &= |\sum_{a \in A} u_{i,\omega}^{\varepsilon^n}(a)(\alpha(a) - \beta(a))| \\ &\leq K \sum_{a \in A} |\alpha(a) - \beta(a)|, \end{aligned} \tag{14.12}$$

where $\alpha(a) = \prod_{i \in N} \alpha_i(a_i)$ denotes the probability that $a \in A$ is played according to the mixed strategy profile α. Let $\beta = (a_i, \alpha_{-i})$ and $\beta^n = (a_i, \alpha_{-i}^n)$ for each $n \in \mathbb{N}$. Inequalities (14.11) and (14.12) imply that for each $n \geq N_\delta$ and $\omega \in \Omega$:

$$\begin{aligned} \delta &\leq u_{i,\omega}^{\varepsilon^n}(a_i, \alpha_{-i}) - u_{i,\omega}^{\varepsilon^n}(\alpha) \\ &\leq u_{i,\omega}^{\varepsilon^n}(a_i, \alpha_{-i}^n) + K \sum_{a \in A} |\beta(a) - \beta^n(a)| \\ &\quad - u_{i,\omega}^{\varepsilon^n}(\alpha^n) + K \sum_{a \in A} |\alpha(a) - \alpha^n(a)|. \end{aligned}$$

As $\alpha^n \to \alpha$ and $\beta^n \to \beta$, this implies that $u_{i,\omega}^{\varepsilon^n}(a_i, \alpha_{-i}^n) - u_{i,\omega}^{\varepsilon^n}(\alpha^n) > 0$ for all $\omega \in \Omega$ and $n \in \mathbb{N}$ sufficiently large. But then $L^{\varepsilon^n}(\alpha^n) = 0$ for large $n \in \mathbb{N}$, in contradiction with Definition 14.10. □

Remark 14.12 In the proof above, the only essential parts of Definitions 14.9 and 14.10 were:

- the fact that ε–perturbations actually exist;
- that payoffs in the perturbed game lie close to those in the original game (part *(ii)* of Definition 14.9);

- that the selected sequence $(\alpha^n)_{n\in\mathbb{N}}$ has a positive likelihood of being a Nash equilibrium in the perturbed games (final part of Definition 14.10).

Consequently, there is still a lot of freedom in redefining the perturbed games without harming the fact that we have a nonempty equilibrium refinement. In particular, considerable strengthenings of condition *(i)* in Definition 14.9 would be possible and would enhance the cutting power of the selection device. ◁

Among the earliest equilibrium refinements is the notion of an essential equilibrium, due to Wu and Jiang (1962). An equilibrium α of a game G is said to be essential if every game with payoffs near those of G has a Nash equilibrium close to α. A game may not have essential equilibria. For instance, in the trivial game where the unique player has a constant payoff function, all strategies are Nash equilibria, but none of them is essential, since for every strategy there exists some slight change in the payoffs that makes no nearby strategy an equilibrium. However, Wu and Jiang (1962) prove that if a game has only finitely many equilibria, at least one of them is essential.

Let $G = \langle N, (A_i)_{i\in N}, (u_i)_{i\in N}\rangle$ and $H = \langle N, (A_i)_{i\in N}, (v_i)_{i\in N}\rangle$ be two finite games with the same set of players and the same strategy space. The distance $d(G, H)$ between G and H is defined to be the maximal distance between its payoffs:

$$d(G, H) = \max_{i\in N, a\in A} |u_i(a) - v_i(a)|.$$

Definition 14.13 Let $G = \langle N, (A_i)_{i\in N}, (u_i)_{i\in N}\rangle$ be a finite strategic game. A strategy profile $\alpha \in NE(G)$ is *essential* if for every $\varepsilon > 0$ there exists a $\delta > 0$ such that every game H within distance δ from the game G has a Nash equilibrium within distance ϵ from α. ◁

Proposition 14.14 *Every essential equilibrium of a finite strategic game* $G = \langle N, (A_i)_{i\in N}, (u_i)_{i\in N}\rangle$ *is robust against randomization.*

Proof. Trivial: construct a sequence of ε–perturbations with only one state, in which the game has distance ε to G. □

14.9 Weakly Strict Equilibria

This section describes the notion of weakly strict equilibria, introduced by Borm *et al.* (1995b). The proofs in this section are mostly simplifications of the original ones. The idea behind weakly strict equilibria is analogous to the principle of robustness against randomization, but attention is restricted to a very simple type of perturbations and to strategies that are (in a sense to be made precise) undominated.

We start by defining the perturbed games that we consider in this section. Let $G = \langle N, (A_i)_{i\in N}, (u_i)_{i\in N}\rangle$ be a finite strategic game and $\varepsilon > 0$. Define the ε-perturbation $G(\varepsilon) = \langle N, (A_i)_{i\in N}, \Omega, \Sigma, P, (G_\omega(\varepsilon))_{\omega\in\Omega}\rangle$ with

- $\Omega = \{-1, 0, 1\}^A, \Sigma = 2^\Omega$, and $P(S) = |S|/|\Omega|$ for each $S \subseteq \Omega$,
- $G_\omega(\varepsilon) = \langle N, (A_i)_{i\in N}, (u^\varepsilon_{i,\omega})_{i\in N}\rangle$ with $u^\varepsilon_{i,\omega} : a \mapsto u_i(a) + \varepsilon\omega(a)$.

This means that there are $|\Omega| = 3^{|A|}$ different payoff perturbed versions of G (one of which coincides with G), each of which is played with equal probability. These perturbed versions are obtained by adding for each $a \in A$ (independently) with probability $\frac{1}{3}$ the amount $-\varepsilon$, zero, or ε to the original payoffs $(u_i(a))_{i\in N}$.

Let $\alpha_i \in \Delta(A_i)$ be a mixed strategy of player i in G and let $\varepsilon > 0$. Strategy α_i is *ε-dominated* if there exists a strategy $\beta_i \in \Delta(A_i)$ such that $u_i(\beta_i, a_{-i}) \geq u_i(\alpha_i, a_{-i}) + \varepsilon$ for each $a_{-i} \in A_{-i}$ with strict inequality for some a_{-i}. A strategy is *ε-undominated* if no player plays an ε-dominated strategy. For each $\varepsilon > 0$ and $\alpha \in \times_{j\in N}\Delta(A_j)$, let

$$U^\varepsilon(\alpha) = P(\{\omega \in \Omega \mid \alpha \text{ is } \varepsilon\text{-undominated in } G(\varepsilon)\})$$

denote the probability that α is ε-undominated in the perturbed game $G(\varepsilon)$. As before,

$$L^\varepsilon(\alpha) = P(\{\omega \in \Omega \mid a \in NE(G_\omega(\varepsilon))\})$$

denotes the likelihood of α yielding a Nash equilibrium in the perturbed game. We can use the measure P instead of the inner measure P_*, since every subset of Ω is measurable.

Instead of considering the limit of a sequence of maximum likelihood equilibria, as was done in the previous section, a weighted objective function is taken into account, where positive weight is assigned to both the probability of being undominated and being a Nash equilibrium.

Definition 14.15 Let $G = \langle N, (A_i)_{i\in N}, (u_i)_{i\in N}\rangle$ be a finite strategic game and $\varepsilon > 0$. Define the function $S^\varepsilon : \times_{j\in N}\Delta(A_j) \to [0, 1]$ by

$$S^\varepsilon : \alpha \mapsto \frac{1}{1+\varepsilon}U^\varepsilon(\alpha) + \frac{\varepsilon}{1+\varepsilon}L^\varepsilon(\alpha).$$

A strategy profile $\alpha \in \times_{j\in N}\Delta(A_j)$ is a *weakly strict equilibrium* if there exist

1. a sequence $(\varepsilon^n)_{n\in\mathbb{N}}$ of positive numbers with $\varepsilon^n \to 0$ and
2. a sequence $(\alpha^n)_{n\in\mathbb{N}}$ of strategy profiles converging to α

such that $S^{\varepsilon^n}(\alpha^n) = \max_\beta S^{\varepsilon^n}(\beta)$ for each $n \in \mathbb{N}$. ◁

Since the state space of each perturbed game $G(\varepsilon)$ is finite, S^ε takes only finitely many values, so the maximum in the definition indeed exists. Notice that α is a strict Nash equilibrium of G if and only if for sufficiently small ε, the strategy profile α is also a strict Nash equilibrium (and hence ε-undominated) in each perturbed game $G_\omega(\varepsilon)$, which in its turn is equivalent with the statement that $S^\varepsilon(\alpha) = 1$.

Proposition 14.16 *The strategy profile $\alpha \in \times_{j\in N}\Delta(A_j)$ is a strict Nash equilibrium of the finite strategic game $\langle N, (A_i)_{i\in N}, (u_i)_{i\in N}\rangle$ if and only if $S^\varepsilon(\alpha) = 1$ for sufficiently small $\varepsilon > 0$.*

This motivates the name 'weakly' strict equilibria in Definition 14.15.

Theorem 14.17 *Let $G = \langle N, (A_i)_{i\in N}, (u_i)_{i\in N}\rangle$ be a finite strategic game.*

(a) G has a weakly strict equilibrium;

(b) Each weakly strict equilibrium of G is an undominated Nash equilibrium of G;

(c) Every strict equilibrium of G is weakly strict;

(d) If G has strict equilibria, these are the only weakly strict ones.

Proof. (a): As in Theorem 14.11(a).
(b): It is well-known that the finite strategic game G has an undominated equilibrium α^*. For $\varepsilon > 0$ small, $S^\varepsilon(\alpha^*) > \frac{1}{1+\varepsilon}$, since $U^\varepsilon(\alpha^*) = 1$ and $L^\varepsilon(\alpha^*) > 0$. Let β be a weakly strict equilibrium of G and suppose β is dominated. As the set of undominated strategy profiles in G is closed, there is an environment U of β such that each $\beta' \in U$ is dominated. Hence for $\varepsilon > 0$ sufficiently small: $U^\varepsilon(\beta') \leq 1 - \frac{1}{|\Omega|}$ for each $\beta' \in U$, so that $S^\varepsilon(\beta) \leq \frac{1}{1+\varepsilon}(1 - \frac{1}{|\Omega|}) + \frac{\varepsilon}{1+\varepsilon} < \frac{1}{1+\varepsilon} < S^\varepsilon(\alpha^*)$, a contradiction. The proof that β is a Nash equilibrium is analogous to Theorem 14.11(b).
(c),(d): Follow easily from Proposition 14.16. □

14.10 Approximate Maximum Likelihood Equilibria

In this section we consider perturbations of a finite strategic game $G = \langle N, (A_i)_{i\in N}, (u_i)_{i\in N}\rangle$ by assuming that, according to some probability distribution, there are trembles in the actions of the players.

Definition 14.18 Let $G = \langle N, (A_i)_{i\in N}, (u_i)_{i\in N}\rangle$ be a finite strategic game and $\varepsilon \in (0,1)$. The *ε-perturbation* of G is the random game $G(\varepsilon) = \langle N, (A_i)_{i\in N}, \Omega, \Sigma, P, (G_\omega(\varepsilon))_{\omega\in\Omega}\rangle$ with

(i) $\Omega = \times_{i\in N}\Delta(A_i)$, Σ is the Borel-σ-algebra on Ω and P is the product measure of the uniform distributions $Un(\Delta(A_i))$ on $\Delta(A_i)$.

(ii) $u^\varepsilon_{i,\omega}(x) = u_i((1-\varepsilon)x + \varepsilon\omega)$ for every $i \in N$, $\omega \in \Omega$, $\varepsilon \in (0,1)$, and $x \in \times_{i\in N}\Delta(A_i)$.

$\triangleleft$

The intuitive idea behind the ε-perturbed games is that players, who choose to act according to some strategies, make small mistakes while implementing these strategies. The approach now is to look for strategy profiles which have a positive probability of being close to some equilibrium of the perturbed games.

Definition 14.19 Let G be a finite game and $\{G(\varepsilon)\}_{\varepsilon\in(0,1)}$ be the corresponding collection of ε-perturbations. For every open set $U \subseteq \times_{i\in N}\Delta(A_i)$ and every $\varepsilon \in (0,1)$ define

$$\gamma(\varepsilon, U) = P_*(\{\omega \in \Omega \mid NE(G_\omega(\varepsilon)) \cap U \neq \emptyset\})$$

and for every open $U \subseteq \times_{i\in N}\Delta(A_i)$ define

$$\mu(U) = \liminf_{\varepsilon\downarrow 0} \gamma(\varepsilon, U).$$

A profile $a \in \times_{i\in N}\Delta(A_i)$ that satisfies

$$\pi(a) = \inf\{\mu(U) \mid U \text{ open}, a \in U\} > 0,$$

is called an *approximate maximum likelihood equilibrium* of G. $\triangleleft$

For an open set U and an $\varepsilon \in (0,1)$ the number $\gamma(\varepsilon, U)$ denotes the probability that U contains a Nash equilibrium of the ε-perturbation $G(\varepsilon)$. An estimate for these probabilities for small ε is given by $\mu(U)$.

A profile a is an approximate maximum likelihood equilibrium if there is a positive constant c such that for ε sufficiently small and any open neighbourhood U of a this probability is at least c.

The concept of approximate maximum likelihood equilibrium is a refinement of the Nash equilibrium concept. In fact we show that every approximate maximum likelihood equilibrium is a perfect Nash equilibrium (Selten, 1975). Recall that a profile a is a perfect equilibrium of the finite game G if there exists a sequence of disturbance vectors $(\eta^k)_{k\in\mathbb{N}}$, $\eta^k = (\eta_i^k)_{i\in N} \in \times_{i\in N}\mathbb{R}_{++}^{A_i}$ for every k, with $\lim_k \eta^k = 0$, and a sequence of profiles $(a^k)_{k\in\mathbb{N}}$, converging to a, such that for every k we have $a^k \in NE(G^{\eta^k})$. Here G^{η^k} is the strategic game which is obtained by restricting the strategy set of every player $i \in N$ to $\{a_i \in \Delta(A_i) \mid a_i \geq \eta_i^k\}$.

Theorem 14.20 *Every approximate maximum likelihood equilibrium of a finite game G is a perfect equilibrium of G.*

Proof. First note that for every $y, \omega \in \times_{i\in N}\Delta(A_i)$ and $\varepsilon \in (0,1)$ we have $y \in NE(G_\omega(\varepsilon))$ if and only if $(1-\varepsilon)y + \varepsilon\omega \in NE(G^{\varepsilon\omega})$.

Now let a be an approximate maximum likelihood equilibrium of G, i.e. $\pi(a) = \alpha > 0$. Let $k \in \mathbb{N}$. Define $U^k = \{x \in \times_{i\in N}\Delta(A_i) \mid \|x-a\| < 1/k\}$, where $\|\cdot\|$ denotes the standard Euclidean norm in $\times_{i\in N}\mathbb{R}_{++}^{A_i}$. Since $\mu(U^k) \geq \alpha$, there exists a $\delta^k \in (0,1)$ such that $\gamma(\varepsilon, U^k) \geq \alpha/2$ for every $\varepsilon \in (0, \delta^k]$. Choose $\varepsilon^k = \min\{1/k, \delta^k\}$. Since $\gamma(\varepsilon^k, U^k) > 0$, there is an $\omega^k \in \times_{i\in N}\text{int}(\Delta(A_i))$ with $NE(G_{\omega^k}(\varepsilon^k)) \cap U^k \neq \emptyset$, so we can choose an $a^k \in NE(G_{\omega^k}(\varepsilon^k)) \cap U^k$. Clearly the sequence $(\varepsilon^k)_{k\in\mathbb{N}}$ converges to 0 and the sequence $(a^k)_{k\in\mathbb{N}}$ converges to a. Moreover, the sequence $(y^k)_{k\in\mathbb{N}}$ defined by $y^k = (1-\varepsilon^k)a^k + \varepsilon^k\omega^k$ also converges to a, and $y^k \in NE(G^{\varepsilon^k\omega^k})$ for every k. Since $\lim_k \varepsilon^k\omega^k = 0$, we conclude that a is a perfect equilibrium of G. □

In the following example we illustrate that the converse of Theorem 14.20 is not true.

Example 14.21 Consider the finite strategic game G given by

$$\begin{array}{c} \\ T \\ B \end{array} \begin{array}{c} \begin{array}{ccc} L & M & R \end{array} \\ \begin{pmatrix} 2,0 & 0,6 & 0,8 \\ 0,8 & 0,6 & 1,0 \end{pmatrix} \end{array}$$

The profile $a = (a_1, a_2)$ given by $a_1(T) = a_1(B) = 1/2$, $a_2(L) = a_2(R) = 0, a_2(M) = 1$ is a Nash equilibrium of G. Consider the sequence of disturbance vectors $(\eta^k)_{k\in\mathbb{N}}$, given by $\eta_1^k(T) = \eta_1^k(B) = \eta_2^k(L) = \eta_2^k(M) =$

$1/(4k)$, $\eta_2^k(R) = 2/(4k)$, and the sequence of profiles $(a^k)_{k\in\mathbb{N}}$, given by $a_1^k(T) = a_1^k(B) = 1/2$, $a_2^k(L) = 1/(4k), a_2^k(M) = 1 - 3/(4k)$, and $a_2^k(R) = 2/(4k)$ for every $k \in \mathbb{N}$. One easily verifies that $a^k \in NE(G^{\eta^k})$ for every k and that $\lim_k a^k = a$. Therefore a is a perfect equilibrium of G. However, a is not an approximate maximum likelihood equilibrium of G. In order to see this consider the set $U = \{x \in \times_{i\in N}\Delta(A_i) \mid |x_i(j) - a_i(j)| < 0.1$ for every $i \in N$ and $j \in A_i\}$, which is an open neighbourhood of a. It suffices to show that for every $\varepsilon \in (0, 0.1)$ we have

$$\{\omega \in \Omega \mid NE(G_\omega(\varepsilon)) \cap U \neq \emptyset\} \subseteq \{\omega \in \Omega \mid \omega_2(R) = 2\omega_2(L)\}.$$

Since then we get, for every $\varepsilon \in (0, 0.1)$,

$$\begin{aligned} 0 \leq \gamma(\varepsilon, U) &= P_*(\{\omega \in \Omega \mid NE(G_\omega(\varepsilon)) \cap U \neq \emptyset\}) \\ &\leq P_*(\{\omega \in \Omega \mid \omega_2(R) = 2\omega_2(L)\}) = 0. \end{aligned}$$

Consequently $\mu(U) = 0$ and hence $\pi(a) = 0$. So, let $\varepsilon \in (0, 0.1)$ and $\omega \in \Omega$ be such that $NE(G_\omega(\varepsilon)) \cap U \neq \emptyset$ and let $y \in NE(G_\omega(\varepsilon)) \cap U$. Then $z = (1-\varepsilon)y + \varepsilon\omega \in NE(G^\eta)$ with $\eta = \varepsilon\omega$. Moreover $z_1(T) \geq (1-\varepsilon)y_1(T) \geq 0.9\cdot 0.4 = 0.36$ and $z_1(B) \geq (1-\varepsilon)y_1(B) \geq 0.9\cdot 0.4 = 0.36$. Consequently $PB_2(z_1) = \{M\}$, where $PB_2(\cdot)$ is the pure best response correspondence of player 2. Hence $z_2(L) = \eta_2(L)$, $z_2(M) = 1 - \eta_2(L) - \eta_2(R)$ and $z_2(R) = \eta_2(R)$. However, since $PB_1(z_2) = \{T, B\}$, we have $\eta_2(R) = 2\eta_2(L)$. Division by ε yields $\omega_2(R) = 2\omega_2(L)$. $\triangleleft$

In the remainder of this section we will show that finite games with two players and finitely many Nash equilibria always possess an approximate maximum likelihood equilibrium. Recall that the *carrier* of strategy $a_i \in \Delta(A_i)$ of player i is defined by $Ca(a_i) = \{j \in A_i \mid a_i(j) > 0\}$.

Lemma 14.22 *Let G be a finite game with two players and let $a = (a_1, a_2) \in NE(G)$. Define*

$$\beta = \min \cup_{i=1}^2 \{a_i(j) \mid j \in Ca(a_i)\}.$$

Let U^a be the open neighbourhood of a defined by

$$\begin{aligned} U^a = \; &\{x \in \Delta(A_1) \times \Delta(A_2) \mid BR(x) \subseteq BR(a), x_i(j) > \beta/2 \\ &\text{for every } i \in \{1,2\}, j \in Ca(x_i)\}. \end{aligned}$$

Finally, let $\eta = (\eta_1, \eta_2)$ be a disturbance vector with $\eta_i(j) < \beta/2$ for every $i \in \{1,2\}$ and $j \in A_i$. Then we have for every $x \in NE(G^\eta) \cap U^a$ and for every $t \in (0,1]$ that $tx + (1-t)a \in NE(G^{t\eta})$.

Proof. First note that if $i \in N$ and $x \in \Delta(A_1) \times \Delta(A_2)$, then for every $j \in PB_i(x_{-i})$ and $k \in A_i \setminus PB_i(x_{-i})$ we have $u_i(j, x_{-i}) > u_i(k, x_{-i})$. By continuity we derive that there is an open neighbourhood V of x such that $PB_i(y_{-i}) \subseteq PB_i(x_{-i})$ for every $y = (y_i, y_{-i}) \in V$. Here $PB_i(\cdot)$ denotes the pure best response correspondence of player i. Consequently the set $\{x \in \Delta(A_1) \times \Delta(A_2) \mid BR(x) \subseteq BR(a)\} = \cap_{i \in N}\{x \in \Delta(A_1) \times \Delta(A_2) \mid PB_i(x_{-i}) \subseteq PB_i(a_{-i})\}$ is an open set and hence U^a is open.

Let $x \in NE(G^\eta) \cap U^a$ and $t \in (0,1]$. Define $y = tx + (1-t)a$. In order to show that $y \in NE(G^{t\eta})$ it suffices to show that $\{j \in A_1 \mid y_1(j) > t\eta_1(j)\} \subseteq PB_1(y_2)$ and $\{j \in A_2 \mid y_2(j) > t\eta_2(j)\} \subseteq PB_2(y_1)$. We only show the first inclusion. Let $j \in A_1$ be such that $y_1(j) > t\eta_1(j)$. Then $x_1(j) > \eta_1(j)$ or $a_1(j) > 0$. If $x_1(j) > \eta_1(j)$, then $j \in PB_1(x_2) \subseteq PB_1(a_2)$, so $j \in PB_1(tx_2 + (1-t)a_2) = PB_1(y_2)$. If $a_1(j) > 0$, then $x_1(j) > \beta/2 > \eta_1(j)$ by definition of U^a, and the same argument can be used. □

In the following lemma we show that if a finite two player game has a Nash equilibrium a and if some state of nature ω yields, for some perturbation level ε, a perturbed game which has a Nash equilibrium close to a, then the same state of nature ω yields for all smaller levels of perturbation a perturbed game with a Nash equilibrium close to a.

Lemma 14.23 *Let G be a finite game with two players and let $a = (a_1, a_2) \in NE(G)$. Let U be a convex, open subset of U^a with $a \in U$. Let $\varepsilon_U \in (0, \beta/2)$ be such that $(1-\varepsilon)x + \varepsilon\omega \in U^a$ for every $x \in U$, $\varepsilon \in (0, \varepsilon_U]$, $\omega \in \Omega$. Let $\varepsilon \in (0, \varepsilon_U]$ and $\omega \in \Omega$ be such that $NE(G_\omega(\varepsilon)) \cap U \neq \emptyset$. Then for every $\varepsilon' \in (0, \varepsilon)$ we have $NE(G_\omega(\varepsilon')) \cap U \neq \emptyset$.*

Proof. Let $x \in NE(G_\omega(\varepsilon)) \cap U$ and let $\varepsilon' \in (0, \varepsilon)$. Write $\varepsilon' = t\varepsilon$ for some $t \in (0,1)$. We have $y = (1-\varepsilon)x + \varepsilon\omega \in NE(G^{\varepsilon\omega}) \cap U^a$ and hence, according to Lemma 14.22, $ty + (1-t)a \in NE(G^{t\varepsilon\omega})$. Consequently,

$$(1-\varepsilon')\left(\frac{t(1-\varepsilon)}{1-\varepsilon'}x + \frac{1-t}{1-\varepsilon'}a\right) + \varepsilon'\omega \in NE(G^{\varepsilon'\omega}).$$

Therefore, by convexity of U, we have
$z = \frac{t(1-\varepsilon)}{1-\varepsilon'}x + \frac{1-t}{1-\varepsilon'}a \in NE(G_\omega(\varepsilon')) \cap U$. □

Corollary 14.24 *Let G be a finite game with two players and let $a = (a_1, a_2) \in NE(G)$. Let U and ε_U be defined as in Lemma 14.22. Then the map $(0, \varepsilon_U] \ni \varepsilon \mapsto \gamma(\varepsilon, U)$ is non-increasing. Consequently, $\mu(U) = \lim_{\varepsilon \downarrow 0} \gamma(\varepsilon, U)$.*

Remark 14.25 The statements above can be generalized in the following way. If $\mathcal{U} = \{U_1, \ldots, U_k\}$ is a collection of open, convex sets, which are mutually disjoint, which is such that every element U_j of this collection contains a Nash equilibrium a_j of the finite two player game G, and which satisfy $U_j \subseteq U^{a_j}$ for every $j \in \{1, \ldots, k\}$, then one can define

$$\gamma(\varepsilon, \mathcal{U}) = P_*(\{\omega \in \Omega \mid NE(G_\omega(\varepsilon)) \cap U \neq \emptyset \text{ for every } U \in \mathcal{U}\}).$$

As a corollary of Lemma 14.23 we get again that the map $(0, \varepsilon_{\mathcal{U}}] \ni \varepsilon \mapsto \gamma(\varepsilon, \mathcal{U})$ is non-increasing. Consequently, $\mu(\mathcal{U}) = \lim_{\varepsilon \downarrow 0} \gamma(\varepsilon, \mathcal{U})$ exists. $\triangleleft$

Now we are able to prove the main theorem of this section.

Theorem 14.26 *Every two player finite game with finitely many Nash equilibria has an approximate maximum likelihood equilibrium.*

Proof. Let G be a two player finite game with finitely many Nash equilibria $a_1, \ldots, a_k$. Let $U_i \subseteq U^{a_i}$ be open, convex sets with $a_i \in U_i$ for every $i \in \{1, \ldots, k\}$. Let $U = \cup_{i=1}^k U_i$. For ε sufficiently small we have $\gamma(\varepsilon, U) = P_*(\{\omega \in \Omega \mid NE(G_\omega(\varepsilon)) \cap U \neq \emptyset\}) = 1$. For, if this statement is not true, there is a sequence $(\varepsilon^k)_{k \in \mathbb{N}}$, converging to 0, a sequence $(\omega^k)_{k \in \mathbb{N}}$ in Ω and a sequence $(x^k)_{k \in \mathbb{N}}$ in $\times_{i \in N} \Delta(A_i)$ with $x^k \in NE(G_{\omega^k}(\varepsilon^k))$ and $x^k \notin U$ for every $k \in \mathbb{N}$. Without loss of generality we may assume that the sequences $(\omega^k)_{k \in \mathbb{N}}$ and $(x^k)_{k \in \mathbb{N}}$ have limits ω and x. Let $i \in N$ and $x'_i \in \Delta(A_i)$. Writing $y = (x'_i, x_{-i})$ and $y^k = (x'_i, x^k_{-i})$ for every $k \in \mathbb{N}$ we have

$$\begin{aligned} u_i(x) &= \lim_k u_i((1 - \varepsilon^k)x^k + \varepsilon^k \omega^k) \\ &= \lim_k u^{\varepsilon^k}_{i,\omega^k}(x^k) \\ &\geq \lim_k u^{\varepsilon^k}_{i,\omega^k}(y^k) \\ &= \lim_k u_i((1 - \varepsilon^k)y^k + \varepsilon^k \omega^k) \\ &= u_i(y). \end{aligned}$$

So, $x \in NE(G)$. However, $x \notin U$ which yields a contradiction. By the rule of inclusion and exclusion we have, for ε sufficiently small,

$$\begin{aligned} 1 &= \gamma(\varepsilon, U) \\ &= \sum_{\mathcal{U}:|\mathcal{U}|=1} \gamma(\varepsilon, \mathcal{U}) - \sum_{\mathcal{U}:|\mathcal{U}|=2} \gamma(\varepsilon, \mathcal{U}) + \cdots + (-1)^{k-1} \sum_{\mathcal{U}:|\mathcal{U}|=k} \gamma(\varepsilon, \mathcal{U}), \end{aligned}$$

and hence, by letting $\varepsilon \downarrow 0$,

$$1 = \sum_{\mathcal{U}:|\mathcal{U}|=1} \mu(\mathcal{U}) - \sum_{\mathcal{U}:|\mathcal{U}|=2} \mu(\mathcal{U}) + \cdots + (-1)^{k-1} \sum_{\mathcal{U}:|\mathcal{U}|=k} \mu(\mathcal{U}).$$

Here all summations are over subsets $\mathcal{U}$ of the collection $\{U_1, \ldots, U_k\}$. If $\pi(a_i) = 0$ for every $i \in \{1, \ldots, k\}$ then the right-hand side of this equality can be made arbitrarily small by choosing an appropriate collection of (small) open convex sets $\{U_1, \ldots, U_k\}$. This yields a contradiction. So, there is at least one equilibrium a with $\pi(a) > 0$. □

References

Aldrich, J. (1997): "R.A. Fisher and the making of maximum likelihood 1912-1922," *Statistical Science*, 12, 162–176.

Aliprantis, C.D., and K.C. Border (1994): *Infinite Dimensional Analysis; A Hitchhiker's Guide.* New York: Springer Verlag.

Borm, P.E.M., R. Cao, and I. García-Jurado (1995a): "Maximum likelihood equilibria of random games," *Optimization*, 35, 77–84.

Borm, P.E.M., R. Cao, I. García-Jurado, and L. Méndez-Naya (1995b): "Weakly strict equilibria in finite normal form games," *OR Spektrum*, 17, 235–238.

Daley, D.J., and D.J. Vere-Jones (1988): *An Introduction to the Theory of Point Processes.* New York: Springer Verlag.

Fagin R., and J.Y. Halpern (1991): "Uncertainty, belief, and probability," *Computational Intelligence*, 7, 160–173.

Gilboa, I., and D. Schmeidler (1999): "Inductive inference: an axiomatic approach," Tel Aviv University.

Harsanyi, J. (1967–1968): "Games with incomplete information played by Bayesian players," *Management Science*, 14, 159–182, 320–334, 486–502.

Milchtaich, I. (1997): "Random-player games," Northwestern University Math Center, discussion paper 1178.

Myerson, R.B. (1998a): "Comparison of scoring rules in Poisson voting games," Northwestern University Math Center, discussion paper 1214.

Myerson, R.B. (1998b): "Extended Poisson games and the Condorcet jury theorem," *Games and Economic Behavior*, 25, 111–131.

Myerson, R.B. (1998c): "Population uncertainty and Poisson games," *International Journal of Game Theory*, 27, 375–392.

Myerson, R.B. (2000): "Large Poisson games," *Journal of Economic Theory*, 94, 7–45.

Selten, R. (1975): "Reexamination of the perfectness concept for equilibrium points in extensive games," *International Journal of Game Theory*, 4, 25–55.

Voorneveld, M. (1999): *Potential Games and Interactive Decisions with Multiple Criteria*, Ph.D. thesis, CentER Dissertation Series 61, Tilburg University.

Voorneveld, M. (2000): "Maximum likelihood equilibria of games with population uncertainty," CentER discussion paper 2000-79.

Wu, W., and J. Jiang (1962): "Essential equilibrium points of n-person non-cooperative games," *Sci. Sinica*, 11, 1307–1322.

THEORY AND DECISION LIBRARY

SERIES C: GAME THEORY, MATHEMATICAL PROGRAMMING AND OPERATIONS RESEARCH

Editor: H. Peters, *Maastricht University, The Netherlands*

1. B.R. Munier and M.F. Shakun (eds.): *Compromise, Negotiation and Group Decision.* 1988 ISBN 90-277-2625-6
2. R. Selten: *Models of Strategic Rationality.* 1988 ISBN 90-277-2663-9
3. T. Driessen: *Cooperative Games, Solutions and Applications.* 1988 ISBN 90-277-2729-5
4. P.P. Wakker: *Additive Representations of Preferences.* A New Foundation of Decision Analysis. 1989 ISBN 0-7923-0050-5
5. A. Rapoport: *Experimental Studies of Interactive Decisions.* 1990 ISBN 0-7923-0685-6
6. K.G. Ramamurthy: *Coherent Structures and Simple Games.* 1990 ISBN 0-7923-0869-7
7. T.E.S. Raghavan, T.S. Ferguson, T. Parthasarathy and O.J. Vrieze (eds.): *Stochastic Games and Related Topics.* In Honor of Professor L.S. Shapley. 1991 ISBN 0-7923-1016-0
8. J. Abdou and H. Keiding: *Effectivity Functions in Social Choice.* 1991 ISBN 0-7923-1147-7
9. H.J.M. Peters: *Axiomatic Bargaining Game Theory.* 1992 ISBN 0-7923-1873-0
10. D. Butnariu and E.P. Klement: *Triangular Norm-Based Measures and Games with Fuzzy Coalitions.* 1993 ISBN 0-7923-2369-6
11. R.P. Gilles and P.H.M. Ruys: *Imperfections and Behavior in Economic Organization.* 1994 ISBN 0-7923-9460-7
12. R.P. Gilles: *Economic Exchange and Social Organization.* The Edgeworthian Foundations of General Equilibrium Theory. 1996 ISBN 0-7923-4200-3
13. P.J.-J. Herings: *Static and Dynamic Aspects of General Disequilibrium Theory.* 1996 ISBN 0-7923-9813-0
14. F. van Dijk: *Social Ties and Economic Performance.* 1997 ISBN 0-7923-9836-X
15. W. Spanjers: *Hierarchically Structured Economies.* Models with Bilateral Exchange Institutions. 1997 ISBN 0-7923-4398-0
16. I. Curiel: *Cooperative Game Theory and Applications.* Cooperative Games Arising from Combinatorial Optimization Problems. 1997 ISBN 0-7923-4476-6
17. O.I. Larichev and H.M. Moshkovich: *Verbal Decision Analysis for Unstructured Problems.* 1997 ISBN 0-7923-4578-9
18. T. Parthasarathy, B. Dutta, J.A.M. Potters, T.E.S. Raghavan, D. Ray and A. Sen (eds.): *Game Theoretical Applications to Economics and Operations Research.* 1997 ISBN 0-7923-4712-9
19. A.M.A. Van Deemen: *Coalition Formation and Social Choice.* 1997 ISBN 0-7923-4750-1
20. M.O.L. Bacharach, L.-A. Gérard-Varet, P. Mongin and H.S. Shin (eds.): *Epistemic Logic and the Theory of Games and Decisions.* 1997 ISBN 0-7923-4804-4
21. Z. Yang (eds.): *Computing Equilibria and Fixed Points.* 1999 ISBN 0-7923-8395-8
22. G. Owen: *Discrete Mathematics and Game Theory.* 1999 ISBN 0-7923-8511-X
23. F. Patrone, I. Garcia-Jurado and S. Tijs (eds.): *Game Practice.* Contributions from Applied Game Theory. 1999 ISBN 0-7923-8661-2

24. J. Suijs: *Cooperative Decision-Making under Risk.* 1999 ISBN 0-7923-8660-4
25. J. Rosenmüller: *Game Theory: Stochastics, Information, Strategies and Cooperation.* 2000 ISBN 0-7923-8673-6
26. J.M. Bilbao: *Cooperative Games on Combinatorial Structures.* 2000 ISBN 0-7923-7782-6
27. M. Slikker and A. van den Nouweland: *Social and Economic Networks in Cooperative Game Theory.* 2000 ISBN 0-7923-7226-3
28. K.J.M. Huisman: *Technology Investment: A Game Theoretic Real Options Approach.* 2001 ISBN 0-7923-7487-8
29. A. Perea: *Rationality in Extensive Form Games.* 2001 ISBN 0-7923-7540-8
30. V. Buskens: *Social Networks and Trust.* 2002 ISBN 1-4020-7010-1

KLUWER ACADEMIC PUBLISHERS – DORDRECHT / BOSTON / LONDON